AF510896

ELIHU BURRITT LIBRARY
CENTRAL CONNECTICUT STATE UNIVERSITY
NEW BRITAIN, CONNECTICUT 06050

Charm Physics

China Center of Advanced Science and Technology (World Laboratory)
Symposium/Workshop Proceedings

Editors-in-Chief

WORLD LABORATORY
A. Zichichi

CCAST
T. D. Lee
K. C. Chou

VOLUME 1 *Lattice Gauge Theory Using Parallel Processors*
Edited by Xiaoyuan Li, Zhaoming Qiu and Hai-cang Ren

VOLUME 2 *Charm Physics*
Edited by Ming-han Ye and Tao Huang

Additional volumes in preparation

This book is part of a series. The publisher will accept continuation orders which may be cancelled at any time and which provide for automatic billing and shipping of each title in the series upon publication. Please write for details.

丛　书　前　言

中国高等科学技术中心（CCAST）于1986年10月在北京成立．它是一个民间研究组织．它参加由意大利政府捐助建立的民间组织世界实验室并获得世界实验室的资助．它和中国的研究机构及大学建立密切的合作关系．中心的目的是在中国创造良好的研究环境，建立中国和世界各国研究机构和科学家之间的联系，以鼓励中国科学家做出有世界水平的研究成果，为全世界科学技术自由交流和发展做出贡献．

中心的主要活动之一是组织国际学术交流；每年将组织三至四次国际学术讨论会，精选与中国现有条件合适的和可发展的前沿科技领域，邀请在这方面有突出贡献的科学家介绍该学科的基础，现状，特别强调存在的问题和困难及将来可盼望的前景，使之有助于中国及世界的物理学家共同努力，推动这一学科的发展．

每一次讨论会将有一本文集，这部丛书就是这些文集的汇合．希望中国的物理学界，尤其是年轻的一代能从这部丛书中得到教益．

李政道

一九八七年五月三十一日

Contents

Preface to the Series vii

Preface ix

List of Invited Contributors xi

An Introduction to Charm and Heavy Quark Physics
Frederick J. Gilman 1

Experimental Review of J/ψ Decays
Walter H. Toki 89

Meson Spectroscopy Viewed from J/ψ Decay
Michael S. Chanowitz 161

Weak Decays of Charmed Quarks: An Experimental Review
David Hitlin 219

On Charm Decays—Present Status and Future Goals
Ikaros Bigi 339

Charm Physics in Fixed Target Experiments
Michael S. Witherell 427

Present Status of Beijing Spectrometer
Ye Minghan and Zheng Zhipeng 475

Charm Physics at BEPC
Tao Huang 503

Panel Discussion 535

List of Participants 555

Index 559

5/16

© 1988 by OPA (Amsterdam) B.V. All rights reserved. Published under license by Gordon and Breach Science Publishers S.A.

QC 793.3 C53 C43 1987

Gordon and Breach Science Publishers

Post Office Box 786
Cooper Station
New York, New York 10276
United States of America

Post Office Box 161
1820 Montreux 2
Switzerland

Post Office Box 197
London WC2E 9PX
England

3-14-9 Okubo
Shinjuku-ku, Tokyo
Japan

58, rue Lhomond
75005 Paris
France

Private Bag 8
Camberwell, Victoria 3124
Australia

Library of Congress Cataloging-in-Publication Data

Charm physics.

(Proceedings of the CCAST (World Laboratory) Symposium/Workshops, ISSN 0894-2536; v. 2)
Includes index.
1. Particles (Nuclear physics)—Charm—Congresses
2. Mesons—Decay—Congresses, I. Ye, Minghan, 1925— . II. Huang, Tao. III. Series.
QC793.3.C53C48 1987 539.7′21 87-19781
ISBN 2-88124-233-2
ISBN 2-88124-275-8 (pbk.)

No part of this book may br reproduced or utilized in any form or by any means, electronic or mechanical, including photocopying and recording, or by any information storage or retrieval system, without permission in writing from the publisher. Printed in Great Britain by Bell and Bain Ltd, Glasgow.

Charm Physics

Edited by

Ming-han Ye
Tao Huang

Institute of High Energy Physics, Beijing, The People's Republic of China

Proceedings of the CCAST (World Laboratory) Symposium/Workshop
held at
The Institute of High Energy Physics, Beijing
The People's Republic of China
June 4–16 1987

GORDON AND BREACH SCIENCE PUBLISHERS
New York London Paris Montreux Tokyo Melbourne

Preface to the Series

The China Center of Advanced Science and Technology (CCAST) was established in Beijing on October 17, 1986 to introduce important frontier areas of science to China and to promote the free exchange of scientific information between China and other nations. It is sponsored by the World Laboratory, with support from the Italian and Chinese Governments.

Every year CCAST (World Laboratory) will organize three or four international symposia/workshops on subjects that are especially selected for their potential as seeds for future development in China. Each symposium brings together about 10 experts from abroad and 60-70 scientists from within China. They work very closely to discuss, in depth, the state of the art of the subject and to explore its future possibilities, with special emphasis on present problems within the area. It is this joint labor of fostering the growth of modern science in an ancient center of civilization that gives these symposia an especially uplifting feeling. This series of proceedings may serve as witness to these efforts on behalf of the younger generation of Chinese physicists.

T. D. Lee

Preface*

From June 4th to June 16th 1987, nearly 100 physicists gathered in Beijing for the Symposium/Workshop on Charm Physics. This was the second of a series of symposia designed to introduce important frontier areas of science to China. The experts invited from abroad were all pioneers in the field. The majority of the participants were Chinese scientists interested in state-of-the-art developments in this area. The discussions not only covered the current status, but also concentrated on present difficulties and future possibilities.

Since the discovery of J/ψ, charm physics has been one of the most productive areas of physics. Recent experimental results, particularly those from Mark III and others, have provided evidence that many important physics problems remain to be solved experimentally. The physics world has accepted quantum chromodynamics (QCD) as the underlying theory of all strong interactions. It is of major interest and novelty in the realm of QCD to observe gluonic bound states. It will be a triumph for QCD if a strongly interacting gluonic bound state is observed. The decays of the charm system provide a unique laboratory in which the subtle interplay of strong and weak interactions is being studied.

China has chosen this highly productive area as a starting point of her high energy physics research. An electron–position collider is being built in Beijing, called the Electron–Positron Collider (BEPC), with 1.5–2.8 GeV as the energy of each beam. Many Chinese participants of this Symposium/Workshop were experimental physicists who are preparing detectors for experiments on charm physics at BEPC.

In this Symposium, we are particularly fortunate to have brought together the leaders from some of the groups devoted to charm physics research and to have the expertise of:

Dr. I. Bigi of the Stanford Linear Accelerator Center,
Professor M. Chanowitz of Lawrence Berkeley Laboratory,
Professor F. J. Gilman of the Stanford Linear Accelerator Center,
Professor D. Hitlin of California Institute of Technology,
Professor T. Huang of the Institute of High Energy Physics in Beijing,
Professor W. H. Toki of the Stanford Linear Accelerator Center,
Professor M. Witherell of University of California, Santa Barbara,
Professor Z. P. Zheng of the Institute of High Energy Physics in Beijing.

Their lectures covered the most recent development in charm

physics and also reviewed the fundamental achievements in this area that have been accomplished through experimental and theoretical studies. Their lectures and the active participation of our Chinese colleagues made this symposium a particularly memorable one. It is difficult to recapture the friendly spirit of such an occasion, the give and take of ideas, and the stimulation that goes far beyond physics *per se*. Much of this will remain in the memory of the participants, but some part may happily be reflected by these proceedings.

T. Huang
M.-h. Ye

*In addition to the China Center of Advanced Science and Technology (World Laboratory), support of this symposium was also provided by China Association for International Personnel Exchange and Institute of High Energy Physics of Academia Sinica.

List of Invited Contributors

Ikaros Bigi	比　基	Stanford Linear Accelerator Center
Michael S. Chanowitz	卡诺维基	Lawrence Berkeley Laboratory
Frederick J. Gilman	吉尔曼	Stanford Linear Accelerator Center
David Hitlin	希特林	California Institute of Technology
Walter H. Toki	托　基	Stanford Linear Accelerator Center
Michael Witherell	韦斯累尔	University of California, Santa Barbara
Ming-han Ye	叶铭汉	Institute of High Energy Physics
Zhi-peng Zheng	郑志鹏	Institute of High Energy Physics
Tao Huang	黄　涛	Institute of High Energy Physics

AN INTRODUCTION TO CHARM AND HEAVY QUARK PHYSICS*

Frederick J. Gilman

Stanford Linear Accelerator Center
Stanford University, Stanford, California 94305

Abstract The physics of the electroweak and strong interactions in the standard model are reviewed, especially as they apply to the tau lepton and to charm and heavy quark physics.

1. INTRODUCTION

Charm physics, as all other high energy physics, is now considered in reference to the framework of the standard model. The strong and electroweak interactions originate in gauge theories with the gauge group being $SU(3)_C \times SU(2) \times U(1)$, together with the assignment of fermions to triplets if they are quarks and singlets if they are leptons under $SU(3)_C$ and to left-handed doublets and right-handed singlets under the electroweak $SU(2)$. Given 18 or so parameters which are put in from outside the standard model, it is in excellent shape. Experiment has again and again confirmed its predictions to whatever accuracy they can be predicted and measured in a given situation. With the beginning of experiments at the Z factories in the near future, the one-loop corrections to the electroweak portion of the standard model will be tested to high accuracy.

* Work supported by the Department of Energy, contract DE–AC03–76SF00515.

While aware of the success of the standard model, we are also not satisfied with it as a final theory of Nature. It is incomplete. Many parameters, such as quark masses and weak mixing angles come, as noted above, from outside the standard model. Nor is the reason for quark and lepton families or generations explained, let alone a connection between quarks and leptons. Why are there 3 (?) generations? And then we have the hierarchy problem—how is a scale as "small" as the weak scale determined if we start at a grand-unified scale or at the Planck scale?

Therefore we are continually looking for evidence of the physics which must exist which is beyond or outside the standard model. We check and recheck with yet higher accuracy the predictions which follow from the standard model, hoping to find some hint of what lies beyond.

While working in what is now comparatively "low" energies, there are possibilities for physics discoveries at BEPC which could point beyond the standard model. Some examples which are of topical interest at the moment are:

- $D^0 - \bar{D}^0$ Mixing - This is a quantity which is small in the standard model, which makes it difficult to accumulate enough data to make a significant measurement, although present experiments are approaching the appropriate level.[1,2] By the same token, this is a good place to look for new physics. Where the standard effects are small (or even better, zero) is precisely where some new effect might stand out.[3]
- Measurement of $c \to d\bar{e}\nu$ as compared to $c \to s\bar{e}\nu$ Decays - Such measurements, if done accurately and combined with theoretical analysis, will permit the extraction of the ratio of

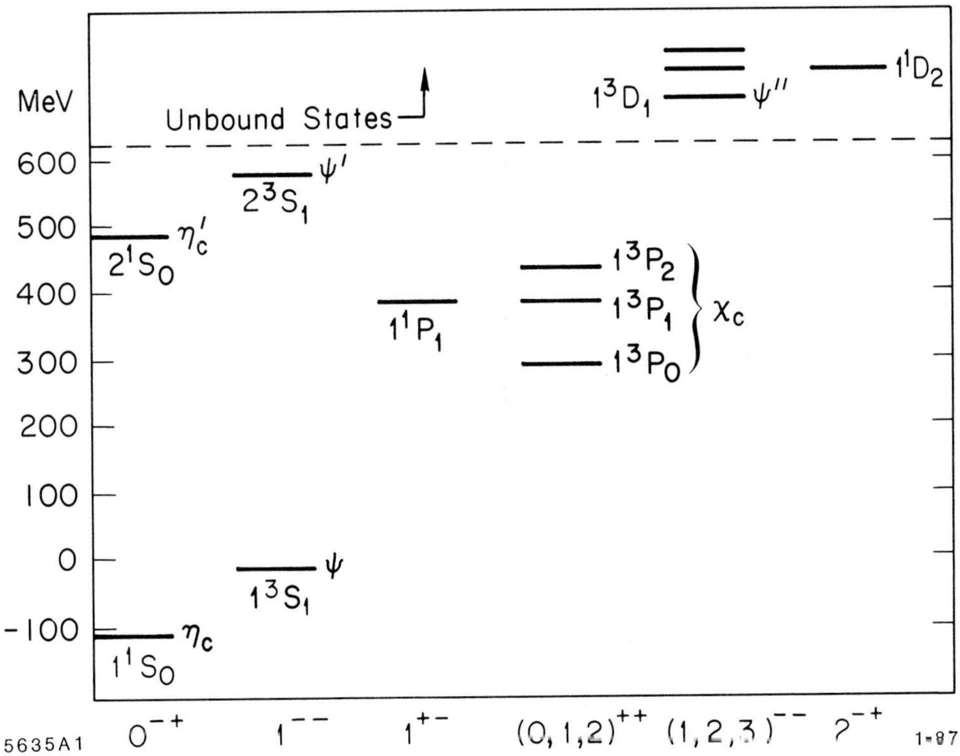

Fig. 1

The spectrum of charmonium.

Kobayashi - Maskawa matrix elements V_{cd}/V_{cs}, if not their separate absolute values. This allows a check of unitarity of the three generation matrix using its second row (see the discussion in Section 2 for the first row).

- τ Decays - The search for "forbidden" (in the standard model) tau decays such as $\tau \to \mu\bar{e}e$, $\tau \to \mu K^0$, and $\tau \to \nu_\tau\eta\pi^-$ needs to be pushed to higher sensitivity. There is an intriguing problem, discussed in Section 4, that has arisen in the past few years as well of accounting for all the exclusive modes that make up the inclusive one-prong decays.

Both open and bound charm serve as a laboratory for standard model physics as well. To cite a few examples:

- Charmonium Spectroscopy - Charmonium presents a rich spectrum of narrow levels (see Figure 1) and has many photon and hadronic transitions which have been studied.[4] It is particularly interesting in comparison to light quark meson spectroscopy of $\bar{q}'q$ states and of bottomonium ($\bar{b}b$ states). It is non-relativistic enough ($v^2/c^2 \sim 0.3$) to apply a Schrodinger equation with an effective two-body potential, and thus to compare to the much more surely non-relativistic bottomonium case (see Figure 2). On the other hand, relativistic corrections are surely important in a number of cases and the charmonium system provides a kind of "bridge" between bottomonium and light quark mesons which are intrinsically relativistic by nature. The comparison of charmonium to other systems, and particularly bottomonium, is something I will stress repeatedly in Section 5, as we learn a great deal from this comparison and it gives us confidence in applying the lessons we learn in one system to another.

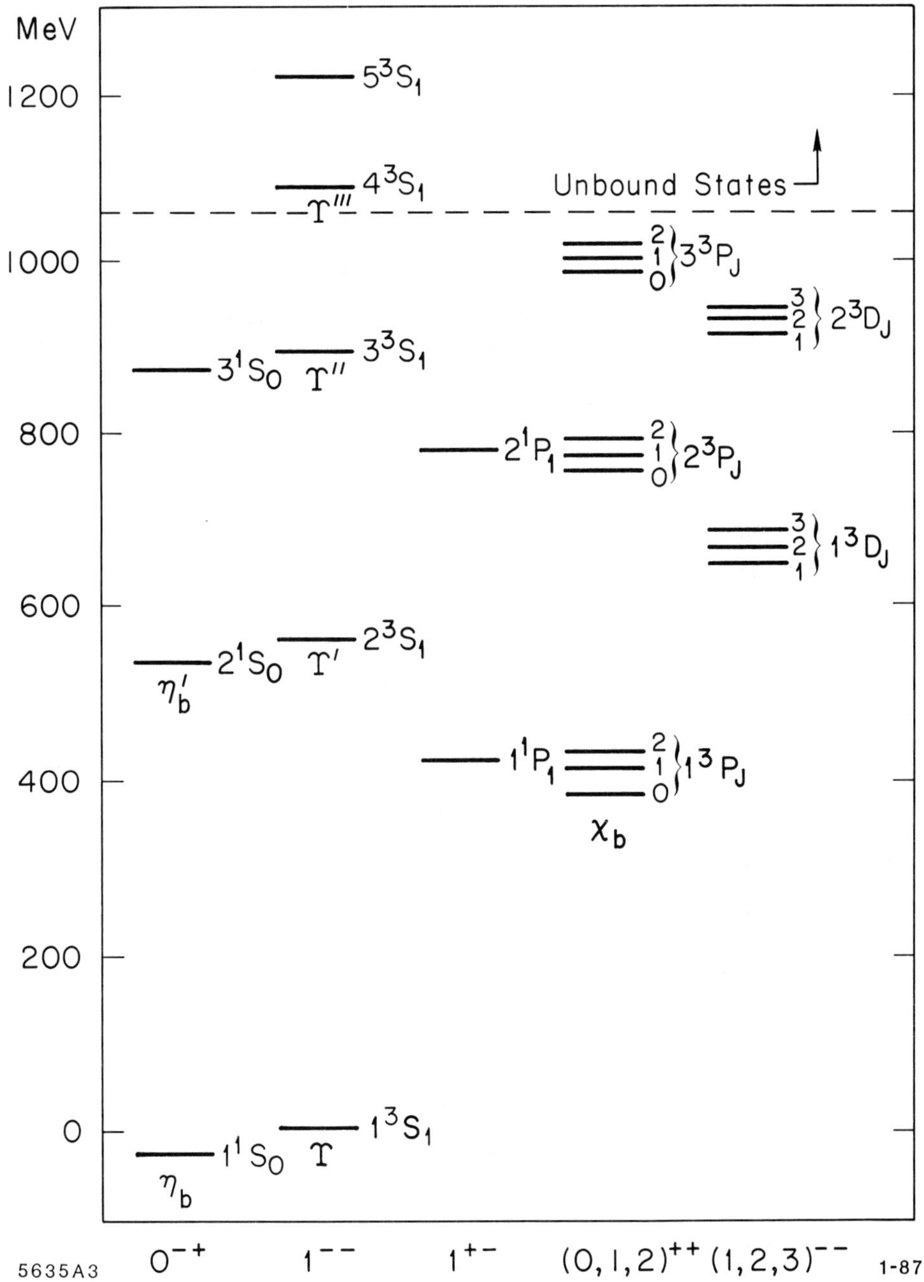

Fig. 2

The spectrum of bottomonium.

- $J/\psi \to \gamma$ + hadrons and $J/\psi \to \omega$ or ϕ + hadrons - These decays permit the high statistics study of light hadron spectroscopy and in particular the production of gluonium and "hybrid" ($\bar{q}qg$) states.[5] It clearly will form an essential part of the BEPC program.

- Weak Decays of D^+, D^0, D_s^+, and Λ_c^+ - Here we can see the interplay of the strong and weak interactions. We are again in a somewhat intermediate situation between K decays and B decays. In the latter case we expect the "spectator model" to work quite well, whereas we already know that there are corrections to it for charm, including the phenomena of final state interactions, color coherence, weak annihilation and Pauli interference which have been extensively discussed.[3]

2. THE STANDARD MODEL

The standard model for strong interactions, Quantum Chromodynamics (QCD), involves an unbroken $SU(3)_C$ gauge group with 8 gauge bosons, the gluons. These gauge bosons are coupled to the color charge carried by the the gluons themselves and by the quarks (and not leptons), which are assigned to the fundamental three dimensional representation of $SU(3)$.

Electroweak physics, on the other hand, is a gauge theory with the group structure $SU(2) \times U(1)$, spontaneously broken so as to have massive W^+, W^-, and Z vector bosons and a massless photon.[6] Within the gauge theory sector itself there are three parameters: g, the $SU(2)$ coupling; g', the $U(1)$ coupling; and the vacuum expectation value, v, of the Higgs field that is associated with spontaneous breaking of the continuous symmetry.

We usually do not work in terms of these three parameters. After defining the weak mixing angle through

$$\cos\theta_W = \frac{g}{(g^2 + g'^2)^{1/2}}, \tag{1}$$

and identifying the electromagnetic coupling

$$e = g\sin\theta_W, \tag{2}$$

and the Fermi effective coupling

$$\frac{G_F}{\sqrt{2}} = \frac{g^2}{8\ M_W^2}, \tag{3}$$

with $M_W = gv/2$, it has been conventional to use $\alpha = e^2/4\pi$, G_F and $\sin^2\theta_W$ as the three parameters of the theory. This is related to the very accurate experimental determinations of the first two, α and G_F, leaving $\sin^2\theta_W$ to be pinned down as the characteristic parameter expressing the unification of weak and electromagnetic interactions. The W and Z boson masses are related by

$$M_W = M_Z\cos\theta_W, \tag{4}$$

and the W mass itself, using Eqs. (2) and (3) in lowest order, is given numerically by

$$M_W = \frac{37.3\ \text{GeV}}{\sin\theta_W}. \tag{5}$$

(Electroweak radiative corrections change this to $M_W = 38.65$ GeV/ $\sin\theta_W$.) While up to the present it is $\sin^2\theta_W$ that is commonly used as the third parameter, Eqs. (4) and (5) make it clear that we could use M_W or M_Z. Until their discovery, this made no sense, but already the experimental uncertainty on their masses gives a comparable accuracy in the determination of $\sin^2\theta_W$ to that from measurements in low energy neutral current experiments.[7] In fact, once we enter the era of Z physics, it is much more appropriate to use α, G_F, and M_Z as the three parameters of the electroweak gauge theory, since M_Z will be fairly easily measured to an accuracy which will far exceed its equivalent in other determinations of $\sin^2\theta_W$.

The fermions in the standard model of electroweak interactions are assigned to left-handed doublets and right-handed singlets, thereby fixing their couplings to the gauge bosons of the electroweak interactions. For example, the charged W is coupled to $\bar{e}\nu_e$ with $g_V = -g_A = g/2\sqrt{2}$ and the Z is coupled to $\bar{\nu}\nu$ with $g_V = g_A = g/4\cos\theta_W$.

For the quark sector, there is the additional complication that the weak and mass eigenstates are not the same. The Kobayashi - Maskawa matrix[8] is defined as the matrix transformation that takes us from the mass eigenstates of the d, s, and b quarks to the weak eigenstates, d', s', and b', the partners in weak doublets of the u, c, and t quarks, respectively, which by convention are unmixed:

$$\begin{pmatrix} d' \\ s' \\ b' \end{pmatrix} = \begin{pmatrix} V_{ud} & V_{us} & V_{ub} \\ V_{cd} & V_{cs} & V_{cb} \\ V_{td} & V_{ts} & V_{tb} \end{pmatrix} \begin{pmatrix} d \\ s \\ b \end{pmatrix} . \tag{6}$$

The 1986 Review of Particle Properties[9] gave the following results for the magnitudes of those matrix elements that can be measured up to the present time:

(1) Nuclear beta decay, when compared to muon decay, gave

$$|V_{ud}| = 0.9729 \pm 0.0012 .$$

which included refinements[10] which had lowered $|V_{ud}|$ by 0.13%.

(2) Analysis[11] of hyperon and K_{e3} decays yielded

$$|V_{us}| = 0.221 \pm 0.002 \ .$$

(3) From ν and $\bar{\nu}$ production of charm, the CDHS group[12] had deduced

$$|V_{cd}| = 0.24 \pm 0.03 \ .$$

(4) By comparing the experimental value[13] for $\Gamma(D \to \overline{K}e^+\nu_e)$ with the expression that follows from the standard weak interaction amplitude, one derives:

$$|f_+^D(0)|^2|V_{cs}|^2 = 0.51 \pm 0.07 \ .$$

where f_+^D is the form factor for $D_{\ell 3}$ decay which is the analogue of f_+ for $K_{\ell 3}$ decay. With the conservative assumption that $|f_+(0)| < 1$,

$$|V_{cs}| > 0.66 \ .$$

(5) The ratio $|V_{ub}/V_{cb}|$ is obtained from the semileptonic decay of B mesons by fitting to the lepton energy spectrum as a sum of contributions involving $b \to u$ and $b \to c$. As more data had accumulated, the inadequacy of previous parametrizations of the lepton spectrum became clear.[6] Using only the lepton momentum region beyond the end-point for $b \to c\ell\overline{\nu}_\ell$ resulted in[6]

$$\Gamma(b \to u\ell\overline{\nu}_\ell)/\Gamma(b \to c\ell\overline{\nu}_\ell) < 0.08 \ ,$$

which translates to

$$|V_{ub}/V_{cb}| < 0.19 \ .$$

(6) The magnitude of V_{cb} itself can be determined if the measured semileptonic bottom hadron partial width is assumed to be that of a b quark decaying through the usual $V - A$ interaction

(which from (5) is $BR(b \to c\ell\overline{\nu}_\ell)$ to within 8%):

$$0.037 < |V_{cb}| < 0.053 \; .$$

One can not prove there are three generations from these data, but only show consistency with that as a hypothesis. The crucial test at present comes from the constraint which unitarity of the 3×3 matrix imposes on the first row:

$$(0.9729 \pm 0.0012)^2 + (0.221 \pm 0.002)^2 + (< 0.01)^2 = 0.9954 \pm 0.0025, \tag{7}$$

with a couple of standard deviations from unity of the right hand side being insufficient to make a fuss about.

Since the Review of Particle Properties went to press, there have been some small shifts in the central values of some of the matrix elements due to reanalysis and/or new data, such as incorporating newer charm semileptonic branching ratios in extracting[14] $|V_{cs}|$. More importantly, a change[15] in the order $Z\alpha^2$ Coulomb corrections brought different experiments into better agreement and raised the value of $|V_{ud}|$:

$$|V_{ud}| = 0.9747 \pm 0.0010,$$

to be compared with a very recent result[16]

$$|V_{ud}| = 0.9755 \pm 0.0017,$$

which also improves on previous analyses of this quantity, primarily in terms of the electron screening correction.[17] The unitarity sum for the first row is now 0.9989 ± 0.0021 or 1.0004 ± 0.0035, depending on which new result one uses. One couldn't ask for better agreement with three generations. Turning this around, and using unitarity to restrict the coupling between the u quark and a new charge $-1/3$

quark results in

$$|V_{ub'}| \leq 0.06. \tag{8}$$

This is not very restrictive and, looking at its primary origin in the error bar on $|V_{ud}|$, it seems unlikely that there will be a very significant improvement upon it in the future.

3. WEAK DECAYS OF HEAVY QUARKS

This is a subject which is comparatively new, but one which is already in a fairly mature state experimentally:[1,2] we have measurements of many D meson branching ratios, including Cabibbo suppressed nonleptonic modes and the decomposition of the semileptonic decays into exclusive channels; excellent lifetime measurements exist for both charm and bottom hadrons; a good beginning has been made on the study of exclusive decays of the D_s, Λ_c and $B^{0,+}$.

On the theoretical side, we have a solid general framework within which to calculate these weak decays. In particular, this means starting with the electroweak interactions and their gauge group, $SU(2) \times U(1)$, and adding the corrections due to the strong interactions through the use of the renormalization group equation (or an equivalent formulation of the same physics), with anomalous dimensions computed from QCD.

These calculations are carried out at the quark level. A first stage in their application to actual hadrons is simply to neglect any other constituent of the decaying hadron aside from the heavy quark. In such a spectator model, as it is called, one directly carries over the quark level calculation to be the hadron level result, with the spectator quarks and gluons assumed to arrange themselves into the final state particles together with the quarks (or leptons) coming from the heavy quark at no cost or benefit to the overall rate.

From the present data on charmed particle lifetimes, it is clear that there are differences of a factor of two or so between different species.[1,2] To understand this theoretically, it is necessary to go beyond the spectator model and to consider not just what happens at the quark level but at the hadron level. In so doing, ideas such as annihilation diagrams, interference, color (mis)matching, and final state interactions have entered the discussion.[18]

We give here only an introduction to the subject of the decays of hadrons containing heavy quarks. We will work at the quark level, where we know quite precisely how to proceed theoretically and the results give a semi-quantitative description of the experimental situation as we know it today. The corrections to the spectator model and the detailed analysis of the approaches to a full understanding of weak charm decays are found elsewhere.[18]

- Semileptonic Decays

It is theoretically simplest to start with semileptonic decays of the form

$$Q \to q + e\bar{\nu}_e$$

(such as $b \to ce\bar{\nu}_e$) or

$$Q \to q + \bar{e}\nu_e$$

(such as $c \to s\bar{e}\nu_e$) which correspond to a Hamiltonian density of the form

$$\mathcal{H} = -V_{Qq} \frac{G_F}{\sqrt{2}} \bar{q}\gamma_\mu(1-\gamma_5)Q \; \bar{e}\gamma^\mu(1-\gamma_5)\nu_e. \tag{9}$$

In Eq. (9) we have used particle names in place of the corresponding spinor operators and introduced the one factor that does not enter into the analogous expression for muon decay, the Kobayashi - Maskawa matrix element V_{Qq}.

The corresponding decay rate can then be easily related to that for muon decay, $G_F^2 m_\mu^5/192\pi^3$:

$$\Gamma(Q \to qe\bar{\nu}_e) = |V_{Qq}|^2 \frac{G_F^2 M_Q^5}{192\pi^3} F\left(\frac{m_q}{M_Q}\right) \tag{10}$$

where

$$F(\Delta) = 1 - 8\Delta^2 + \Delta^6 - \Delta^8 - 24\Delta^4 \ell n \Delta. \tag{11}$$

Note again the extra Kobayashi - Maskawa factor in front and the phase space factor, $F(\Delta)$, which is unity for a massless final quark ($\Delta = 0$). This factor drops off rather quickly, so that $F(0.3) = 0.52$, a value relevant approximately for the $c \to s$ and the $b \to c$ transitions.

The electron (positron) energy spectrum is different in the two cases. For $b \to ce\bar{\nu}_e$ it is like that in muon decay and gives rise to a "hard" spectrum that does not vanish at the high energy end:

$$\frac{1}{\Gamma}\frac{d\Gamma}{dx} = \frac{12}{5} x^2(2-x), \tag{12}$$

while for $c \to s\bar{e}\nu_e$ (and for $t \to b\bar{e}\nu_e$) it vanishes at the two ends

$$\frac{1}{\Gamma}\frac{d\Gamma}{dx} = 12x^2(1-x) \tag{13}$$

where the scaled energy variable is

$$x = \frac{2E_e}{M_Q} \leq 1 - \frac{m_q^2}{M_Q^2}. \tag{14}$$

Similar results of course hold for decays involving muons or taus.

- Nonleptonic Decays

The Hamiltonian density for nonleptonic decays such as $Q \to q + u\bar{d}$ has the same basic form,

$$\mathcal{H} = -V_{Qq} \frac{G_F}{\sqrt{2}} \bar{q}_\alpha \gamma_\mu (1 - \gamma_5) Q_\alpha \; \bar{u}_\beta \gamma^\mu (1 - \gamma_5) d_\beta \tag{15}$$

as for semileptonic decays, aside from the addition of the color indices α and β which are summed over the three colors to form color singlet currents. The decay rate

$$\Gamma(Q \to qu\bar{d}) = 3 \;\; |V_{Qq}|^2 \, \frac{G_F^2 M_Q^5}{192\pi^3} \; F\left(\frac{m_q}{M_Q}\right) \tag{16}$$

is also identical to the semileptonic case except for the factor of three on the right hand side due to color (we are neglecting the masses of the u and d quarks, just as we neglected those of the e and ν_e previously).

Now let us rewrite the Hamiltonian in a slightly different form:

$$\begin{aligned}\mathcal{H} = -V_{Qq}\,\frac{G_F}{2\sqrt{2}}\cdot\\ \Big[c_+\Big[\bar{q}_\alpha\gamma_\mu(1-\gamma_5)Q_\alpha\bar{u}_\beta\gamma^\mu(1-\gamma_5)d_\beta\\ +\bar{q}_\alpha\gamma_\mu(1-\gamma_5)d_\alpha\bar{u}_\beta\gamma^\mu(1-\gamma_5)Q_\beta\Big]\\ +c_-\Big[\bar{q}_\alpha\gamma_\mu(1-\gamma_5)Q_\alpha\bar{u}_\beta\gamma^\mu(1-\gamma_5)d_\beta\\ -\bar{q}_\alpha\gamma_\mu(1-\gamma_5)d_\alpha\bar{u}_\beta\gamma^\mu(1-\gamma_5)Q_\beta\Big]\Big]\end{aligned} \tag{17}$$

with $c_+ = c_- = 1$ initially. All we have done is to add and subtract a term which is nothing but the original expression with $Q \leftrightarrow d$. Moreover, this term would be identical to the original one if it were not for the presence of the color indices; without them the interchange $Q \leftrightarrow d$ is a Fierz transformation under which $V - A$ interactions go into themselves. In the decay rate, the three on the right-hand side of Eq. (16) is replaced by $2c_+^2 + c_-^2$, which again is no change at all when $c_+ = c_- = 1$.

So why make a more complicated expression out of something simple? The answer lies in what happens when we turn on the strong interactions and add the effects of QCD to the purely weak interactions that we have considered up to this point. The weak interaction will be modified by the presence of strong interaction effects and c_+ and c_- will be renormalized. But they have been carefully chosen in this regard, for they only go into themselves under this renormalization, i. e. the corresponding operators, which are even or odd under interchange of color indices, do not mix through QCD corrections. Not only do the strong interactions modify c_+ and c_- from their initial value of unity, but they introduce new operators into the effective Hamiltonian. These so-called "penguin" operators, which come in beginning at the one loop level, have a different space-time structure than the $V - A \times V - A$ structure we have had up to now. We proceed to consider each of these effects and their magnitude in turn.

- The Calculation of c_+ and c_-

Calculating what happens to c_+ and c_- under renormalization due to QCD is equivalent to studying their behavior as one moves from one momentum scale to another. More specifically, at the momentum scale corresponding to M_W the weak interactions are characterized by the "bare" Hamiltonian density of Eq. (15) and $c_+ = c_- = 1$. We are interested in what happens when we move down to a momentum scale of relevance to a particular hadron, *i.e.*, roughly the mass of the decaying heavy quark.

The study of what happens when one moves from one momentum scale to another is directly formulated through a renormalization group equation. In the case of c_+ and c_-, they satisfy such an equation of the form:

$$\left[\mu\frac{\partial}{\partial\mu} + \beta(g)\frac{\partial}{\partial g} - \gamma_\pm(g)\right] c_\pm(q/\mu, g) = 0, \tag{18}$$

where μ is some reference scale of momentum (the renormalization point) and q is a second scale at which we wish to calculate the effective weak Hamiltonian. In this equation, $\beta(g)$ is the standard beta function of the theory,

$$\beta \equiv \mu\frac{\partial g}{\partial\mu},$$

which characterizes how the coupling changes with a change of scale. For QCD, it has the perturbation theory expansion,

$$\beta(g) = \frac{g^3}{48\pi^2}(2n_f - 33) + ... \tag{19}$$

where n_f is the number of quark flavors. Notice that the coefficient of g^3 is negative as long as $33 > 2n_f$. In other words, the coupling decreases as we increase the scale of momentum at which we are looking. This is just the property of asymptotic freedom; the theory of QCD becomes more and more like a free field theory as we increase the momentum scale. The quantities $\gamma_\pm$ are the anomalous dimensions associated with the operators $c_\pm$, respectively. They also can be calculated in a perturbative expansion, starting in order g^2, where they originate in graphs where a single gluon is exchanged between fermion lines in the basic four-fermion weak interaction:[19]

$$\gamma_+ = +\frac{g^2}{4\pi^2} + ... \tag{20a}$$

$$\gamma_- = -\frac{g^2}{2\pi^2} + ... \tag{20b}$$

Note that if $\gamma_\pm = 0$, then the combination of derivatives on the left-hand side of the renormalization group equation can be rewritten as a total derivative:

$$\left[\mu\frac{\partial}{\partial\mu} + \beta(g)\frac{\partial}{\partial g}\right] c_\pm(q/\mu, g) = \left[\mu\frac{d}{d\mu}\right] c_\pm = 0, \tag{21}$$

simply expressing the fact that $c_\pm$ does not change under a change of momentum scale if the anomalous dimensions are zero. In the case at hand, as we have just seen, the anomalous dimensions are non-zero and the operator coefficients $c_\pm$ change with scale.

We now proceed to solve this renormalization group equation.[20] The method of solution that follows looks like it is pulled out of the hat, but bear with me.

We begin by defining the quantity $\bar{g}$ through an integral:

$$\ln(q/\mu) \equiv \int_{g}^{\bar{g}(q/\mu,g)} \frac{dx}{\beta(x)} \tag{22}$$

with $\bar{g}(1,g) = g$. The quantity $\bar{g}$, which is dimensionless, can only be a function of the ratio of the momentum scales q and μ and the coupling g at the reference scale μ; it is just the "running coupling" that is familiar to all of us. To see this, let us put it in a more familiar form by looking at the situation when $\bar{g}$ is small, so that we can use the perturbative result for $\beta(x) = \frac{x^3}{48\pi^2}(2n_f - 33) + \ldots$ under the integral in Eq. (22). If we take the first term in this expansion we obtain on performing the integral,

$$\ln q/\mu = \frac{48\pi^2}{2n_f - 33}\frac{1}{2}\left(\frac{1}{g^2} - \frac{1}{\bar{g}^2}\right) \tag{23}$$

or

$$\alpha_s(q^2) \equiv \frac{\bar{g}^2}{4\pi} = \frac{\alpha_s(\mu^2)}{1 + \frac{33-2n_f}{12\pi}\alpha_s(\mu^2)\ln q^2/\mu^2}. \tag{24}$$

This is the standard expression for the running of α_s when it is small.

The definition of $\bar{g}$ in Eq. (22) is perfectly general; it simply reduces to the standard form in the small coupling region upon using lowest order perturbation theory for $\beta(x)$. Moreover, it is relevant to solving our equation since it exactly satisfies the renormalization group equation with zero anomalous dimensions:

$$\left(\mu\frac{\partial}{\partial\mu}+\beta(g)\frac{\partial}{\partial g}\right)\bar{g}(q/\mu,g)=0, \tag{25}$$

as may be seen by applying the differential operator on the left-hand side of this equation to both sides of Eq. (22).

Now we are ready to solve the full equations for $c_\pm$. The solution of Eq. (18) is:

$$c_\pm(q/\mu,g)=c_\pm(1,\bar{g})\exp\left(\int_g^{\bar{g}}-\frac{\gamma_\pm(x)dx}{\beta(x)}\right), \tag{26}$$

as can be seen directly by substitution and employing Eq. (25) together with the fact that the derivative of an integral with respect to its upper limit of integration is just the integrand evaluated at that point.

We go again to perturbation theory to evaluate the integral in the exponent of Eq. (26):

$$\int_g^{\bar{g}}-\frac{\gamma_+(x)dx}{\beta(x)}=\frac{6}{33-2n_f}\ln\frac{\bar{g}^2}{g^2}. \tag{27}$$

Therefore,

$$c_+(q/\mu,g)=c_+(1,\bar{g})\left(\frac{\bar{g}^2}{g^2}\right)^{\frac{6}{33-2n_f}}, \tag{28}$$

and similarly,

$$c_-(q/\mu,g)=c_-(1,\bar{g})\left(\frac{\bar{g}^2}{g^2}\right)^{\frac{-12}{33-2n_f}}. \tag{29}$$

We are interested in what happens between a momentum scale

which is characteristic of the decaying hadron (which we take to be μ) and the weak scale, M_W (which we take to be q). Moreover, if we had also taken our reference momentum scale μ to be the weak scale, our coefficients should make the Hamiltonian density correspond to the "bare" density in Eq. (17), *i.e.*, $c_\pm(1,\bar{g}) = 1$. Therefore,

$$c_+(M_W/\mu, g) = \left(\frac{\alpha_s(M_W^2)}{\alpha_s(\mu^2)}\right)^{\frac{6}{33-2n_f}} \tag{30}$$

and

$$c_-(M_W/\mu, g) = \left(\frac{\alpha_s(M_W^2)}{\alpha_s(\mu^2)}\right)^{\frac{-12}{33-2n_f}} . \tag{31}$$

This, finally, is our long-sought result for the coefficients $c_\pm(M_W/\mu, g)$.

When we recall that $\alpha_s(q^2)$ runs down as the momentum scale goes up, we see that $c_+(M_W/\mu, g) < 1$ and $c_-(M_W/\mu, g) > 1$. In fact, there is the simple relation

$$c_+^2 c_- = 1 \tag{32}$$

(which is traceable to the factor of -2 between the anomalous dimensions of the corresponding operators), so that one of the corresponding terms in the weak Hamiltonian is necessarily suppressed if the other is enhanced by the effects of QCD.

- Penguins

Before using these results to look at the overall picture of decays in the spectator model, let us take a brief look at the additional operators introduced by QCD, the "penguins". A set of lowest order graphs which contribute to the existence of "penguin" operators relevant to various quark decays is shown in Figure 3.

On the upper left is a one loop, "penguin" graph relevant to strange quark decay (and in particular, to neutral K decay). Once the loop integral is performed this diagram contributes to an effective operator whose space-time structure is $(V-A)\times V$, or equivalently a mixture of $(V-A)\times(V-A)$ and $(V-A)\times(V+A)$. The latter operator has a structure that is not in the original weak Hamiltonian density. Arguments have been made that although its relative coefficient is small, the corresponding operator has a big matrix element in K decays and that it contributes a large part of the experimentally observed amplitude.[21] This is a subject still very much under debate.[22]

The diagram on the upper right shows a potential "penguin" in Cabibbo suppressed charm decays. Estimates generally put its strength well below that from ordinary graphs which contribute to the same process.

In bottom decay, however, it may be possible to have processes (Cabibbo suppressed to be sure) where "penguin" diagrams give rise to contributions comparable to, or maybe even larger than, those of ordinary tree level graphs.[23] The bottom portion of Figure 3 shows a possible example. The "penguin" diagram on the lower left contributes an effective Hamiltonian density:

$$\mathcal{H} = \frac{G_F}{\sqrt{2}}\frac{\alpha_s}{3\pi}\, V_{tb}\, V_{ts}^*\, \ln(m_t^2/m_c^2)\; \bar{s}\gamma_\mu(1-\gamma_5)b\bar{u}\gamma^\mu u, \qquad (33)$$

whereas the usual spectator diagram (aside from factors of $c_\pm$, which are close to unity) corresponds to

$$\mathcal{H} = \frac{G_F}{\sqrt{2}}\, V_{ub}\, V_{us}\; \bar{u}\gamma_\mu(1-\gamma_5)b\; \bar{s}\gamma^\mu(1-\gamma_5)u. \qquad (34)$$

The "penguin" loses to the spectator graph because of the

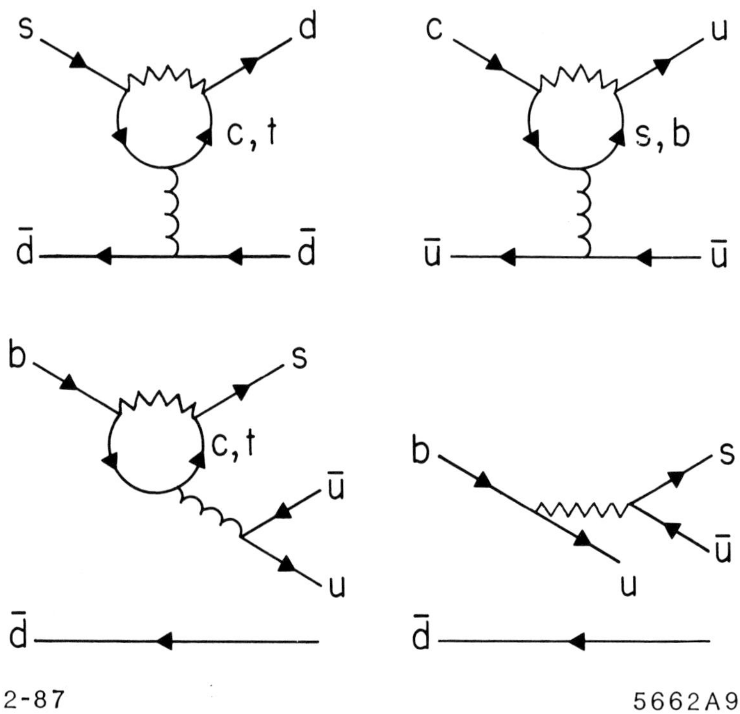

Fig. 3

Set of lowest order "penguin" graphs contributing to strange quark decay (upper left), Cabibbo suppressed charm quark decay (upper right), and Cabibbo suppressed bottom quark decay (lower left). Also shown is a spectator graph which also contributes to Cabibbo suppressed bottom quark decays (lower right).

$\frac{\alpha_s}{3\pi}\ln(m_t^2/m_c^2)$ that arises from having one loop and the presence of the gluon, but it wins because of the Cabibbo (or more exactly, Kobayashi-Maskawa) factor $V_{tb}V_{ts}^*$, which involves zero and one generation jumps, as compared to $V_{ub}V_{us}$, which involves two and one generation jumps, respectively. Depending in part on how small V_{ub} is (something still not known), it could well be that the spectator graph gives the lesser of the two contributions. Then, for example, in the decays $B_d \to K^+\pi^-$ or $B_s \to \phi K^0$ the "penguin" contribution may be dominant.[24]

- Decays in the Spectator Model

What does all this mean numerically for the decay of the various quark flavors? First consider the strange quark. The statement that $c_- > 1$ corresponds to the enhancement of the $\Delta I = 1/2$ amplitude in strange particle decay, which is what one desires in order to be in accord with experiment. However, it already requires some stretching to get a factor of 3 or 4 in the amplitude, while what is needed is something like a factor of 20. Another piece of physics, perhaps "penguins" (see the discussion above), is required in addition to the QCD enhancement of c_-.

For the charm quark, if we set $\mu = m_c$, we find $c_- \sim 2$ and $c_+ \sim 1/\sqrt{2}$. At the quark level the Cabibbo allowed decay channels are $c \to s\bar{e}\nu_e$, $c \to s\bar{\mu}\nu_\mu$, and $c \to s\bar{d}u$. In the spectator model, all charmed hadrons would have the same lifetime and the same semileptonic branching fraction, which would be identified with that for the charm quark as if it decayed in isolation from other hadronic constituents:

$$BR_{\text{semileptonic}} \sim \frac{1}{2 + 2c_+^2 + c_-^2} \sim 14\%. \tag{35}$$

For the bottom quark, with $\mu = m_b$, the QCD enhancement (suppression) of c_- (c_+) is less than that for charm: $c_- \sim 1.5$ and $c_+ \sim 0.8$. In this case we have an expanded list of decay channels at the quark level: $b \to ce\bar{\nu}_e$, $b \to c\mu\bar{\nu}_\mu$, $b \to c\tau\bar{\nu}_\tau$, $b \to cd\bar{u}$, and $b \to cs\bar{c}$. We have neglected decays where the final c quark is replaced by a u quark (using the experimental result[1] that $b \to u/b \to c$ is small). The corresponding semileptonic branching ratio is

$$BR_{\text{semileptonic}} \sim \frac{1}{2.2 + 1.2(2c_+^2 + c_-^2)} \sim 15\%, \tag{36}$$

where the semileptonic decays involving $c\bar{\tau}\nu_\tau$ and $c\bar{c}s$ have been given an approximate phase space weight which is 0.2 times that for $c\bar{e}\nu_e$.

On the one hand, as noted at the beginning of this Section, these results do not agree with experiment, e.g., the D^0 and the D^+ lifetimes differ by a factor of two or so, the average B semileptonic branching ratio is about 12%, *etc.*[1,2] On the other hand, before fixing up the shortcomings of the spectator model,[18] I emphasize that this is not so bad – I only wish that I was able to calculate so simply everything else involving strong interactions to a factor of two or better in the rate! The spectator model provides a semi-quantitative basis for calculating the weak decays of heavy quarks, and is the starting point for the improvements needed to obtain a detailed understanding when we take into account the effects of the other quarks and gluons present in the initial or final state.

4. THE TAU LEPTON

All the properties of the tau are consistent with its being a third generation lepton, *i.e.*, just another copy of the electron and muon, albeit much heavier.[25] The front-back asymmetry measurements in $e^+e^- \to \mu^+\mu^-$ determine the g_A coupling of the Z to $\bar{\tau}\tau$ so as to make the tau the lower component of a weak doublet. The other member of the $SU(2)$ doublet, the tau neutrino, must be distinct from the electron and muon neutrinos. The upper bound on the tau neutrino mass has been lowered to 70 MeV, below the mass of the charged lepton of the previous generation.[26]

The recent measurement of the charged lepton momentum spectrum in leptonic tau decay by the MAC collaboration gives a new world average[26] for the Michel ρ parameter for τ decay of 0.73 ± 0.07, to be compared with the 0.75 expected for a pure $V - A$ coupling of the W at the $\bar{\tau}\nu_\tau$ vertex. While much improved over the situation of a few years ago, one can exclude only $> 47\%$ $V + A$ at 95% confidence level.

The clean separation of τ events at PEP and PETRA has allowed the accurate measurement of the distribution of charged-prong multiplicity in its decay. The present world average numbers for the inclusive topological branching ratios[26] are $86.8 \pm 0.3\%$ for one-prong decays and $13.1 \pm 0.3\%$ for three-prong decays of the tau. The five -prong branching ratio is $\sim 0.14\%$, divided approximately equally between decays with and without extra neutrals.

The advent of vertex detectors has given us accurate measurements of the tau lifetime. Until this year the best measurements gave numbers near 0.29 picoseconds. With new measurements from ARGUS, CLEO, HRS, and MAC, this number has moved up somewhat to[26] 0.305 ± 0.009 picoseconds, with the error obtained from combining statistical and systematic errors in each measurement and

ELIHU BURRITT LIBRARY
CENTRAL CONNECTICUT STATE UNIVERSITY
NEW BRITAIN, CONNECTICUT 06050

then averaging. For later reference we note that

$$\begin{aligned}\tau_\tau &= BR(\tau \to \nu_\tau e^- \bar{\nu}_e)\left(\frac{M_\mu}{M_\tau}\right)^5 \tau_\mu \\ &= BR(\tau \to \nu_\tau e^- \bar{\nu}_e)\ (1.595 \text{ picoseconds}).\end{aligned} \tag{37}$$

The present world average lifetime of the tau then corresponds to an electronic branching fraction of $19.1 \pm 0.6\%$, about 2σ from the direct measurement (see below).

- Leptonic Decays

Neglecting the mass of the electron, the width for the decay $\tau \to \nu_\tau e^- \bar{\nu}_e$ is

$$\Gamma(\tau \to \nu_\tau e^- \bar{\nu}_e) = \frac{G_F^2 M_\tau^5}{192\pi^3} = \frac{1}{1.595 \text{ picoseconds}} \tag{38}$$

Taking account of the mass of the muon, we have

$$\Gamma(\tau \to \nu_\tau \mu^- \bar{\nu}_\mu) = \frac{G_F^2 M_\tau^5}{192\pi^3}\, F(M_\mu/M_\tau), \tag{39}$$

where $F(M_\mu/M_\tau) = 0.97$. The present world average values for $\text{BR}(\tau \to \nu_\tau e^- \bar{\nu}_e)$ and $\text{BR}(\tau \to \nu_\tau \mu^- \bar{\nu}_\mu)$ among experiments where some have assumed, and others not, the factor of 0.97 between them, are[26] $17.9 \pm 0.4\%$ and $17.5 \pm 0.3\%$, respectively.

Note that leptonic decays like $\tau \to e\bar{e}e$, $\tau \to e\bar{\mu}e$, $\tau \to \mu\bar{e}e$, *etc.*, are forbidden in the standard model, as they involve lepton flavor changing neutral currents. The present upper limits on branching ratios for such processes have been reduced to the level of a few times 10^{-5} by the ARGUS collaboration.[27] As these are decays which involve only charged particles which must reconstruct to the beam energy and the tau mass, the searches (up to now) have been straightforward ones which have turned up no candidates in the correct mass and energy bin. The limit essentially depends on the number of tau's produced, a very large sample of which BEPC should get for "free" when running above open charm (and therefore tau) threshold near the design luminosity.

- Semi-Hadronic Decays Through the Vector Current

We are now interested in decays of the form $\tau \to \nu_\tau + W^-_{\text{virtual}}$, where the virtual W^- couples through a vector current to quarks, and thence to hadrons. Here we may make essential use the fact that the vector current in strangeness non-changing weak processes is an isospin rotation of the isovector part of the (vector) current of electromagnetism (*i.e.*, the conserved vector current hypothesis, CVC). There is consequentially a relation between the tau decay rate into ν_τ and a final hadronic state, f^-, and an integral over electron-positron annihilation cross sections for production of the corresponding hadronic final state, f^0. More precisely, the relation is:

$$\frac{\Gamma(\tau^- \to \nu_\tau f^-)}{\Gamma(\tau^- \to \nu_\tau e^- \bar{\nu}_e)} =$$

$$\cos^2\theta_c \, \frac{3}{2\pi\alpha^2 M_\tau^8} \int_0^{M_\tau^2} dQ^2 Q^2 (M_\tau^2 - Q^2)^2 (M_\tau^2 + 2Q^2)\sigma_{e^+e^- \to f^0}(Q^2), \tag{40}$$

or

$$\frac{\Gamma(\tau^- \to \nu_\tau f^-)}{\Gamma(\tau^- \to \nu_\tau e^- \bar{\nu}_e)} = 2\cos^2\theta_c \int_0^1 dx (1-x)^2 (1+2x) \frac{\sigma_{e^+e^- \to f^0}}{\sigma_{pt}}, \tag{41}$$

in terms of the scaled variable $x = Q^2/M_\tau^2$ and the point cross section, $\sigma_{pt}(Q^2) = 4\pi\alpha^2/3Q^2$. As the vector, strangeness non-changing current is even under G parity, the final state f contains an even number of pions.

The simplest and most important example of such a decay is $\tau^- \to \nu_\tau \pi^- \pi^0$. Here the strength of the weak vector current coupling to $\pi^-\pi^0$ can be related, using CVC, to that of the electromagnetic current to $\pi^+\pi^-$.

The $\pi\pi$ system is very much dominated by the ρ resonance and an approximate result can be obtained[28] from computing $\tau^- \to \nu_\tau \rho^-$ in the narrow resonance approximation, with the coupling to the vector current extracted from $e^+e^- \to \rho^0$ experiments. A more accurate result is obtained by integrating directly over the $e^+e^- \to \pi\pi$ cross section using Eq. (41); one predicts[29]

$$\frac{\Gamma(\tau^- \to \nu_\tau \pi^- \pi^0)}{\Gamma(\tau^- \to \nu_\tau e^- \bar{\nu}_e)} = 1.23, \tag{42}$$

with an error which is due principally to the possible overall normalization error in the measurement of the e^+e^- cross sections. The agreement of Eq. (42) with experiment[26] provides a successful test of CVC.

A closely related result follows for the Cabibbo suppressed decay $\tau^- \to \nu_\tau (K\pi)^-$. Just as the $\pi\pi$ system is dominated by the ρ, we expect this decay to be dominated by the $K^*(890)$, as is indeed observed experimentally. The rate can be obtained from that for $\tau^- \to \nu_{/tau} \rho^-$ by multiplying by $\tan^2\theta_c$ due to the strangeness-changing current, a factor to correct for phase space, and a correction for SU(3) breaking, yielding[30]

$$\frac{\Gamma(\tau^- \to \nu_\tau K^{*-})}{\Gamma(\tau^- \to \nu_\tau e^- \bar{\nu}_e)} = 0.064. \tag{43}$$

Because of the decay $K^0_s \to \pi^+\pi^-$, two ninths of these will appear with as three-prong decays.

A more complicated situation is presented by $\tau^- \to \nu_\tau (4\pi)^-$. While this decay proceeds through the vector current and we can again relate the decay rate to an integral over e^+e^- cross sections, there are now two possible charge configurations in e^+e^- annihilation, $2\pi^- 2\pi^+$ and $\pi^-\pi^+ 2\pi^0$, and two in τ decay as well, $\nu_\tau 2\pi^-\pi^+\pi^0$ and $\nu_\tau \pi^- 3\pi^0$. The constraint of being produced by different I_3 components of the same $I = 1$ weak current forces one linear relation between the rates for producing the respective charge states at each 4π invariant mass. As a result, not only can we write the $\tau^- \to \nu_\tau (4\pi)^-$ total decay rate as an integral over $\sigma_{e^+e^- \to 4\pi}$, but we have separately:

$$\frac{\Gamma(\tau^- \to \nu_\tau \pi^- 3\pi^0)}{\Gamma(\tau^- \to \nu_\tau e^- \bar{\nu}_e)} = \frac{3}{2\pi\, \alpha^2 M_\tau^8} \int_0^{M_\tau^2} dQ^2 Q^2 (M_\tau^2 - Q^2)^2 \times (M_\tau^2 + 2Q^2) \left[\frac{1}{2} \sigma_{e^+e^- \to 2\pi^- 2\pi^+}(Q^2) \right] \tag{43a}$$

and

$$\frac{\Gamma(\tau^- \to \nu_\tau 2\pi^-\pi^+\pi^0)}{\Gamma(\tau^- \to \nu_\tau e^- \bar{\nu}_e)} = \frac{3}{2\pi\, \alpha^2 M_\tau^8} \int_0^{M_\tau^2} dQ^2 Q^2 (M_\tau^2 - Q^2)^2 (M_\tau^2 + 2Q^2) \times \left[\frac{1}{2} \sigma_{e^+e^- \to 2\pi^- 2\pi^+}(Q^2) + \sigma_{e^+e^- \to \pi^+\pi^- 2\pi^0}(Q^2) \right] \tag{43b}$$

The e^+e^- cross sections are dominated by the ρ' resonance. Very rough results may be obtained by approximating the integrand using a single narrow resonance, but this is somewhat dangerous in that the mass ($\sim$ 1550 MeV) and width ($\sim$ 300 MeV) of the ρ' make the factor $(M_\tau^2 - Q^2)^2(M_\tau^2 + 2Q^2)$ vary strongly over the resonance, considerably distorting its shape in τ decay.[30] Integrating directly over the e^+e^- cross sections gives:

$$\frac{\Gamma(\tau^- \to \nu_\tau \pi^- 3\pi^0)}{\Gamma(\tau^- \to \nu_\tau e^- \bar{\nu}_e)} = 0.055, \tag{44a}$$

$$\frac{\Gamma(\tau^- \to \nu_\tau 2\pi^- \pi^+ \pi^0)}{\Gamma(\tau^- \to \nu_\tau e^- \bar{\nu}_e)} = 0.275, \tag{44b}$$

and the sum,

$$\frac{\Gamma(\tau^- \to \nu_\tau (4\pi)^-)}{\Gamma(\tau^- \to \nu_\tau e^- \bar{\nu}_e)} = 0.33. \tag{44c}$$

The result for the one charged-prong decay $\tau^- \to \nu_\tau \pi^- 3\pi^0$ in Eq. (44a) is rather certain as it depends only on the integral over the well measured cross section for $e^+e^- \to 2\pi^+ 2\pi^-$. The numerical result in Eq. (44b) is more uncertain because of the poorly measured cross section for $e^+e^- \to \pi^+\pi^- 2\pi^0$ above 1.4 GeV, although it is in excellent accord with the direct measurement[26] of the branching ratio for this process in tau decay. Moreover, a subset of these decays arise from the process $\tau^- \to \nu_\tau \omega \pi^-$. This has been observed in tau decay,[31] permitting a test of CVC by comparing this data to that for $e^+e^- \to \omega\pi^0$ as a function of invariant $\omega\pi$ mass. CVC passes the test successfully.[31]

Decays involving more than four pions are very small, as is seen in particular for $\tau^- \to \nu_\tau (6\pi)^-$. In this case we can use the constraints that follow from the six pion system having total isospin one:

$$0 \leq f_1 = \frac{\pi^- 5\pi^0}{\text{all } (6\pi)^-} \leq 9/35 \ , \tag{45a}$$

$$1/5 \le f_3 = \frac{2\pi^-\pi^+3\pi^0}{\text{all}\,(6\pi)^-} \le 4/5 \ , \tag{45b}$$

$$1/5 \le f_5 = \frac{3\pi^-2\pi^+\pi^0}{\text{all}\,(6\pi)^-} \le 4/5 \ . \tag{45c}$$

From Eq. (45c),

$$BR(\tau^- \to \nu_\tau(6\pi)^-) \le 5BR(\tau^- \to \nu_\tau 3\pi^- 2\pi^+\pi^0) \le 0.005 \tag{46}$$

and from Eqs. (45a) and (45c),

$$BR(\tau \to \nu_\tau \pi^- 5\pi^0) \le (9/7)BR(\tau^- \to \nu_\tau 3\pi^- 2\pi^+\pi^0) \le 0.001 \tag{47}$$

Nor can decays involving eta mesons be appreciable. The simplest example proceeding through the vector current in the standard model is $\tau \to \nu_\tau \eta 2\pi$. Only within the past year have data for $\sigma_{e^+e^- \to \eta\pi^+\pi^-}$ been published which cover the full energy region needed for performing the integral in Eq. (41). The cross sections are at most a few nanobarns, and carrying out the calculation[32] gives a branching ratio for $\tau^- \to \nu_\tau \eta \pi^- \pi^0$ of 0.15%. Even taking 5 nb as a generous upper limit on the $e^+e^- \to \eta\pi^+\pi^-$ cross section above 1.4 GeV produces an upper limit of 0.24%.

- Semi-Hadronic Decays Through the Axial-Vector Current

The simplest decay that proceeds through the axial-vector current is $\tau^- \to \nu_\tau \pi^-$, which can be precisely calculated because the strength of the pion's coupling to the axial-vector current is directly determined in the leptonic decay of the pion. Exactly the same quantity, $f_\pi \cos\theta_c$, is relevant in $\tau^- \to \nu_\tau \pi^-$:

$$\frac{\Gamma(\tau^- \to \nu_\tau \pi^-)}{\Gamma(\tau^- \to \nu_\tau e^- \bar{\nu}_e)} = \frac{(f_\pi \cos\theta_c)^2}{M_\tau^2} 12\pi^2 \left[1 - \frac{m_\pi^2}{M_\tau^2}\right]^2 = 0.607 \; . \qquad (48)$$

The related, Cabibbo suppressed decay, $\tau^- \to \nu_\tau K^-$, also has the quantity of relevance, $f_K \sin\theta_c$, directly measured elsewhere (in the dominant decay of the charged kaon), yielding

$$\frac{\Gamma(\tau^- \to \nu_\tau K^-)}{\Gamma(\tau^- \to \nu_\tau e^- \bar{\nu}_e)} = \frac{(f_K \sin\theta_c)^2}{M_\tau^2} 12\pi^2 \left[1 - \frac{m_K^2}{M_\tau^2}\right]^2 = 0.0395 \; . \qquad (49)$$

The decay $\tau \to \nu_\tau (3\pi)^-$ can not be calculated with great accuracy, although it was expected[28] to be dominated by the A_1 resonance with a branching ratio of order 10%. Excellent measurements of the decay $\tau^- \to \nu_\tau 2\pi^-\pi^+$ now exist with a branching ratio[26] of $6.6 \pm 0.4\%$ and showing a clear A_1 resonance in the $J^P = 1^+$ partial wave, decaying almost entirely through $\pi\rho$.

Measurements of the other charge configuration, $\tau^- \to \nu_\tau \pi^- 2\pi^0$ are not nearly so good.[26] We may gain information on it, however, from the constraint of having total isospin one for the $(3\pi)^-$ system; it follow that independent of the dynamics

$$1/5 \leq f_1 = \frac{\pi^- 2\pi^0}{\text{all}\,(3\pi)^-} \leq 1/2 \; , \qquad (50a)$$

and

$$1/2 \leq f_3 = \frac{2\pi^-\pi^+}{\text{all}\,(3\pi)^-} \leq 4/5. \qquad (50b)$$

Thus the number of three-prong tau decays must be greater than that of one-prong decays for $\tau^- \to \nu_\tau (3\pi)^-$. The equality of the two charge configurations occurs when two of the pions are in an isospin one state, *i. e.*, precisely the case of relevance where the quasi-two body $\pi\rho$ channel dominates. The branching ratio for $\tau^- \to \nu_\tau \pi^- 2\pi^0$ is, if anything, measured[26] to be slightly bigger than that for $\tau^- \to \nu_\tau 2\pi^- \pi^+$, but is consistent with equality within the statistical and especially the systematic errors of background subtractions.

Other decays through the axial-vector current are measured to be small, like the Cabibbo suppressed decay $\tau \to \nu_\tau (K\pi\pi)^-$, or can be bounded, like $\tau^- \to \nu_\tau (5\pi)^-$, from the very small number of five-prong decays using isospin.[30]

- The "Missing" One-Prongs

Having done all the work of enumerating the decay modes of the tau in the standard model, we come to the one outstanding problem. We have discussed the decays $\tau \to \nu_\tau e \bar{\nu}_e$, $\tau \to \nu_\tau \mu \bar{\nu}_\mu$, $\tau \to \nu_\tau \pi$, $\tau \to \nu_\tau 2\pi$, $\tau \to \nu_\tau 3\pi$, and $\tau \to \nu_\tau 4\pi$, plus a number of smaller modes. They occur at the expected rates[26,30] and the sum of their exclusive branching ratios accounts for almost 90% of tau decays. However, there are no modes thought to be of consequence which haven't been included and there is a nagging problem in accounting for all tau decays.[26,30,32] In particular, it is difficult to account for all the one-prong decays of the tau.

In more detail, the problem arises as follows. Consider first three-prong decays, which are shown in Table I. All the theoretical calculations of decay branching ratios are taken from the discussion above and are normalized to that for $\tau \to \nu_\tau e \bar{\nu}_e$, for which we take the world average value of 17.9%. Where theory and experiment can be compared, they are in excellent agreement. Moreover, the sum of all the exclusive decays measurements is in agreement with the inclusive three-prong branching ratio.

The agreement between theory (where there is a prediction of some accuracy) and experiment (where there is a definite measurement) is also very good in the case of tau decays involving one charged prong, as shown in Table II. Note in particular that aside from the leptonic decays (whose branching ratio is used as an input), the excellent agreement between theory and experiment for the semi-hadronic decays $\tau \to \nu_\tau \pi$, $\tau \to \nu_\tau K$, and $\tau^- \to \nu_\tau \pi^- \pi^0$.

The theoretical upper bound on the branching ratio for $\tau^- \to \nu_\tau \pi^- 2\pi^0$ comes from Eq. (50) and the measured rate for $\tau^- \to \nu_\tau 2\pi^- \pi^+$ in Table I. Adding one percent for the sum of small modes not explicitly written in Table II, the sum of the one-prong branching ratios from theory is $< 80\%$, to be compared with the $86.8 \pm 0.3\%$ inclusive one-prong branching ratio measured experimentally. We have accounted for all the purely leptonic modes and all the modes of the form $\tau^- \to \nu_\tau (n\pi)^-$ of any substance, as well as Cabibbo suppressed modes. Where are the remaining of one-prong decays?

There is no reason to suspect that they come from the decays in Table II for which there are no experimental numbers entered. The only two expected to be substantial are $\tau^- \to \nu_\tau \pi^- 2\pi^0$ and $\tau^- \to \nu_\tau \pi^- 3\pi^0$. The former should occur at the same rate ($6.6 \pm 0.4\%$) as $\tau^- \to \nu_\tau 2\pi^- \pi^+$ (see above). The latter is predicted theoretically on the basis of CVC and well measured e^+e^- cross sections; it should be solid. Adding these theoretical branching ratios (and that for $\tau \to \nu_\tau \eta \pi^- \pi^0$) to the second column of Table II gives a sum of $78 \pm 2\%$ for the branching ratios of the exclusive channels. Although with bigger experimental error bars, this is the situation that was already

TABLE I

THREE CHARGED PRONG DECAYS OF THE TAU

Decay Mode	Branching Ratio (%) Theory	Experiment[26]
$\tau^- \rightarrow \nu_\tau 2\pi^-\pi^+$		6.6 ± 0.4
$\tau^- \rightarrow \nu_\tau 2\pi^-\pi^+\pi^0$	4.9	5.2 ± 0.5
$\tau^- \rightarrow \nu_\tau (K\pi)^-$	0.3	0.3 ± 0.1
$\tau^- \rightarrow \nu_\tau K^-\pi^-\pi^+(\pi^0)$		0.22 ± 0.14
$\tau^- \rightarrow \nu_\tau K^-K^+\pi^-$		0.22 ± 0.14
$\tau^- \rightarrow \nu_\tau 2\pi^-\pi^+3\pi^0$	< 0.4	
TOTAL EXCLUSIVE		12.5 ± 0.7
TOTAL INCLUSIVE		13.1 ± 0.3

TABLE II

ONE CHARGED PRONG DECAYS OF THE TAU

Decay Mode	Branching Ratio (%) Theory	Experiment[26]
$\tau^- \rightarrow \nu_\tau e^-\bar{\nu}_e$	17.9 (Input)	17.9 ± 0.4
$\tau^- \rightarrow \nu_\tau \mu^-\bar{\nu}_\mu$	17.4	17.5 ± 0.3
$\tau^- \rightarrow \nu_\tau \pi^-$	10.9	10.9 ± 0.6
$\tau^- \rightarrow \nu_\tau \pi^-\pi^0$	22.0	22.1 ± 1.1
$\tau \rightarrow \nu_\tau \pi\ 2\pi^0$	≤ 7.0	
$\tau^- \rightarrow \nu_\tau \pi^- 3\pi^0$	1.0	
$\tau^- \rightarrow \nu_\tau \pi^- 4\pi^0$	< 0.1	
$\tau^- \rightarrow \nu_\tau \pi^- 5\pi^0$	< 0.1	
$\tau^- \rightarrow \nu_\tau (K\overline{K})^-$		< 0.26
$\tau^- \rightarrow \nu_\tau (K\overline{K}\pi)^-$	< 0.5	
$\tau^- \rightarrow \nu_\tau \eta\pi^-\pi^0$	0.15	
$\tau^- \rightarrow \nu_\tau K^-$	0.7	0.7 ± 0.2
$\tau^- \rightarrow \nu_\tau (K\pi)^-$	0.9	1.1 ± 0.3
TOTAL EXCLUSIVE	< 80	
TOTAL INCLUSIVE		86.8 ± 0.3

noted over two years ago.[30]

A sharpening of the problem has come from the increased accuracy with which many of the main branching ratios are known and from measurements of the branching ratio for $\tau \rightarrow \pi^-$+multineutrals, which includes $\tau^- \rightarrow \nu_\tau \pi^- 2\pi^0$, $\tau^- \rightarrow \nu_\tau \pi^- 3\pi^0$, $\tau^- \rightarrow \nu_\tau \eta \pi^- \pi^0$), *etc.* This measurement gives consistently higher values[33] than is expected theoretically from the sum of the components. (It is difficult to reconstruct the individual multi-neutral decay modes.) For a time it was suspected that modes involving eta's were the "missing" modes,[34] but this was recently shown to be untenable[32] in the standard model.

In the past few months, the HRS collaboration claimed that there were decays involving the eta at the 5% level in exactly the mode, $\tau^- \rightarrow \nu_\tau \pi^- \eta$, that is not expected in the standard model.[35] (The $\eta\pi$ system, which is G odd, has natural spin-parity and in the standard model it must come from the vector current, which is G even; we have by definition a process that involves a second class current.) Within the standard model this should happen at a level of roughly α^2 in the rate when compared to processes arising through the usual first class currents and such a decay would be completely negligible.[36] Other experiments have since contradicted this result.[37]

With other conventional (or even unconventional) modes that contribute to one-prong tau decays severely limited, there are two possible ways out of the problem. The first is that the branching ratio for $\tau \rightarrow \nu_\tau e \bar{\nu}_e$, which we took to be 17.9% (and to which we normalized all our theoretical predictions), should be $\approx$ 19%. This would scale up all the predicted branching ratios by $\approx$ 6% and make the sum for theory agree with the measurement of the one-prong inclusive branching ratio. Of course it is one thing to scale up the theory by a common factor, as all the predictions are normalized to the single mode $\tau^- \to \nu_\tau e^- \bar{\nu}_e$, and another to get all the individually measured experimental branching ratios to change, all in the same direction! Nevertheless this is what the present world average value of the tau lifetime indicates (by 2σ) is the solution, and it is the easiest (purely from the point of view of not straining the standard model) way out.

Second, there is the possibility of something new. This is the direction in which the directly measured branching ratios, and particularly that for $\tau \to \pi^- +$ multineutrals, points. But what? It is hard to find a scenario for this situation which is not very contrived (especially if it is not to be in conflict with other existing experiments).

The experiments necessary to decide between these two possibilities are hard as well. I suspect that although the puzzle has been sharpened considerably in the past year, it will take some additional measurements over the next several years by detectors well-instrumented for detection of neutrals to resolve it.

5. HEAVY QUARK SPECTROSCOPY

The spectroscopy of heavy quark systems is the showcase of our understanding of hadron physics. It is sometimes even advertised as the "hydrogen atom of strong interactions".

We do indeed have a fundamental gauge theory of the strong interactions in Quantum Chromodynamics (QCD). This theory in principle explains the vast body of data that has been accumulated over the past dozen years.[38] However, as we will soon see, the connection between the fundamental theory and experimental observables is not (yet) as it is for the electroweak gauge theory, $SU(2) \times U(1)$. The situation we confront is essentially non-perturbative, and the underlying gauge theory is one-step-removed from detailed numerical confrontation with experiment.

What we have at present to accompany the data is more like a phenomenology, inspired or backed-up by QCD. At times it gives us an asymptotic form. At other times it gives an expression for the general structure of some quantity, with free parameters or hadronic matrix elements contained within it. While these latter are determined by QCD in principle, for the moment they are often only approximately calculable (at best). So we take a peek at the data and 'adjust' the parameters, thereby learning something about the nature of the solution of QCD. Then we predict additional quantities and iterate the whole process again.

This is then a place where theory and experiment intertwine; basic theory, models inspired by theory, and experiment meet and influence one another. It is quite different from the situation in the electroweak theory where there is a well-defined and clean set of perturbative predictions to compare with experiment. In one sense this is frustrating, as one would like clean and decisive tests of the underlying theory. In another sense, this is what makes it exciting and makes the subject still worth pursuing: the interplay between theory and experiment is interesting in itself, and we often learn things which are applicable either as techniques or as results in other areas as well.

In fact, progress has been made and continues to be made.[39] Eventually, one has every reason to believe that we will be able to calculate the "potential" from first principles, presumably using lattice techniques. Everything then will be predicted starting from the QCD Lagrangian. We have come a long way in this direction already,[40] and perhaps in a Charm Symposium of a few(?) years hence we may well no longer need to give such a talk on this subject.

- The Spin Independent Potential

Let's start with the nonrelativistic, spin independent potential. Even the use of the word potential is a bit loose for we are starting with a strong interaction bound state problem and extracting from it an effective two-body, non-relativistic potential. The problem at hand is intrinsically a relativistic field-theoretic one in which the $q\bar{q}$ sector, for example, is coupled to what happens in the $qq\bar{q}\bar{q}$, $q\bar{q}$ + gluon, *etc.*, sectors as well. Some justification for the success of the "naive," non-relativistic approach have recently been given,[41] but simultaneously questions have been raised as to the effect of what is being neglected, and how it changes the relationship between parameters in the underlying theory and the effective potential.[42] There is even a whole, well-developed approach to understanding some of the same body of data through QCD sum rules.[43]

With these questions in mind, we shall proceed to think in terms of a two-body potential obtained by expanding in powers of v^2/c^2. Indeed, such an expansion does make some sense, for the smallness of v^2/c^2 in a system composed of a heavy quark and heavy antiquark encourages us to think in terms of a non-relativistic potential with spin-dependent terms which arise first in order v^2/c^2 and give rise to splittings which are smaller than those between levels of the spin independent potential. In the cases at hand v^2/c^2 is ~ 0.4 for the low-lying charmonium states and $\lesssim 0.1$ for the low-lying bottomonium states. Still, when we go to compare with the calculated energy levels with experiment we need to bear in mind that predictions from alternate potentials that differ by 10 or 20 MeV are not necessarily significant in favoring one potential over another. We may in any case be making (especially for charmonium) approximations to the exact theory in "deriving" a non-relativistic potential which render the resulting model incapable of discriminating differences at this level.

We have good theoretical guidance in two opposite regimes. At short distances, or equivalently large momentum scales, there is the property of asymptotic freedom. The running coupling becomes smaller as we decrease the distance scale at which we work, and the effective potential approaches the lowest order one gluon exchange result,

$$V(r) \to -\frac{4}{3}\frac{\alpha_s}{r} \tag{51}$$

as $r \to 0$.

Note the additional factor of $\frac{4}{3}$ compared to the usual Coulomb interaction; this arises from color. It is so pervasive that it is worth a short derivation, so here it is again.[44] We want the additional factor due to color. It arises from a normalized color singlet quark-antiquark wave function, $\delta_{ij}/\sqrt{3}$ in the initial and final state, a color SU(3) matrix $\lambda^a_{ij}/2$ at each quark-gluon vertex, and a color sum over the (eight) gluons, δ^{ab}, in the gluon propagator. The sum over indices gives a trace:

$$\sum_{a,b=1}^{8}\sum_{i,j,k,l=1}^{3}\frac{\delta_{il}}{\sqrt{3}}\frac{\lambda^a_{ij}}{2}\frac{\delta_{jk}}{\sqrt{3}}\frac{\lambda^b_{kl}}{2}\delta^{ab} = \frac{1}{3}\sum_{a=1}^{8} Tr\left[\frac{\lambda^a}{2}\frac{\lambda^a}{2}\right] = \frac{1}{3}(8)\frac{1}{2} = \frac{4}{3}. \tag{52}$$

The trace of $\frac{\lambda^a}{2}\frac{\lambda^a}{2}$ is just $\frac{1}{2}$, as befits the generators of a Lie algebra (or, as may be checked for the case of the λ matrices of $SU(3)$ directly); thus the ubiquitous factor of 4/3.

That's one regime. The second regime where we have very solid theoretical input is at the other extreme, as $r \to \infty$, and we have confinement of the quarks. From relatively general theoretical arguments we know that the potential behaves as a linear function of the distance:

$$V(r) \to kr. \tag{53}$$

There is a corresponding physical picture of a "color-electric flux tube" joining the quarks. As you pull the quarks apart, the flux tube is increased in length, at the cost of an increase in energy per unit length given by the constant k. The value of k is about 0.2 GeV^2.

Given those two regimes we might hope to construct the full potential. The simplest possibility is to simply add together terms with the correct functional dependence in the two asymptotic regimes. This is basically the Cornell potential,[45]

$$V(r) = \frac{-0.48}{r} + \frac{r}{(2.34\ \text{GeV}^{-1})^2}, \tag{54}$$

with the two coefficients having been adjusted to fit the charmonium spectrum, although the model does a quite adequate job in describing bottomonium as well.

In the late 70's Richardson combined the two behaviors in one form.[46] Here is his potential in configuration space,

$$V(r) = \frac{8\pi}{33 - 2n_f}\,\Lambda\left(\Lambda r - \frac{f(\Lambda r)}{\Lambda r}\right), \tag{55}$$

with

$$f(t) = 1 - 4 \int_1^\infty \frac{dq}{q} \frac{e^{-qt}}{\ell n^2(q^2-1)+\pi^2}, \tag{56}$$

where it looks like two terms. What's going on is more transparent in momentum space where it can be written as one term:

$$\widetilde{V}(q^2) = -\frac{4}{3}\frac{12\pi}{33-2n_f}\frac{1}{q^2 \ell n(1+q^2/\Lambda^2)}. \tag{57}$$

As q^2 goes to infinity, this expression becomes precisely $\frac{4}{3}\alpha_s/q^2$, as required from one gluon exchange. In the other limit of $q^2 \to 0$, one obtains something proportional to $1/q^4$. This may be an unfamiliar behavior in momentum space, but if you Fourier transform back to configuration space, this is just a potential which is linear in r. It is by no means guaranteed that you will get the "right" coefficient to fit the data. Richardson, along with others[47] who proposed modified versions of this potential, showed that you do in fact get a very reasonable, even excellent, description of the data, especially for bottomonium.

Finally, Martin has proposed the potential[48]

$$V(r) = (5.82 \text{ GeV}) \left(\frac{r}{1 \text{ GeV}^{-1}}\right)^{0.104} . \tag{58}$$

This potential, with the absurd power of 0.104, lacks fundamental motivation (as Martin knew very well). We will use it as a kind of straw man, for it also does quite a credible job of fitting the charmonium and bottomonium data. But why?

The reason can be seen in Figure 4. Here you can see the various potentials for comparison purposes. In particular, aside from

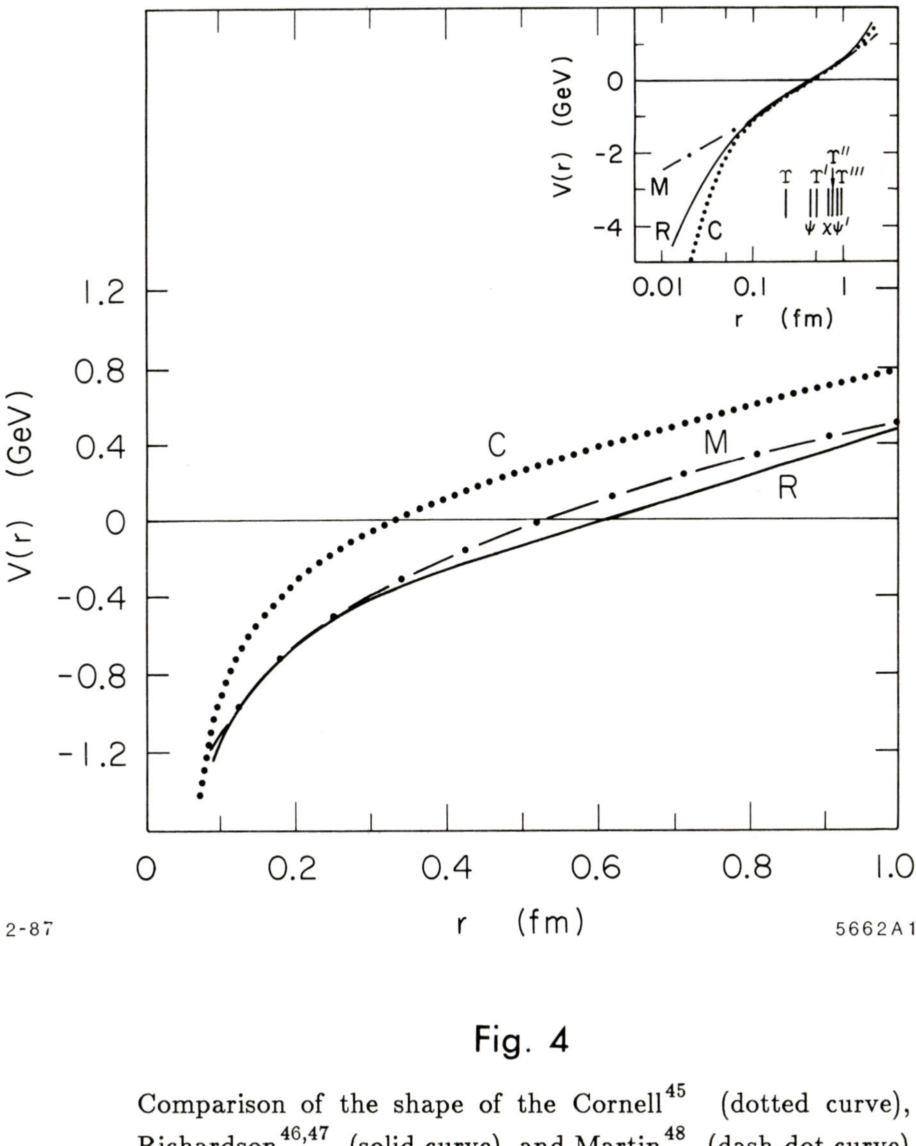

Fig. 4

Comparison of the shape of the Cornell[45] (dotted curve), Richardson[46,47] (solid curve), and Martin[48] (dash-dot curve) potentials. The inset shows the same comparison with the potentials displaced slightly on the vertical scale and a logarithmic horizontal scale, along with the mean radii of some charmonium and bottomonium states.

being displaced vertically from one another a little bit (which you are free to remove by adjusting the quark mass), they all have about the same behavior between 0.1 and 1 fermi. This can be seen even better looking at the inset, where r is given on a logarithmic scale and the potentials have been shifted slightly relative to one another vertically, as discussed above. Also shown are the mean radii of the psi, upsilon, *etc.* These are all between 0.1 and 1 fermi, and that's why the different potentials all can fit the data; the wave functions for these states mostly (but not entirely) live in this region where the potentials coincide.

Thus, where our theoretical insight is best and tells us something very well-defined for the behavior of the spin independent potential, it is mostly irrelevant to the present data. Conversely, the experiments up to now mostly tell us about a region where theory does not have much to say about the spin independent potential. In fact, one can invert the data to obtain a potential[49] which describes what happens from 0.1 to 1 fermi. Within errors, it coincides with what we have just seen in Figure 4.

Even without a particular potential and detailed calculation, we can get a good qualitative idea of what the spectrum of states will look like. In Figure 5a is the familiar spectrum due to a Coulomb potential, which is what we have at short distances. The ground state with $\ell = 0$ is labelled 1S; its radial excitation (labelled 2S) is degenerate in the case of a Coulomb potential with the first set of $\ell = 1$ states (labelled 1P), and so on. As an example of a confining potential which we want at large distance, Figure 5b shows the levels of a three dimensional harmonic oscillator, which is more familiar than a linear potential and turns out also to be a special "boundary"

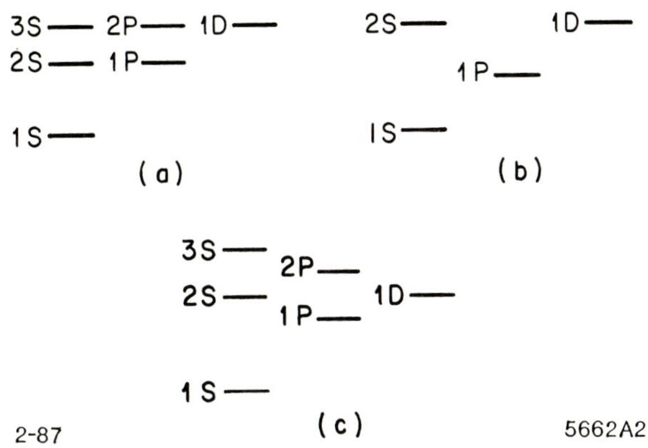

Fig. 5

The spectrum of energy levels in the case of the Coulomb potential (a), the three dimensional harmonic oscillator (b), and a hybrid of the two (c).

case from the point of view of the ordering of levels. What will happen when we combine the two? For the energy levels we will naturally get something in between Figures 5a and 5b. This is shown in Figure 5c. The ordering, starting at the bottom, is 1S, 1P (between the 1S and 2S as for the harmonic oscillator, but closer to the 2S, as it would be degenerate with it for the Coulomb potential), 2S, 1D (above 2S as for Coulomb, but close to it, as it would be degenerate for the harmonic oscillator), *etc.* You can therefore get a qualitative understanding of the spectrum from quite general considerations.

There is a theorem[50] which is quite useful in this regard and puts the qualitative ordering discussed above on a rigorous footing. It states that if $\nabla^2 V(r) > 0$ for all r, something which is true for all suggested potentials, then $E_{nS} > E_{(n-1)P}$. Related theorems are provable for the ordering of levels with other angular momenta.[51]

Each of the potentials discussed above can give a quantitative understanding of the levels of charmonium (Figure 1) and bottomonium (Figure 2) to 30 MeV or better. Even the statement that one flavor independent potential can fit both systems is nontrivial. The agreement between theory and experiment, which is shown in Schindler's lectures,[38] I regard as quite spectacular. It includes not just energy levels, but wave functions at the origin for the nS states as well. Where there is a disagreement, it is difficult to know whether to blame it on the potential or on corrections due to relativistic or other effects which have been left out.

When and how will we be able to distinguish between potentials? The answer appears to be that toponium will provide the crucial system. In Figure 6 is shown the spectrum of toponium[52] corresponding to m_t in the range of 40 to 50 GeV. There are 10 or more nS states

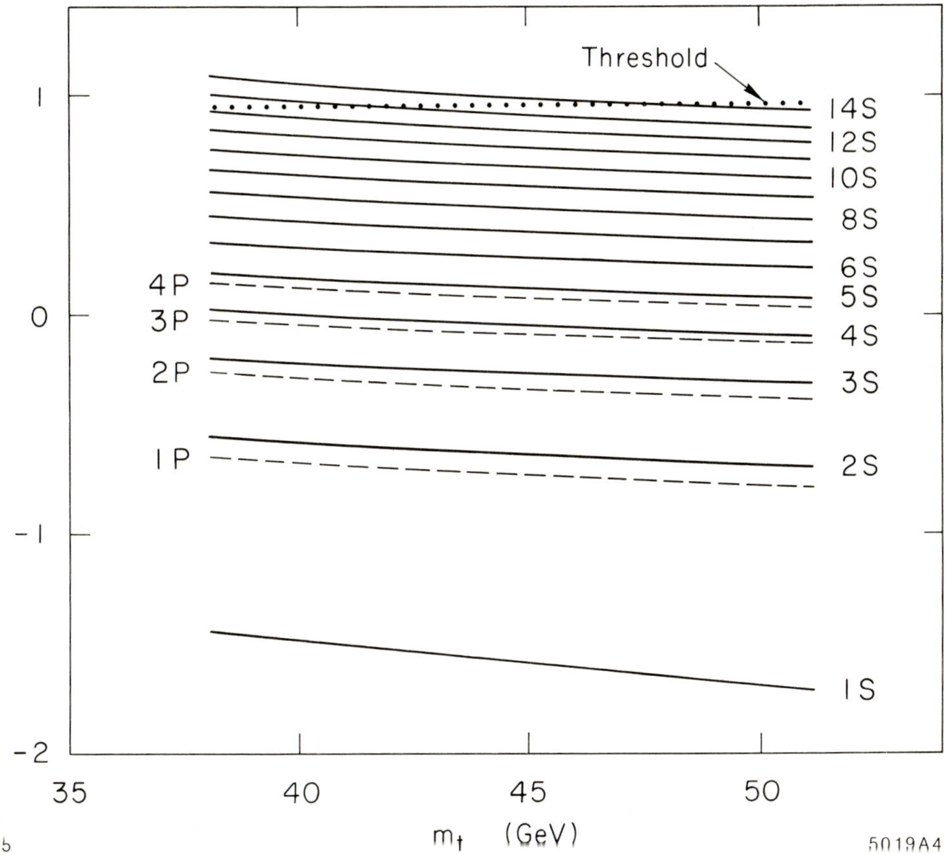

Fig. 6

The spectrum of nS states of toponium obtained[52] from the Richardson potential with m_t in the range of 40 to 50 GeV.

below open top threshold; near that threshold there is one state per 100 MeV.

More important for the physics at hand, aspects of the spectrum of states and of the wave functions at the origin are now sensitive to the behavior of the potential at short distances. The values of the wave function at the origin are shown in Figure 7, with that for the ground state corresponding to a width into electron-positron pairs, which is proportional to the square of the wave function at the origin, of about 9 keV (from the one photon intermediate state alone). This is larger than one would expect from a naive extrapolation from the psi and the upsilon by about a factor of two. We are beginning to see the effect of the $1/r$ term in the potential pulling in the wave function. Higher levels are affected less, as seen in Figure 7, for on average they live at larger distances.

The same physical effect is shown in Table III, with the t quark mass assumed to be 50 GeV. Notice in particular how much the energy of the 1S level is pulled down by the Cornell potential (3 GeV below 2 m_t). This is to be compared with 1.7 GeV for the Richardson and 1.4 GeV for the Martin potentials. Correspondingly the radius of the 1S state is much smaller for the Cornell potential and the 2S to 1S difference much bigger. Even more dramatic is the comparison of the wave function at the origin for the 1S state, where the Cornell result is about 3 times that for Richardson and 9 times that for Martin. Remember, the predicted electron-positron width goes like these numbers squared!

Before leaving this subject, we should note that this same property makes toponium a fairly sensitive place to look for extra short range forces. A good example is the presence of an extra term in the

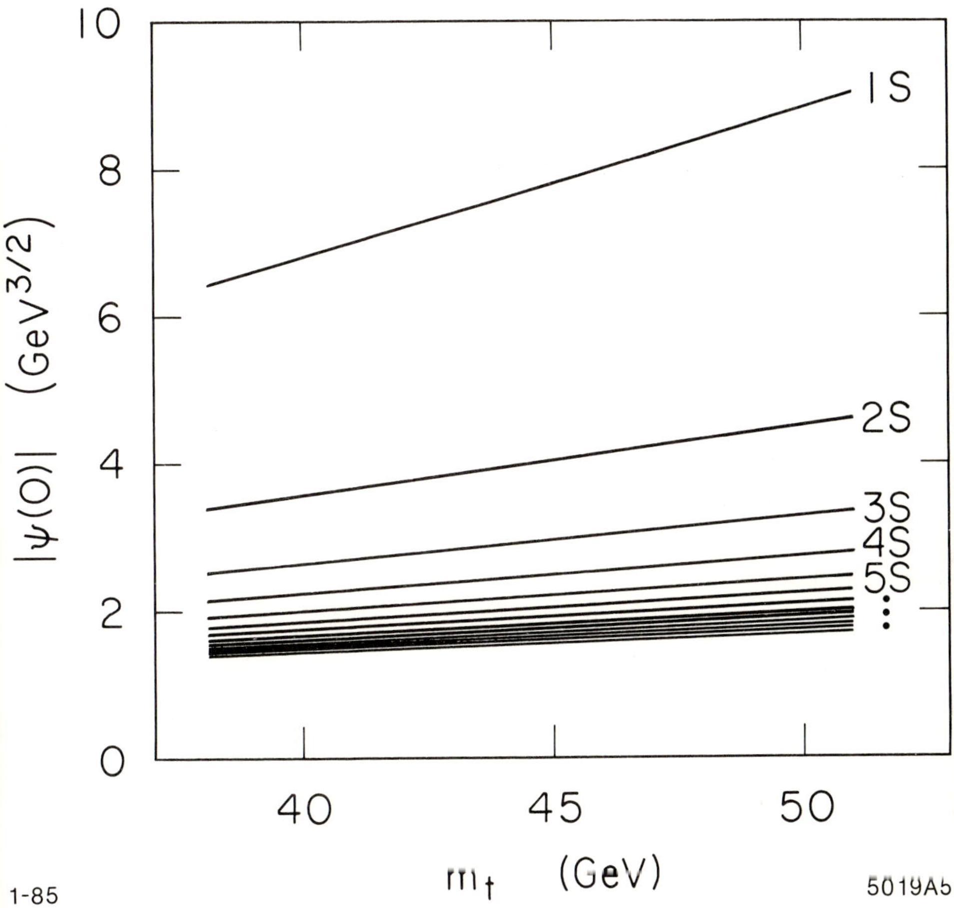

Fig. 7

The value of the wave function at the origin for toponium nS states obtained[52] with the Richardson potential and m_t in the range of 40 to 50 GeV.

TABLE III

CHARACTERISTICS OF TOPONIUM STATES FOR VARIOUS POTENTIALS

Potential	E_{1S} (GeV)	$< r_{1S} >$ (fermi)	$E_{2S} - E_{1S}$ (GeV)	$\Psi(0)_{1S}$ ($\mathrm{GeV}^{3/2}$)
"Cornell"	97.1	0.028	2.2	23.3
"Richardson"	98.3	0.048	1.0	8.5
"Martin"	98.6	0.084	0.5	2.7

potential due to neutral Higgs exchange with enhanced couplings.[53] This changes both the wave functions and the ordering of the energy levels in a characteristic fashion, and allows it to be distinguished from a simple change in the strength of the $1/r$ piece of the strong interaction potential.

- The Spin Dependent Potential

Now we turn to the spin dependent potential. In its full glory it has the form:

$$
\begin{aligned}
V_{SD}(r) = & \left(\frac{\vec{S}_1 \cdot \vec{L}}{2m_1^2} + \frac{\vec{S}_2 \cdot \vec{L}}{2m_2^2}\right) \left(\frac{dV(r)}{rdr} + 2\frac{dV_1(r)}{rdr}\right) \\
& + \frac{(\vec{S}_1 + \vec{S}_2) \cdot \vec{L}}{m_1 m_2} \frac{dV_2(r)}{rdr} \\
& + \frac{1}{6m_1 m_2}\left(6\vec{S}_1 \cdot \hat{r}\vec{S}_2 \cdot \hat{r} - 2\vec{S}_1 \cdot \vec{S}_2\right) V_3(r) \\
& + \frac{2}{3m_1 m_2}\vec{S}_1 \cdot \vec{S}_2 \, V_4(r)
\end{aligned}
\tag{59}
$$

as given by Eichten and Feinberg[54] and discussed pedagogically by

Eichten[55] and by Peskin.[56] The term $V(r)$ is the spin independent potential we discussed previously. The other terms involving V_1, V_2, V_3, and V_4 are not necessarily simply related to $V(r)$. As can be seen particularly clearly in Michael Peskin's lectures,[56] these extra terms originate in expectation values of color electric and magnet fields which are different than those that enter in the spin independent potential; they are new objects.

Although the situation is more complicated than one might have hoped, at least initially it was possible to entertain the idea that all the new spin dependent terms are of short range. This hope was dashed when it was shown that[57]

$$V(r) + V_1(r) = V_2(r). \tag{60}$$

Since V has a long range confining part, so must either V_1 or V_2.

Let us use Eq. (60) to eliminate V_1 from the spin dependent potential. It now reads:

$$\begin{aligned} V_{SD}(r) = & \left(\frac{\vec{S}_1 \cdot \vec{L}}{2m_1^2} + \frac{\vec{S}_2 \cdot \vec{L}}{2m_2^2}\right) \left(\frac{-dV(r)}{rdr} + 2\frac{dV_2(r)}{rdr}\right) \\ & + \frac{(\vec{S}_1 + \vec{S}_2) \cdot \vec{L}}{m_1 m_2} \frac{dV_2(r)}{rdr} \\ & + \frac{1}{6m_1 m_2}\left(6\vec{S}_1 \cdot \hat{r}\vec{S}_2 \cdot \hat{r} - 2\vec{S}_1 \cdot \vec{S}_2\right) V_3(r) \\ & + \frac{2}{3m_1 m_2}\vec{S}_1 \cdot \vec{S}_2 \, V_4(r). \end{aligned} \tag{61}$$

Could it now be that the remaining new potentials V_2, V_3, and V_4 are short range?

Not only is there no information to contradict this possibility, but it is supported by the results of recent lattice gauge theory calculations,[58,59] the results of some of which[22] are shown in Figures 8, 9, 10, and 11. We see that V_1 (which we have eliminated from Eq. (61)) is not short range, but V_2 looks completely different; it is very short range, and similarly for V_3 and V_4. All of this is done on a $16^3 \times 32$ lattice. It should be regarded as a qualitative result, but an important step toward the more quantitative results we can expect in the future.

Let us now go back to the spin dependent potential in the equal mass case relevant to quarkonium. We rewrite it a little bit, combining the first two terms:

$$\begin{aligned} V_{SD}(r) = & \frac{\vec{S}\cdot\vec{L}}{2m^2}\left(\frac{-dV(r)}{rdr} + 4\frac{dV_2(r)}{rdr}\right) \\ & + \frac{1}{12m^2}\left(6\vec{S}\cdot\hat{r}\vec{S}\cdot\hat{r} - 2\vec{S}\cdot\vec{S}\right)\ V_3(r) \\ & + \frac{1}{6m^2}\left(2\ \vec{S}\cdot\vec{S} - 3\right)\ V_4(r). \end{aligned} \tag{62}$$

Now, to get a simple physical picture of what is happening, let us forget for a moment the previous discussion about the spin dependent and spin independent potentials being independent entities. Let us consider what we would obtain from a (relativistic) four-fermion interaction arising from the exchange of a vector and a scalar between a quark and the antiquark of equal mass. In momentum space this is represented by an interaction:

$$L_{int} = \tilde{s}(q^2)\ \bar{u}u\bar{v}v + \tilde{v}(q^2)\ \bar{u}\gamma_\mu u\bar{v}\gamma^\mu v. \tag{63}$$

If we do an expansion in powers of v^2/c^2, the static limit is the spin

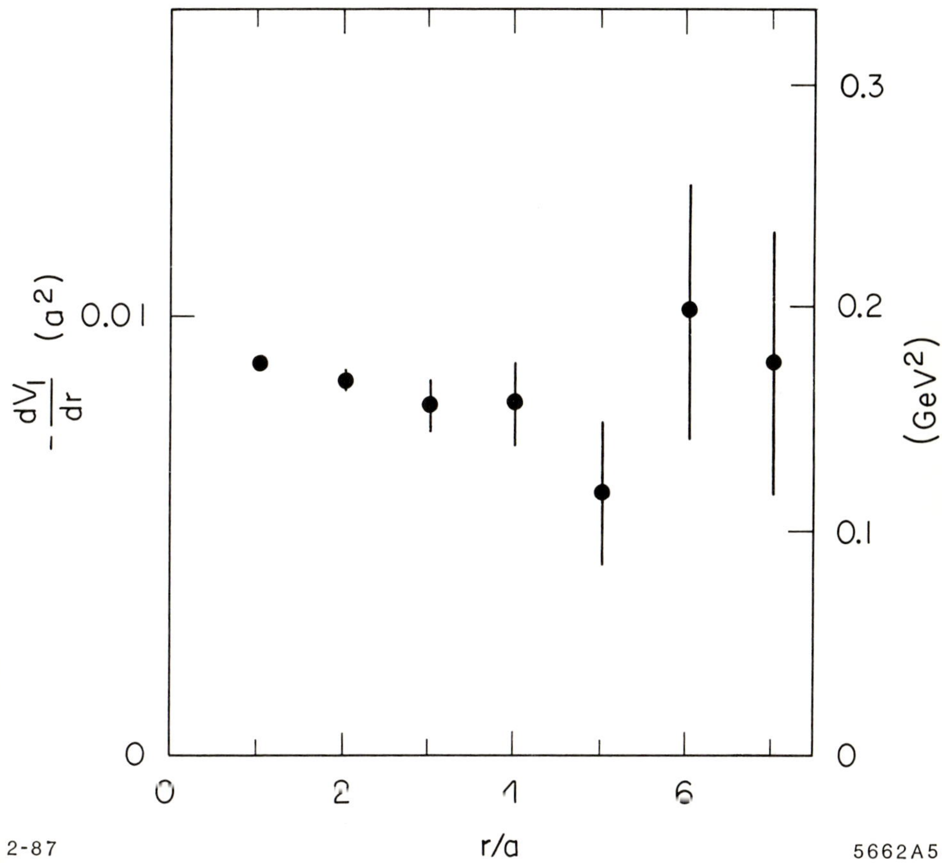

Fig. 8

Results of a lattice Monte Carlo calculation[59] of the spin-dependent potential $-dV_1/dr$ as a function of radial distance in units of the lattice spacing, a.

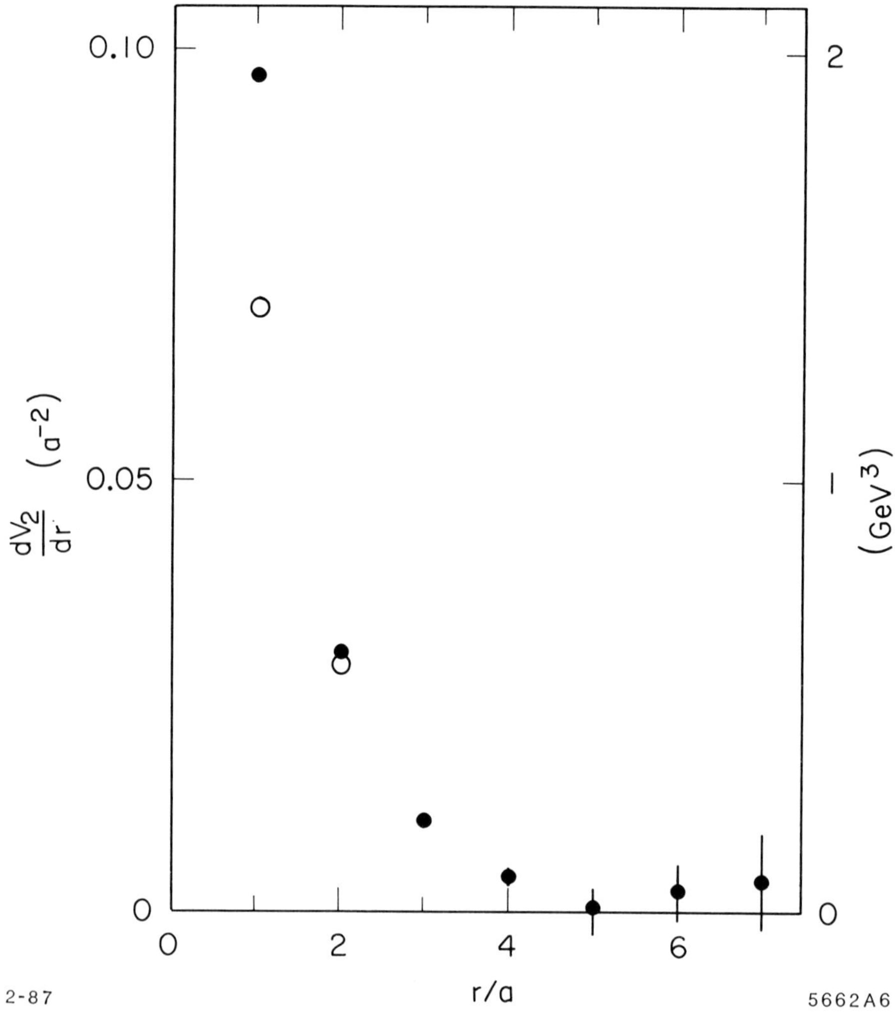

Fig. 9

Results of a lattice Monte Carlo calculation[59] of the spin-dependent potential dV_2/dr as a function of radial distance in units of the lattice spacing, a. The solid points are before, and the open points after a correction for lattice artifacts described in Ref. 59.

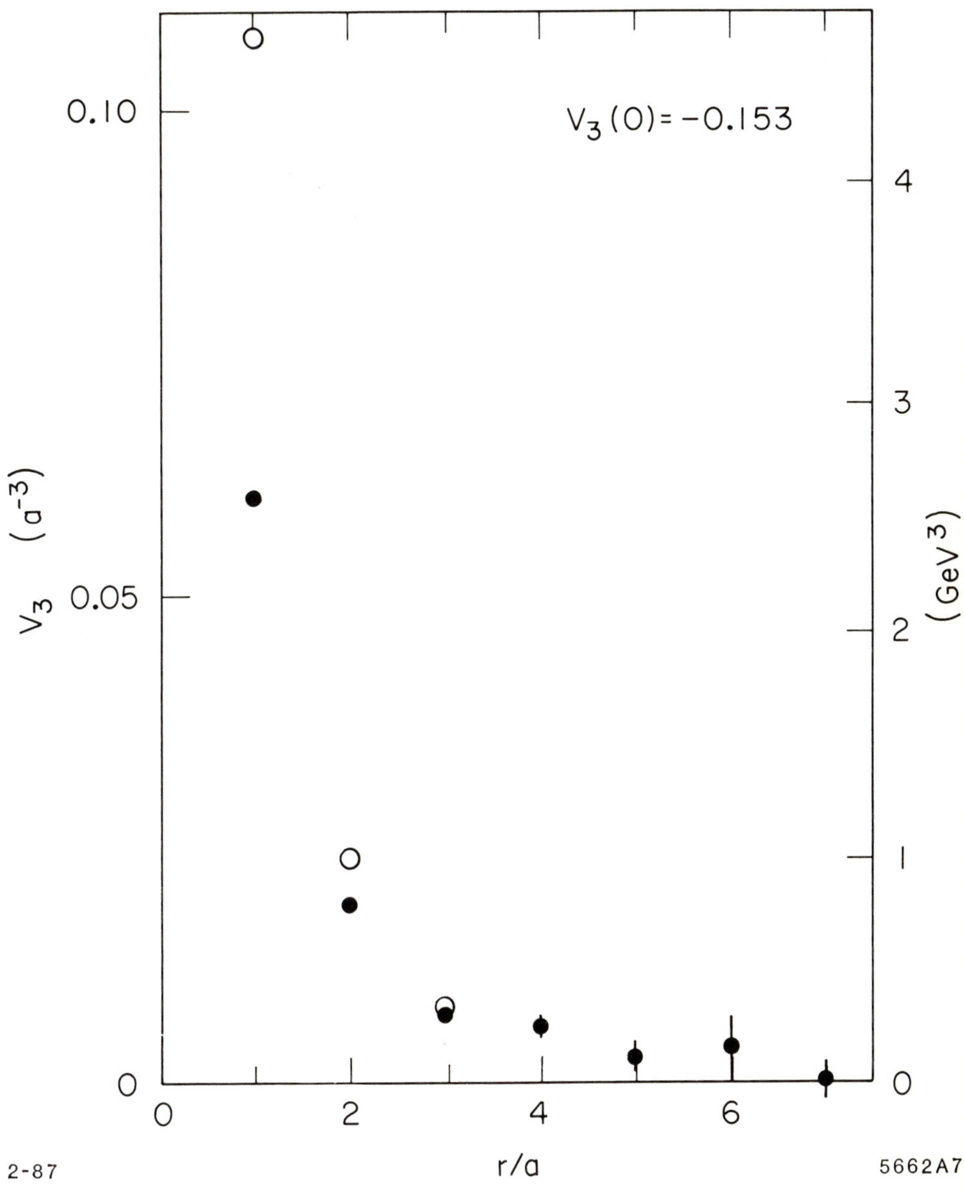

Fig. 10

Results of a lattice Monte Carlo calculation[59] of the spin-dependent potential V_3 as a function of radial distance in units of the lattice spacing, a. The solid points are before, and the open points after a correction for lattice artifacts described in Ref. 59.

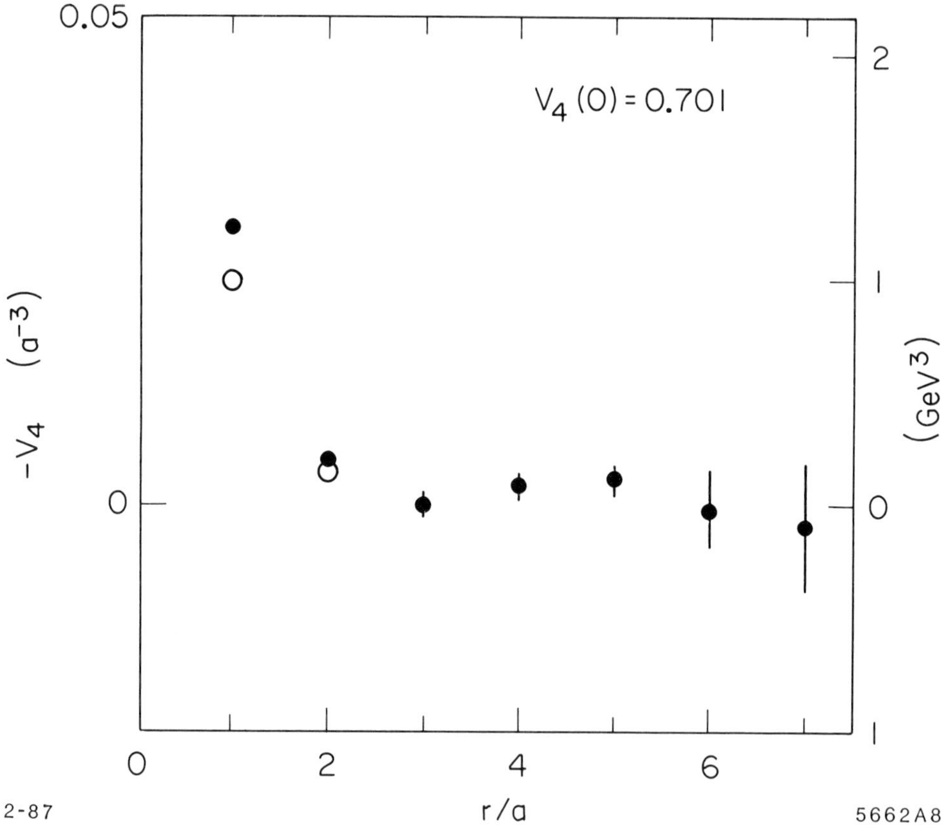

Fig. 11

Results of a lattice Monte Carlo calculation[59] of the spin-dependent potential $-V_4$ as a function of radial distance in units of the lattice spacing, a. The solid points are before, and the open points after a correction for lattice artifacts described in Ref. 59.

independent potential $v + s$, and the spin dependent terms give the Breit-Fermi potential, which in configuration space is:

$$
\begin{aligned}
V_{SD}(r) = {} & \frac{\vec{S}\cdot\vec{L}}{2m^2}\left(-\frac{dv(r)+ds(r)}{rdr} + 4\frac{dv(r)}{rdr}\right) \\
& + \frac{1}{12m^2}\left(6\vec{S}\cdot\hat{r}\vec{S}\cdot\hat{r} - 2\vec{S}\cdot\vec{S}\right)\left(\frac{dv(r)}{rdr} - \frac{d^2v(r)}{d^2r}\right) \qquad (64) \\
& + \frac{1}{6m^2}\left(2\,\vec{S}\cdot\vec{S} - 3\right)\,\nabla^2\,v(r).
\end{aligned}
$$

The term $-(dv(r) + ds(r))/rdr$ in the first line is due to the familiar Thomas precession, and it is followed by usual spin-orbit, tensor (on the second line), and spin-spin (on the third line) interactions, each with a coefficient related to $v(r)$ or $s(r)$.

Now we are in a position to compare what is in Eq. (64) to the generic decomposition in Eq. (62) involving V_1, V_2, and V_3. First, the spin independent potential V is here given by the sum of the vector and scalar potentials, $v+s$. Second, the spin dependent potentials V_2. V_3, and V_4 are all expressible in terms of derivatives of only the vector part of the potential, v. Hence, if v is related to gluon exchange and its associated $1/r$ behavior, then the potentials V_2, V_3, and V_4 are all short range in character

This encourages us to make the following division: the scalar term is long range and associated with quark confinement, while the vector term is short range (we include $1/r$ behavior as short range) and associated with gluon exchange. From the short range Coulomb-like piece one obtains the spin dependent terms we are long accustomed to in atomic physics: a spin-orbit interaction (minus the piece due to Thomas precession), a tensor interaction, and a spin-spin interaction. As you go to long range, the confining interaction, which is Lorentz scalar in character, becomes dominant. The associated physical picture[60] has a color flux tube that connects the quark and antiquark, and as they rotate around each other the flux tube rotates along with them. Consequently there are no spin dependent forces generated from this part of the potential, aside from the Thomas term which comes in with a minus sign and is generated from the spin rotation associated with Lorentz transforming from the center-of-mass to the quark or antiquark rest frame. So we get a simple way of understanding all the terms in Eq. (64). From now on we will take this identification of v and s seriously. Occasionally we will slip over to the stronger assumption that $s(r) \propto r$ and $v(r) \propto 1/r$, even to the point of thinking that we know the respective constants of proportionality.

- The Spin-Spin Interaction

The spin-spin interaction, which in the equal mass case takes the form

$$V_{SS} = \frac{1}{6m^2} \left(2\, \vec{S} \cdot \vec{S} - 3\right) \nabla^2 v(r), \tag{65}$$

is the analogue for the color forces of QCD of the interaction which gives rise to the hyperfine splittings between atomic levels. If we are brave enough to follow this analogy further and insert a $1/r$ behavior for $v(r)$, then since $\nabla^2(1/r) = -4\pi\delta^3(\vec{r})$, the spin-spin interaction is of very short range!

This delta function at the origin can be tested by noting that for quarkonium p-wave states, whose wave function at the origin vanishes, the expectation value of the spin-spin interaction should be zero. Therefore the center-of-gravity of the three states with total quark spin one and $J = 0$, 1, and 2 should be the same as the mass of the $J = 1$ state with quark spin zero:

$$\frac{5M_2 + 3M_1 + M_0}{9} = M_{\text{spin singlet}}. \tag{66}$$

(The p-wave states with total quark spin one are split in mass by the spin-orbit and tensor interactions, and the weighted average is just such as to cancel out these contributions).

For charmonium, the left-hand side of Eq. (66) is 3525.38 MeV, and an experiment in the last days of the ISR found a few candidate events with an average mass of 3525.4 ± 0.8 MeV.[61] For the bottomonium system, the corresponding values for the center-of-gravity are 9900.2 MeV for the 1P states and 10,261.6 MeV for the 2P states.[38] It would be very interesting to measure the mass of the corresponding singlet p wave states for bottomonium. There is a little bit of evidence from the CLEO experiment, studying $\pi\pi$ transitions from the 3S resonance, for a state a little below the 1P center-of-gravity.[62] As the $\bar{b}b$ system is more non-relativistic than $\bar{c}c$, the agreement with Eq. (66) should be excellent. Otherwise, the agreement in the charm case was an accident, and we had better take a close look at our assumptions on the short range nature of the spin-spin interaction.

Let us specialize to a system that consists of one heavy and one light quark. The assumption that $v(r)$ behaves as $1/r$ still gives a delta function at the origin in the part of the potential that gives the spin-spin interaction. Furthermore, the physical origin of this term in a quark color magnetic moment interacting with an antiquark color magnetic moment is still correct, and so it still depends inversely on the product of the quark mass and the antiquark mass (see the coefficient of V_4 in Eq. (59)). For example, the mass difference of the ground state vector and pseudoscalar states should behave as

$$M(^3S_1) - M(^1S_0) \propto \frac{|\Psi(0)|^2}{m_i m_j}. \tag{67}$$

If we use the fact that the spin-spin splitting is small and that in terms of constituent masses,

$$M(^3S_1) \sim M(^1S_0) \sim m_i + m_j, \tag{68}$$

then we can rewrite Eq. (17) in terms of mass squared,

$$M^2(^3S_1) - M^2(^1S_0) \propto \frac{m_i + m_j}{m_i m_j}|\Psi(0)|^2 \propto |\Psi(0)|^2/\mu_{ij} \tag{69}$$

and get a result that depends on the reduced mass of the quark-antiquark system.

One the other hand, in a system composed of a heavy and a light quark we have a atomic hydrogen-like situation with the heavy quark playing the role of the nucleus and the light quark primarily living at "large" distances. The corresponding wave function is determined by the long distance part of the potential which behaves as kr. However, for a potential which behaves as r^β, the Schrodinger equation yields a scaling law that makes[63]

$$|\Psi(0)|^2 \propto \mu_{ij}^{\frac{3}{2+\beta}}.$$

Therefore, corresponding to the case at hand with $\beta = 1$,

$$|\Psi(0)|^2 \propto \mu_{ij},$$

and substituting this into Eq. (69), one finds[64,65]

$$M^2(^3S_1) - M^2(^1S_0) \sim \text{const.} \tag{70}$$

This relation is compared to experiment in Table IV. The input masses come from the Particle Data Tables[66] except for the $F - F^*$ mass splitting where the new result from the Mark III experiment is used.[67]

TABLE IV

GROUND STATE VECTOR –

PSEUDOSCALAR MASS² DIFFERENCES

Mass² Difference	Experimental value[66,67] in GeV²
$M_\rho^2 - M_\pi^2$	0.57
$M_{K^*}^2 - M_K^2$	0.56
$M_{D^*}^2 - M_D^2$	0.55
$M_{F^*}^2 - M_F^2$	0.55
$M_{B^*}^2 - M_B^2$	0.55

The $\rho - \pi$ difference is thrown in for good measure, even though it involves only light quarks. Even the $K^* - K$ case should not be in Table IV, for the strange quark is not all that heavy. Of course, they are in Table IV because they all agree magnificently with each other, so much the more so now that we have the new data on the $F^* - F$ mass difference. Equation (70) works far better than it should, as not only are the "heavy" quarks involved not all that heavy, but even the statement that the wave function at the origin squared is proportional to the reduced mass is only approximate. Such superb agreement must be an accident.

Now let us return to systems with two heavy quarks. There the wave functions are not determined by the linear part of the potential and Eq. (70) should not hold. (It doesn't!) But here we can be braver yet and insert $v(r) = -4\pi\alpha_s/r$ into Eq. (65) and sandwich it between ground state vector and pseudoscalar meson wave functions to obtain[68]

$$M(^3S_1) - M(^1S_0) = \frac{32\pi\alpha_s}{9}\frac{|\Psi(0)|^2}{m^2}\left(1 + \mathcal{O}(\frac{\alpha_s}{\pi})\right), \tag{71}$$

where even the next order QCD corrections have been calculated. If we take the measured splitting between the ψ and η_c and invert Eq. (71) to find α_s, the result[zz31] is 0.3 to 0.4. This is perhaps a little bit too big, not to be regarded as very significant at this time.

- The Spin-Orbit and Tensor Interactions

Spin-orbit terms give rise to the fine structure in the old atomic physics terminology. In the case of equal constituent masses they take the form

$$V_{S.O.} = \frac{\vec{S}\cdot\vec{L}}{2m^2}\left(-\frac{ds(r)}{rdr} + 3\frac{dv(r)}{rdr}\right), \tag{72a}$$

and

$$V_{Tensor} = \frac{1}{12m^2}\left(6\vec{S}\cdot\hat{r}\vec{S}\cdot\hat{r} - 2\vec{S}\cdot\vec{S}\right)\left(\frac{dv(r)}{rdr} - \frac{d^2v(r)}{d^2r}\right). \tag{72b}$$

If we take the spin-orbit and tensor interactions and calculate their contributions to the ^3P_J state masses, we get[69]

$$M(^3P_2) = \vec{M} + a - 2b/5 \tag{73a}$$

$$M(^3P_1) = \vec{M} - a + 2b \tag{73b}$$

$$M(^3P_0) = \vec{M} - 2a - 4b, \tag{73c}$$

where the matrix elements a and b are defined as

$$a = \frac{1}{2m^2}\left\langle -\frac{ds}{rdr} + 3\frac{dv}{rdr}\right\rangle, \tag{74a}$$

$$b = \frac{1}{12m^2}\left\langle \frac{dv}{rdr} - \frac{d^2v}{dr^2}\right\rangle. \tag{74b}$$

We can summarize the relative values of the matrix elements in terms of one number by forming the ratio

$$r = \frac{M(^3P_2) - M(^3P)_1}{M(^3P_1) - M(^3P_0)} = \frac{2a - \frac{12}{5}b}{a + 6b}. \tag{75}$$

If only the spin-orbit term contributes, $r = 2$, while if the Coulomb-like vector part of the potential $v(r)$ is present, $r = 0.8$. As one turns on the scalar term, $s(r)$, the matrix element a decreases, as does r.

If you look at the experimental numbers,[38] updated with recent data, particularly from CUSB,[70] one finds[38] for charmonium $r_{\chi_c} = 0.50 \pm 0.02$, and for bottomonium $r_{\chi_b} = 0.67 \pm 0.05$ and $r_{\chi_b'} = 0.70 \pm 0.20$, for the 1P and 2P states, respectively. All those values are smaller than would result from solely a Coulomb-like vector term, and point toward a non-negligible scalar term. Moreover, the detailed predictions from taking the vector term as $-\frac{4}{3}\alpha_s/r$ and the scalar term as kr give quite good agreement[71] with the data, particularly for bottomonium (recall that one expects some corrections, particularly for charmonium). The case is getting fairly good for a substantial part of the long range, confining part of the potential being scalar rather than vector.

For mesons composed of a heavy quark and a light antiquark or *vice versa* the physical situation is different, as we discussed previously in considering the spin-spin interaction. The light quark lives at larger distances, so that the Thomas term, $-ds(r)/rdr$, can "beat" the net vector term, $3dv(r)/rdr$, and the effect of the spin orbit interaction, $\langle V_{S.O.}\rangle$, can be reversed in sign. This would result in an inversion of spin multiplets[72] compared to atomic physics, charmonium, and bottomonium where the higher spin state lies higher: the ordering would now be $M(^3P_0) > M(^3P_1) > M(^3P_2)$. This idea might be testable in the ^3P charm meson states, labelled here D^{**}'s. A candidate state, the $D^{**}(2420)$, already has been found and must be $J = 1$ or 2, as it decays to $D^*\pi$. If this multiplet is inverted, the $J = 0$ state, which decays to $D\pi$, will lie at a higher mass than the $D^{**}(2420)$.

- Conclusion

In this brief and incomplete review of the spectroscopy of heavy quark systems, we have seen that we have a good qualitative picture of the nature of the spin-independent forces. That, plus some phenomenological potentials inspired by fundamental theory, carry us a long way. For the spin-dependent effects we even have a semi-quantitative understanding, as they are more sensitive to the short distance part of the potential and we have more insight and more tools to help us sort things out.

Eventually, we want a quantitative calculation of both the spin-independent potential $V(r)$ and the spin-dependent potentials $V_2(r)$, $V_3(r)$, and $V_4(r)$, from QCD. This will likely come in due time from improvements in computer power and in technique over the present lattice calculations.

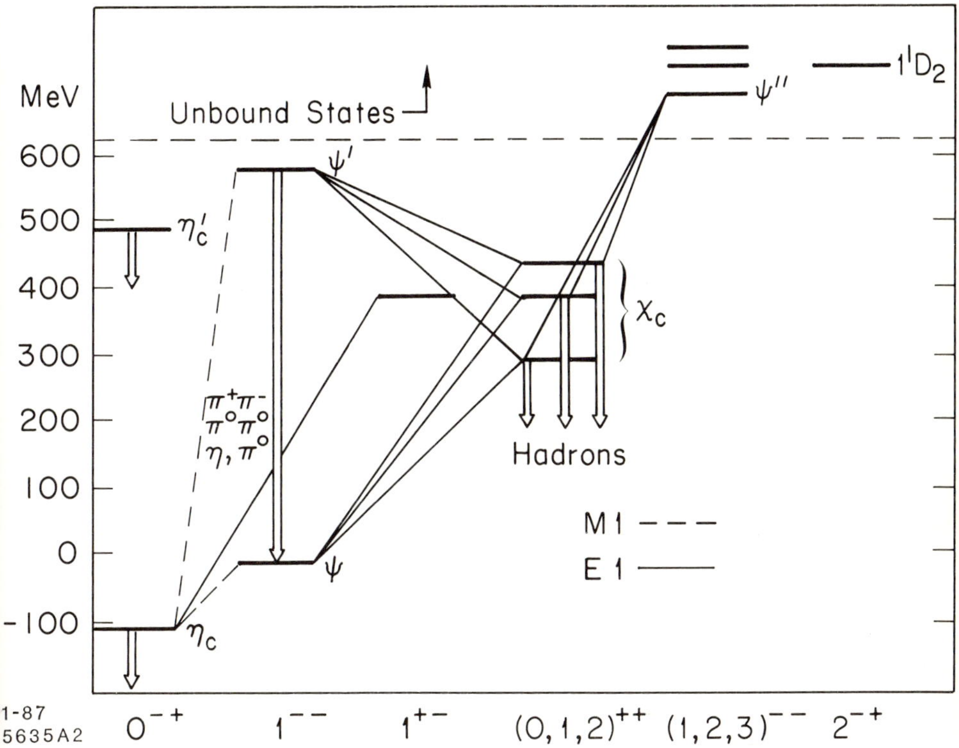

Fig. 12

Radiative and hadronic transitions between charmonium levels.

In the meantime, to clarify the emerging picture, we need more data. We need to find or confirm the $^1P_1$ states of charmonium and bottomonium. We need to find the η_b. We need to find the other D^{**}'s. And maybe best of all, we need to see the spectrum of toponium.

6. RADIATIVE TRANSITIONS

As in our earlier discussion of charmonium spectroscopy, we assume that we are dealing with heavy quarks for which the non-relativistic, two-body bound state regime is a good approximation of reality. Radiative transitions between levels then occur preferentially through the emission of photons in the lowest multipole consistent with the spin and parity of the initial and final hadronic states.

The various radiative transitions for charmonium are shown in Figure 12 and those for bottomonium in Figure 13. Transitions of the form $^3P \to {}^3S + \gamma$ or $^3S \to {}^3P + \gamma$ are electric dipole in character, while $^3S \to {}^1S + \gamma$ transitions are magnetic dipole. The corresponding rates behave as

$$\Gamma_{E1} \propto e_Q^2 (2J_f + 1)\ k_\gamma^3\ | < f|\vec{r}|i > |^2, \tag{76}$$

and

$$\Gamma_{M1} \propto e_Q^2 (2J_f +!)\ k_\gamma^3\ | < f|\vec{\sigma}/2M_Q|i > |^2, \tag{77}$$

respectively.

Table V summarizes[38] the situation for electric dipole transitions of the type $2^3S_1 \to \gamma\, 1^3P_J$, *i. e.*, $\psi' \to \gamma\, \chi_J$. One sees that the simple

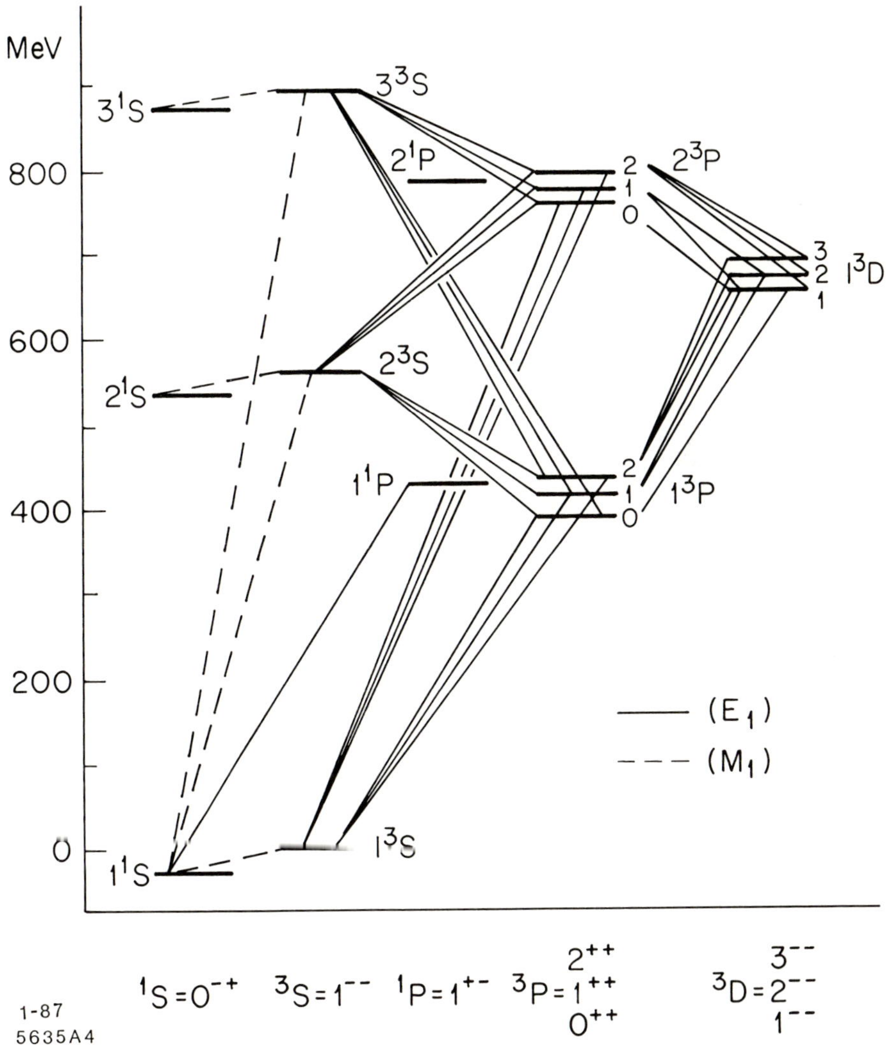

Fig. 13

Radiative transitions for bottomonium.

TABLE V

RADIATIVE TRANSITIONS[38] FROM THE ψ' TO THE χ STATES

Transition	$\Gamma_{Experiment}$ (keV)	Γ_{NR} (keV)	Γ_{CC} (keV)	Γ_{REL} (keV)
$\psi' \to \gamma\chi_0$	20.1 ± 4	45	16	19
$\psi' \to \gamma\chi_1$	18.6 ± 4	40	23	31
$\psi' \to \gamma\chi_2$	16.8 ± 4	27	22	27

non-relativistic model predicts widths (Γ_{NR}) which are consistently about a factor of two bigger than the experimental widths.

There are two explanations for this discrepancy, both of which put theory and experiment in rough accord. One is in terms of coupled-channel effects (of the charmonium bound states to the open charm states above threshold), whose predictions[45] are indicated as Γ_{CC} in Table V. A second explanation comes from relativistic effects, an example[73] of which is given in Table V as Γ_{REL}. Both effects should be considerably less in the bottomonium system. This is indeed the case, as shown in Table VI, where the experimental widths for the transitions $\Upsilon' \to \gamma\ \chi_b$ are compared to the predictions, scaled to the correct photon energies, of one specific (relativistic) model,[74] although a number of theoretical models all give comparable results for these widths.[38]

Within the experimental errors, there is good agreement between theory and experiment - certainly there is no factor of two discrepancy here. A similar conclusion holds for the $\Upsilon'' \to \gamma\ \chi_b'$ transitions.[70]

TABLE VI

RADIATIVE TRANSITIONS[38] FROM THE Υ' TO THE χ_b STATES

Transition	$\Gamma_{Experiment}$ (keV)	$\Gamma_{Moxhay-Rosner}$ (keV)
$\Upsilon' \to \gamma\chi_{b,0}$	1.3 ± 0.4	1.0
$\Upsilon' \to \gamma\chi_{b,1}$	2.0 ± 0.5	2.1
$\Upsilon' \to \gamma\chi_{b,2}$	2.0 ± 0.5	2.2

This agreement for the more non-relativistic bottomonium case allows us to conclude that we are on the right track for charmonium. Another indication of this comes from the $1^3P_J \to \gamma\ 1^3S_1$, *i.e.*, $\chi \to \gamma\psi$ transitions in charmonium shown in Table VII.

TABLE VII

RADIATIVE TRANSITIONS[38] FROM THE χ STATES TO THE ψ

Transition	$\Gamma_{Experiment}$ (keV)	Γ_{NR} (keV)	Γ_{CC} (keV)	Γ_{REL} (keV)
$\chi_0 \to \gamma\psi$	95 ± 37	121	117	128
$\chi_1 \to \gamma\psi$	≤ 252	250	240	270
$\chi_2 \to \gamma\psi$	429^{+270}_{-169}	362	305	347

Although the experimental widths in Table VII are not pinned down with high accuracy (because the total widths of the χ states are not accurately known), all the calculations are in rough agreement with each other and with the data. The reason that the relativistic corrections have such a small effect here, but a large effect in the case of $\psi' \to \gamma\,\chi_J$, is shown in Figure 14. We see that the matrix element for $2S \to 1P$ electric dipole transitions involves a cancellation which is sensitive to the position of the node in the $2S$ wave function, something which is shifted by the relativistic corrections.[73] This does not occur for matrix elements for $2P \to 1S$ transitions, which are relatively insensitive to corrections.

The situation for magnetic dipole transitions between charmonium levels is shown in Table VIII. We recall that the $\psi \to \gamma\eta_c$ and $\psi' \to \gamma\eta_c'$ transitions involve a flip of the quark spin, but the same spatial wavefunction in the initial and final states in a non-relativistic picture. This makes the calculation rather unambiguous, aside from deciding what value to assign to the charm quark mass in the expression for the magnetic dipole matrix element. Within their respective errors, theory and experiment are in rough agreement.

The transition $\psi' \to \gamma\eta_c$ on the other hand is of the form $2^3S_1 \to 1^1S_0$, and involves a $2S-1S$ wavefunction overlap which is zero in the non-relativistic limit. Indeed, there is a factor of 175 from the k_γ^3 factor favoring this transition over $\psi \to \gamma\eta_c$; yet Table VIII shows they have roughly the same width. There must be an order of magnitude suppression in the matrix element. As such, the absolute prediction of the rate is very sensitive to relativistic and other effects; the theoretical prediction for this transition in Table VIII has intentionally been left blank, although it is possible to obtain agreement[45,73] with

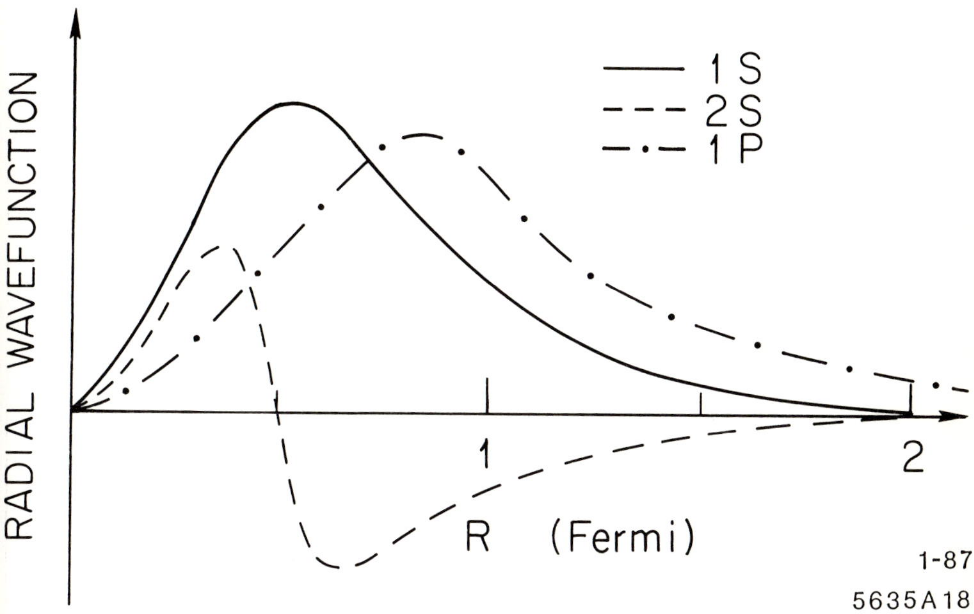

Fig. 14

The 1S, 2S, and 1P wave functions showing the relative position of their nodes.

TABLE VIII

MAGNETIC DIPOLE RADIATIVE TRANSITIONS FOR CHARMONIUM

Transition	$\Gamma_{Experiment}$ (keV)	Γ_{Theory} (keV)
$\psi \to \gamma\eta_c$	0.8 ± 0.24	~ 2
$\psi' \to \gamma\eta_c'$	$0.4 - 2.9$	~ 1
$\psi' \to \gamma\eta_c$	0.6 ± 0.13	

the experimentally measured decay rate. Nevertheless, the strong suppression of the matrix element observed experimentally should be taken as one more indication that the underlying picture of a bound, non-relativistic $\bar{c}c$ system is a good first approximation to the physics of charmonium.

7. HADRONIC TRANSITIONS

The most remarkable fact about hadronic transitions between onium states is that they have such small widths. After all, the ordinary hadrons (pions, etas) emitted in such decays have a low combined invariant mass; these are strong interaction processes in a regime where the running coupling is large. The reason[56] for the small widths emerges when we consider these transitions as occurring through the emission of gluons which later manifest themselves as ordinary hadrons. Each gluon "sees" the color carried by the quark and antiquark in a heavy $\bar{Q}Q$ system. However, the system as a whole is a color singlet. As its size shrinks to zero, a gluon will "see" no net color; the amplitude for the transition will also go to zero.

Not only does this give us a simple understanding of the small rates for such processes, but it suggests an expansion in the size of the system, i. e., a multipole expansion,[75–79] in which the leading term (proportional to the total color charge) is zero. The first non-zero term involves the emission of two gluons and is of dipole - dipole character, with the form[78]

$$M_{E1-E1} \propto g_s^2 < \Phi_f | \vec{r} \cdot \vec{E}^{\alpha} \frac{1}{E_i - H_8 - iD} \vec{r} \cdot \vec{E}^{\alpha} | \Phi_i >, \quad (78)$$

where the index α runs over the eight gluons, the Hamiltonian H_8 describes the intermediate state where the $\bar{Q}Q$ system is in a color octet state, and D is the time component of the covariant gauge theory derivative. The explicit occurence of two powers of r in Eq. (78) shows how the small size of heavy onia enters the amplitude and makes it small. For $M1 - M1$ transitions, the transition amplitudes are small because $1/M_Q^2$ is small.

It is very difficult to calculate absolute rates in this situation, as we know neither the spectrum of H_8 nor the amplitude for the two gluons to turn into specific final hadrons. We can still maintain predictive power, but avoid these issues, by taking ratios of amplitudes. A prime example is provided by the processes $\psi' \to \psi\pi\pi$ and $\Upsilon' \to \Upsilon\pi\pi$. We have the additional piece of luck that the phase space available for the $\pi\pi$ system is almost the same in these two processes. Thus everything in the expressions for the respective amplitudes is nearly the same except the expectation value of the two powers of r that occur in the dipole - dipole matrix element. Therefore, one expects[75,79]

$$\frac{\Gamma(\psi' \to \psi\pi\pi)}{\Gamma(\Upsilon' \to \Upsilon\pi\pi)} \sim \left[\frac{< r^2 >_{\psi'}}{< r^2 >_{\Upsilon'}}\right]^2, \tag{79}$$

with the right-hand-side found[79] to be ~ 16 from calculations using explicit bound state wave functions for the ψ' and Υ'. The present data for the left-hand-side are in good accord with this number.[38]

This agreement is an important verification of the basic idea that these hadronic transitions are small because of the small size of the onium system; the $\Upsilon' \to \Upsilon$ transition is smaller because the bottomonium system is smaller.

Bottomonium hadronic transitions are shown in Figure 15 (the charmonium transitions were contained in Figure 12). Table IX contains an abreviated selection of predictions for hadronic transitions in the bottomonium system, many of them obtained[79] by using the corresponding charmonium transition rate together with the multipole expansion to relate their amplitudes.

The $\pi\pi$ transitions among the Υ family have been partly dealt with above, and are in general accord with experiment. The transition $\Upsilon' \to \Upsilon\eta$ is particularly interesting because it is of $M1 - M1$ type, scales like $1/M_Q^2$, and should be detectable at the predicted rate. There are 9 possible transitions between the $2P$ and $1P$ states, but only 3 independent amplitudes in the multipole expansion. Thus there are 6 relations among the rates, which, however, are disappointingly small for experimental testing. The last transition in the Table is the best way to find the ${}^1P_1$ state, and may have already been observed (see Section 4).

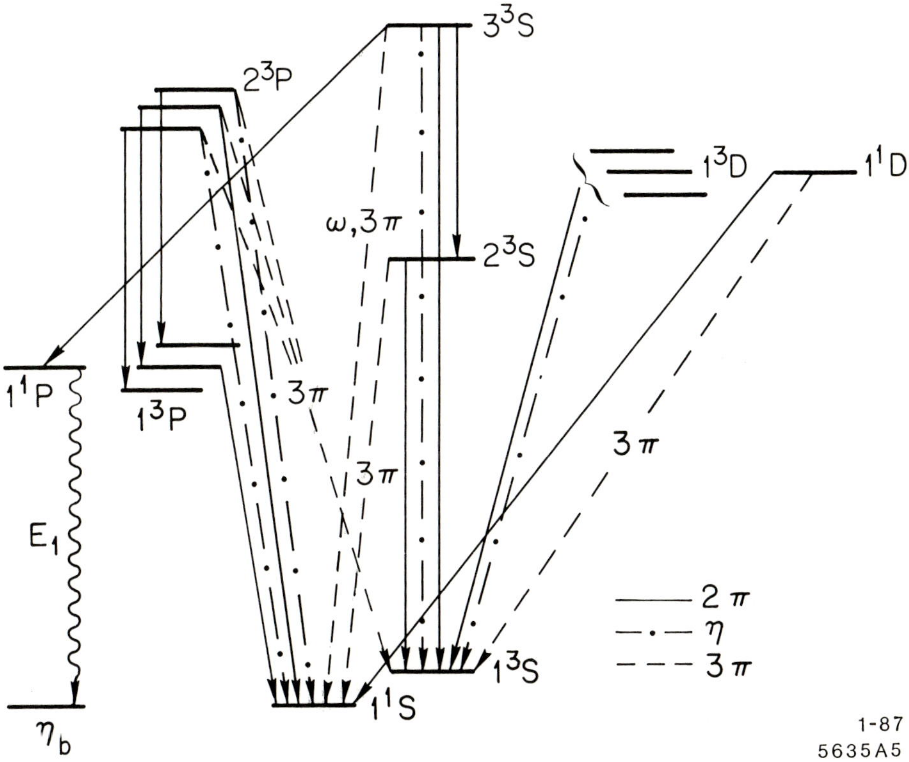

Fig. 15

Hadronic transitions between bottomonium levels.

TABLE IX

PREDICTIONS FOR HADRONIC TRANSITIONS BETWEEN BOTTOMONIUM LEVELS

Hadronic Transition	Branching Ratio (%) Theory[78,79]	Experiment[38]
$\Upsilon' \to \Upsilon\ \pi\pi$	$25 - 30$	27.4 ± 1.5
$\Upsilon' \to \Upsilon\ \eta$	0.04	< 0.2
$\Upsilon'' \to \Upsilon\ \pi^+\pi^-$	$2 - 5$	4.5 ± 0.8
$\Upsilon'' \to \Upsilon'\ \pi^+\pi^-$	$2 - 3$	3.1 ± 2.0
$2^3P_J \to 1^3P_J\ \pi\pi$	$0.1 - 0.3$	
$2^3P_J \to 1^3P_{J'}\ \pi\pi$	~ 0.01	
$1^3D_1 \to \Upsilon\ \pi\pi$	20 keV$/\Gamma$	
$\Upsilon'' \to {}^1P_1\ \pi\pi$	< 1	

There are also a number of hadronic transitions in the charmonium system that have small branching ratios, but may well be observable. Among them are: $\psi'' \to \eta_c\ \omega$ involving three gluons (and very small?); $\psi'' \to \psi\pi\pi$, which must occur at least through the 2^3S_1 component of the ψ''; $1^3D_2 \to \psi\pi\pi$ and $1^1D_2 \to \eta_c\pi\pi$, which involve narrow charmonium states above $\bar{D}D$ (into which they can not decay) but not $\bar{D}D^*$ threshold; and $\eta_c' \to \eta_c\ \pi\pi$, which should have approximately the same width ($\sim$ 100 keV) as $\psi' \to \psi\pi\pi$, but a much smaller ($\sim$ 2%) branching ratio because of the much larger width of the η_c'.

8. CHARM PRODUCTION ABOVE OPEN CHARM THRESHOLD

Once we cross open charm threhold at 3.73 GeV, a wealth of interesting new phenomena occur in e^+e^- annihilation. We still have the continuation of the charmonium levels, the 3^3S_1, $4^3S_1, \ldots$ and

1^3D_1, 2^3D_1,... states with $J^P = 1^-$. In addition, there may be "mixed" states[80] of the form $\bar{c}cg$ with these quantum numbers. These have been actively looked for in the bottom[81] system without success, but they may well be lurking in the region above 4 GeV in the center-of-mass for charm.

In addition to this there may be "charm molecules" with the quark content of $\bar{c}c\bar{q}q$, or more accurately, $(\bar{c}q)(\bar{q}c)$, as these are quasi-bound states of $\bar{D}D$, $\bar{D}D^*$, *etc.*, just above their thresholds.[82] Even without resonant states there may be enhancements in various channels just above their respective thresholds.

The continuum itself rapidly becomes fairly complicated. Even restricting ourselves to production of two-body or quasi-two-body final states, we have:

- $\bar{D}D$, $\bar{D}D^* + \bar{D}^*D$, $\bar{D}^*D^*$, $\bar{D}D^{**}$,... where D^{**} indicates an $L = 1$ charm state which decays to πD and/or πD^*;
- $\bar{D}_sD_s$, $\bar{D}_sD_s^* + \bar{D}_s^*D_s$, $\bar{D}_s^*D_s^*$, $\bar{D}_sD_s^{**}$,... where the D_s^{**} indicates and $L = 1$ charm-strange state that decays to $\pi\pi D_s$ (a suppressed strong interaction decay) or, if allowed, to $D\ K$ and/or $D^*\ K$;
- $\bar{\Lambda}_c\Lambda_c$, $\bar{\Sigma}_c\Sigma_c$, $\bar{\Sigma}_c\Sigma_c^*$,... including the production of a lowest $L = 1$ charmed baryon with a mass of $\sim$ 2500 MeV, which may be forbidden to decay strongly,[83] and therefore has the dominant decay $\Lambda_c' \to \gamma\ \Lambda_c$;
- $\bar{\Xi}_c\Xi_c$, $\bar{\Xi}_c\Xi_c^*$,...

The interplay of the continuum and the resonances is a whole subject in itself.[84] A representative decomposition from one calculation[45] of the part of the charm production cross section that involves D

mesons is shown in Figure 16. Note in particular the valleys in particular channels produced by overlaps of wave functions with radial nodes. For comparison, one set of experimental measurements[85] of the total e^+e^- annihilation cross section into hadrons is shown in Figure 17.

With all this in mind, let us start from open charm threshold and make our way upward in energy, highlighting some of the open questions:

- The $\psi''(3.77)$ is a "D factory" which dominantly decays to $\bar{D}D$ and is dominantly a ${}^3D_1$ state with a small admixture of ${}^3S_1$. How big are decays like $\psi'' \rightarrow \psi\pi\pi$ and $\psi'' \rightarrow \gamma\ \chi$ of the ψ'' into other charmonium states?

- As noted before, the ${}^3D_2$ and ${}^1D_2$ states should be narrow as they likely lie near the ψ'', are forbidden to decay into $\bar{D}D$, and are probably below threshold for decay into $\bar{D}D^*$. They are not directly produced in e^+e^-, but could be found in $\bar{p}p$ experiments and have interesting decays into other charmonium states.

- There is a local maximum in the cross section around ~ 3.96 GeV, not far above $\bar{D}D^*$ threshold (see Figure 17). Is this indeed some sort of enhancement associated with $\bar{D}D^*$, or is it associated with $\bar{D}_sD_s$ threshold, which we now know is also very nearby?

- The $\psi(4.03)$ is a clear peak (see Figure 17) with a major decay into $\bar{D}^*D^*$ even though this channel has just opened, but its character as a $3S$ charmonium level, charm molecule, or something else remains to be settled.

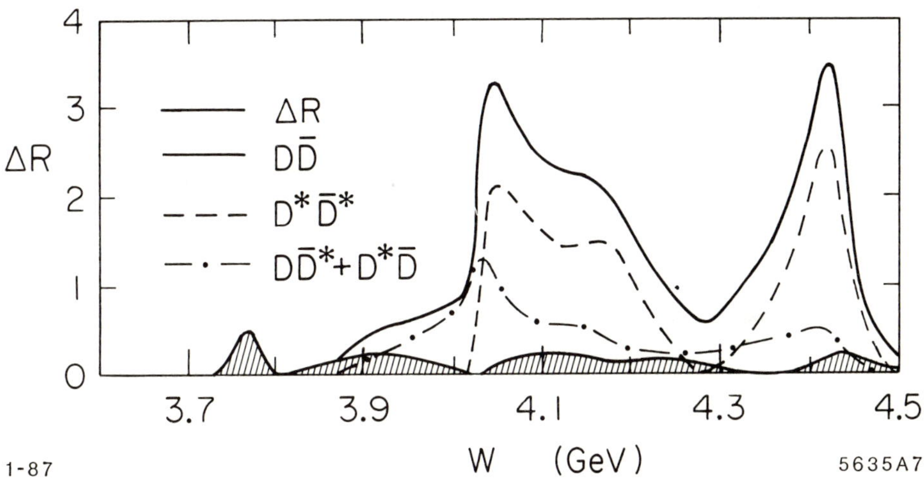

Fig. 16

Decomposition of the charm cross section involving D mesons into its component channels according to Ref. 45.

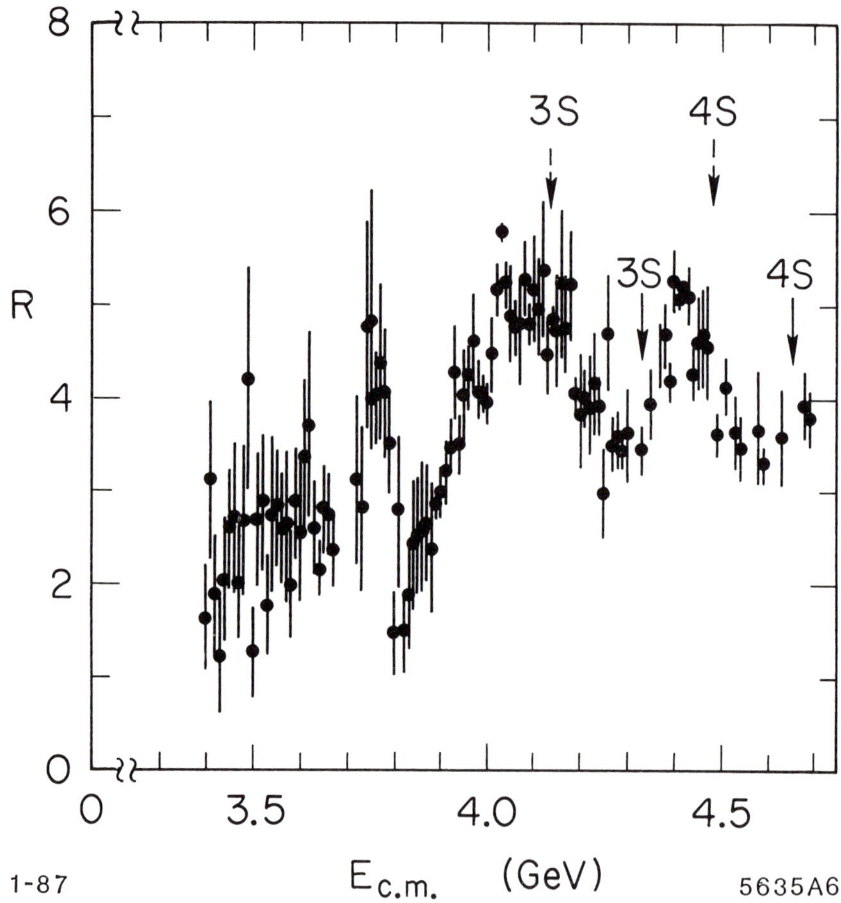

Fig. 17

The total cross section for e^+e^- annihilation into hadrons in units of the point cross section from Ref. 85. The arrows point to the positions of the $3S$ and $4S$ charmonium states in two theoretical calculations.

- The $\psi(4.16)$ is another peak. It is at too low an energy to be the $4S$ state. Does it have anything to do with $\bar{D}_s D_s^*$ threshold which is just below it?
- There is a long valley in the cross section with a minimum at ~ 4.3 GeV. Is there no enhancement above $\bar{D}_s^* D_s^*$ threshold, which lies in this region?
- The $\psi(4.415)$ would seem to be a good candidate for the $4S$ state of charmonium. Does it decay to $\bar{D}D^{**}$?
- Charmed baryon threshold is at ~ 4.57 GeV, and experiments just above this energy have the capability of making an absolute branching ratio measurement for the Λ_c if enough decays can be "tagged". At energies above ~ 5 GeV, the charmed-strange baryons (Ξ_c) should be produced, as well as Σ_c, Σ_c^*, and other excited charmed baryons. The production and weak decays of charmed baryons is a subject that has barely been touched.

REFERENCES

1. M. Witherell, these Proceedings.
2. D. G. Hitlin, these Proceedings.
3. I. Bigi, these Proceedings.
4. W. Toki, these Proceedings.
5. M. S. Chanowitz, these Proceedings.
6. S. Weinberg, Phys. Rev. Lett. 19, 1264 (1967); A. Salam, *Proceedings of the Eighth Nobel Symposium*, edited by N. Svartholm (Almqvist and Wiksell, Stockholm, 1968), p. 367.
7. See the excellent recent review by P. Langacker, in *Proceedings of the 1985 Int. Symp. on Lepton and Photon Interactions at High Energies*, edited by M. Konuma and K. Takahashi (Kyoto University, Kyoto, 1986), p. 186. The latest review is found in U. Almaldi *et al.*, University of Pennsylvania preprint UPR-0331T, 1987 (unpublished).
8. M. Kobayashi and T. Maskawa, Prog. Theor. Phys. 49, 652 (1973).
9. Particle Data Group, Phys. Lett. 170B, 74 (1986).
10. W. J. Marciano and A. Sirlin, Phys. Rev. Lett. 56, 22 (1986).
11. H. Leutwyler and M. Roos, Z. Phys. C25, 91 (1984).
12. H. Abramowicz *et al.*, Z. Phys. C15, 19 (1982).
13. The c and b decay data at that time were reviewed by E. Thorndike, in *Proceedings of the 1985 International Symposium on Lepton and Photon Interactions at High Energies*, edited by M. Konuma and K. Takahashi (Kyoto University, Kyoto, 1986), p. 406. For a review of the experimental situation as of the summer of 1986 see M. G. D. Gilchriese, rapporteur talk in *Proceedings of the XXIII International Conference on High Energy Physics*, edited by S. C. Loken (World Scientific, Singapore, 1987), p. 140.
14. K. Kleinknecht and B. Renk, Mainz preprint MZ-ETAP/86-4, 1986 (unpublished).
15. A. Sirlin and R. Zucchini, Phys. Rev. Lett. 57, 1994 (1986).

16. E. G. Drukarev and M. I. Strikman, Leningrad Institute of Nuclear Physics preprint, No. 1221, 1986 (unpublished).

17. I thank Bill Marciano for a clarifying discussion of the comparison of the two new results.

18. Some recent reviews from various viewpoints are found in A. J. Buras, Max Planck Institute preprint MPI-PAE/PTH 40/86, 1986 (unpublished); R. Ruckl, in *Proceedings of the XXIII International Conference on High Energy Physics*, edited by S. C. Loken (World Scientific, Singapore, 1987), p. 797; I. Bigi, Phys. Lett. 169B, 101 (1986) and Ref. 3.

19. M. K. Gaillard and B. W. Lee, Phys. Rev. Lett. 33, 108 (1974); G. Altarelli and L. Maiani, Phys. Lett. 52B, 351 (1974).

20. The discussion which follows parallels that of F. J. Gilman and M. B. Wise, Phys. Rev. D20, 2392 (1979).

21. M. A. Shifman *et al.*, Zh. Eksp. Teor. Fiz. Pis'ma Red. 22, 123 (1975) [JETP Lett. 22, 55 (1975)]; M. A. Shifman *et al.*, Nucl. Phys. B120, 316 (1977).

22. See, for example, F. J. Gilman, in the *Second Aspen Winter Particle Physics Conference*, Annals of the New York Academy of Sciences, Vol. 490, edited by L. Durand (The New York Academy of Sciences, New York, 1987), p. 1; and A. J. Buras, W. A. Bardeen, and J. M. Gerard, Phys. Lett. 180B, 133 (1980) and Max Planck Institute preprints MPI-PAE/Pth-45/86 and MPI-PAE/Pth-55/86, 1986 (unpublished).

23. B. Guberina, R. D. Peccei, and R. Ruckl, Phys. Lett. 90B, 169 (1980).

24. For a recent calculation of "penguin" effects in exclusive B decays see M. B. Gavela *et al.*, Phys. Lett. 154B, 425 (1985).

25. M. L. Perl, Ann. Rev. Nucl. Part. Sci. 30, 299 (1980).

26. In the following we take the properties of the tau, unless otherwise specified, from the review of J. G. Smith, invited talk at the 1987 Spring Meeting of the American Physical Society, Bulletin of the American Physical Society, Series II, 32, No. 4, p. 1010, 1987 (unpublished). The situation as of the summer of 1986 is found in M. G. D. Gilchriese, Ref. 13.

27. H. Albrecht *et al.*, Phys. Lett. 185B, 228 (1987).

28. Y. S. Tsai, Phys. Rev. D4, 2821 (1971); H. B. Thacker and J. J. Sakurai, Phys. Lett. 36B, 103 (1971).

29. F. J. Gilman and D. H. Miller, Phys. Rev. D17, 1846 (1978) and unpublished calculations with the present value of M_τ.

30. F. J. Gilman and S. H. Rhie, Phys. Rev. D31, 1066 (1985) and references to other previous work therein.

31. H. Albrecht *et al.*, Phys. Lett. 185B, 223 (1987).

32. F. J. Gilman, SLAC preprint SLAC-PUB-4365, 1987 and Phys. Rev., in press.

33. P. R. Burchat *et al.*, Phys. Rev. D35, 27 (1987); H. Aihara *et al.*, Phys. Rev. Lett. 57, 1836 (1986); K. K. Gan *et al.*, SLAC preprint SLAC-PUB-4110, 1987 (unpublished).

34. S. Keh *et al.* and S. Abachi *et al.*, contributions to the Berkeley Conference quoted in M. G. D. Gilchriese, Ref. 13.

35. M. Derrick *et al.*, Phys. Lett. 189B, 260 (1987).

36. E. L. Berger and H. J. Lipkin, Phys. Lett. 189B, 226 (1987); Y. Meurice and C. K. Zachos, Mod. Phys. Lett. A 2, 247 (1987).

37. The ARGUS collaboration has set an upper limit of 1.3 % at 95% CL on the branching fraction for $\tau^- \to \nu_\tau \pi^- \eta$ (talk at the XXIInd Rencontre de Moriond, Les Arcs, France, March 8 - 16, 1987). The TPC collaboration set an upper limit on the branching fraction of the $SU(3)$ related decay, $\tau^- \to \nu_\tau K^- K^0$, of 0.26% (H. Aihara *et al.*, LBL preprint LBL-23176, 1987 (unpublished)). Also, the Mark III collaboration, D. Coffman *et al.*, SLAC preprint SLAC-PUB-4314, 1987 (unpublished), set a limit of 2.5%.

38. R. Schindler, in *Proceedings of the 14th SLAC Summer Institute on Particle Physics*, SLAC Report No. 312, edited by Eileen Brennan (SLAC, Stanford, 1987), p. 239 reviews much of the data relevant to the discussion here. See also W. Toki, Ref. 4.

39. Recent reviews are to be found in A. Martin, Comm. Nucl. Part. Phys. 16, 249 (1986) and W. Kwong, J. L. Rosner, and C. Quigg, University of Chicago preprint EFI 86-60, 1986 (to be published in Ann. Rev. Nucl. Part. Sci 37, 1987.

40. The basics and earlier accomplishments of lattice QCD calculations are set out in M. Creutz, *Quarks, Gluons, and Lattices* (Cambridge University Press, Cambridge, 1983).

41. S. Capstick, N. Isgur, and J. Paton, Phys. Lett. 175B, 457 (1986).

42. Relativistic corrections have been considered, for example, by P. Moxhay and J. L. Rosner, Phys. Rev. D28, 1132 (1983) and by R. McClary and N. Byers, Phys. Rev. D28, 1692 (1983). The effects of light quark pairs are studied by N. Byers and V. Zambetakis, invited talk at the Second Conference on Interactions Between Particle and Nuclear Physics, Lake Louise, Canada, May 23-32, 1986 and UCLA preprint UCLA/86/TEP/24, 1986 (unpublished).

43. The sum rule approach is reviewed by L. J. Reinders, H. Rubinstein, and S. Yazaki, Phys. Rep. 127, 1 (1985); and by M. A. Shifman, Ann. Rev. Nucl. Part. Sci 33, 199 (1984).

44. J. D. Jackson, *Proceedings of the Summer Institute on Particle Physics*, SLAC Report No. 198, edited by M. C. Zipf (SLAC, Stanford, 1976), p. 147.

45. E. Eichten *et al.*, Phys. Rev. D17, 3090 (1978) and D21, 203 (1980).

46. J. L. Richardson, Phys. Lett 82B, 272 (1979).

47. W. Buchmuller, G. Grunberg, and S. H. H. Tye, Phys. Rev. Lett. 45, 103 (1980); W. Buchmuller and S. H. H. Tye, Phys. Rev. D24, 132 (1981).

48. A. Martin, Phys. Lett. 93B, 338 (1980).

49. H. B. Thacker, C. Quigg and J. L. Rosner, Phys. Rev. D18 274 (1978) and D21, 234 (1980); C. Quigg and J. L. Rosner, Phys. Rev. D23, 2625 (1981).

50. B. Baumgartner, H. Grosse, and A. Martin, Phys. Lett. 146B, 363 (1984).

51. B. Baumgartner, H. Grosse, and A. Martin, Nucl. Phys. B254, 528 (1985).

52. P. J. Franzini and F. J. Gilman, Phys. Rev. D32, 237 (1985).

53. G. G. Athanasiu, P. J. Franzini, and F. J. Gilman, Phys. Rev. D32, 3010 (1985).

54. E. Eichten and F. Feinberg, Phys. Rev. D23, 2724 (1981).

55. E. Eichten, *Proceedings of the 11th SLAC Summer Institute on Particle Physics*, SLAC Report No. 267, edited by P. M. McDonough (SLAC, Stanford, 1983), p. 497.

56. M. Peskin, *Proceedings of the 11th SLAC Summer Institute on Particle Physics*, SLAC Report No. 267, edited by P. M. McDonough (SLAC, Stanford, 1983), p. 151.

57. D. Gromes, Z. Phys. C26, 401 (1984).

58. P. de Forcrand and J. D. Stack, Phys. Rev. Lett. 55, 1254 (1985); C. Michael, Phys. Rev. Lett. 56, 1219 (1985).

59. M. Campostrini, K. Moriarty, and C. Rebbi, Phys. Rev. Lett. 57, 44 (1986).

60. W. Buchmuller, Phys. Lett. 112B, 479 (1982). See also J. L. Rosner, J. de Phys. 46, C2, Supp. 2 (1985); R. D. Pisarski and J. D. Stack, Fermilab preprint FERMILAB-PUB -86/122-T, 1986 (unpublished), and M. G. Olsson and C. J. Suchyta, University of Wisconsin preprint MAD/PH/312, 1986 (unpublished).

61. C. Baglin *et al.*, Phys. Lett. 171B, 135 (1986).

62. T. Bowcock *et al.*, Phys. Rev. Lett. 58, 307 (1987).

63. C. Quigg and J. L. Rosner, Comm. Nucl. Part. Phys. 8, 11 (1978).

64. M. Frank and P. J. O'Donnell, Phys. Lett. 157B, 174 (1985).

65. H. J. Schnitzer, Phys. Lett. 134B, 253 (1984) and Brandeis preprint, 1985 (unpublished).

66. Particle Data Group, Phys. Lett. 170B, 1 (1986).

67. W. Toki, in *Proceedings of the 14th SLAC Summer Institute on Particle Physics*, SLAC Report No. 312, edited by Eileen Brennan (SLAC, Stanford, 1987), p. 495.

68. W. Buchmuller, Y. J. Ng, and S. H. H. Tye, Phys. Rev. D24, 3003 (1981); A. Martin and J. M. Richard, Phys. Lett. 115B, 323 (1982).

69. J. L. Rosner, *Proceedings of the 1985 Symposium on Lepton and Photon Interactions at High Energies*, edited by M. Konuma and K. Takahashi (Kyoto University, Kyoto, 1986), p. 448.

70. P. M. Tuts, in *Proceedings of the 14th SLAC Summer Institute on Particle Physics*, SLAC Report No. 312, edited by Eileen Brennan (SLAC, Stanford, 1987), p. 507.

71. See Table X of Ref. 1 and references to the original calculations there.

72. H. J. Schnitzer, Phys. Lett. 76B, 461 (1978). See also, J. L. Rosner, Comm. Nucl. Part. Phys. 16, 109 (1986).

73. R. McClary and N. Byers, Ref. 42.

74. P. Moxhay and J. L. Rosner, Ref. 42.

75. K. Gottfried, Phys. Rev. Lett. 40, 598 (1978).

76. G. Bhanot, W. Fischler, and S. Rudaz, Nucl. Phys. B155, 208 (1979).

77. M. E. Peskin, Nucl. Phys. B156, 365 (1979); G. Bhanot and M. E. Peskin, Nucl. Phys. B156, 391 (1979).

78. T.-M. Yan, Phys. Rev. D22, 1652 (1980).

79. Y.-P. Kuang and T.-M. Yan, Phys. Rev. D24, 2874 (1981).

80. R. C. Giles and S.-H. H. Tye, Phys. Rev. Lett. 37, 1175 (1976); Phys. Rev. D16, 1079 (1977); Phys. Lett. 73B, 30 (1978).

81. W. Buchmuller and S.-H. H. Tye, Phys. Rev. Lett. 44, 850 (1980).

82. A. De Rujula, H. Georgi, and S. L. Glashow, Phys. Rev. Lett. 38, 317 (1977).

83. L. A. Copley, N. Isgur, and G. Karl, Phys. Rev. D20, 768 (1979).

84. E. Eichten *et al.*, Phys. Rev. Lett. 36, 500 (1976) and Ref. 45.

85. J. Siegrist, SLAC Report No. 225, 1979.

Experimental Review of J/ψ Decays*

WALTER H. TOKI

Stanford Linear Accelerator Center
Stanford University, Stanford, California, 94305

Abstract

This is a review of J/ψ physics from e^+e^- colliders presented at the Charm Workshop in Beijing, China. The review includes a brief historical summary of J/ψ physics, a general discussion of theoretical models and detailed results on radiative and hadronic J/ψ decays from the MARK III , DM2, Crystal Ball and MARK II groups.

Invited Talk at the Charm Workshop
Beijing, Peoples Republic of China
June 4-16, 1987

* **This work was supported in part by the Department of Energy, under contracts DE-AC03-76SF00515, DE-AC02-76ER01195, DE-AC03-81ER40050, DE-AM03-76SF0034, and by the National Science Foundation.**

1. Introduction

The study of J/ψ decays has produced many new results. With high statistics J/ψ data, the MARK III and DM2 groups have continued a study that begin with the MARK II and Crystal Ball groups in the early 1980's. The main focus of research has been the search for gluonium states. This progressed in the direct search for these states in radiative decays and in a study of hadronic decays, to understand light quark spectroscopy. In order to understand the evidence for gluonium states, the mesons predicted by the quark model must be thoroughly understood and high statistics J/ψ decays is providing an SU(3) flavor singlet source of mesons to probe the quark content of different mesons.

In this review we begin with a brief review of (1) the J/ψ discovery and (2) the early measurements of its quantum numbers and a short discussion of the high statistics J/ψ e^+e^- detectors. Next (3) a theoretical review is presented to provide guidelines of how to understand the J/ψ decays. The remaining sections cover major experimental results; (4) η_c decays, (5) two body hadronic decays, (6) inclusive radiative decays, (7) radiative decays into conventional mesons, π^0, η, η', f and f', and finally radiative decays into the new states, (8) the $\iota/\eta(1440)$ and the $E/f_1(1420)$, (9) the $\theta/f_2(1720)$, (10) the vector-vector resonance and (11) the $\xi(2.2)$.

1.1 J/ψ Discovery

The J/ψ discovery was announced in November 1974 by the MIT-BNL group at Brookhaven and the SLAC-LBL group at SLAC. The MIT-BNL group[1] was studying the reaction,

$$p + Be \to e^+e^- + X$$

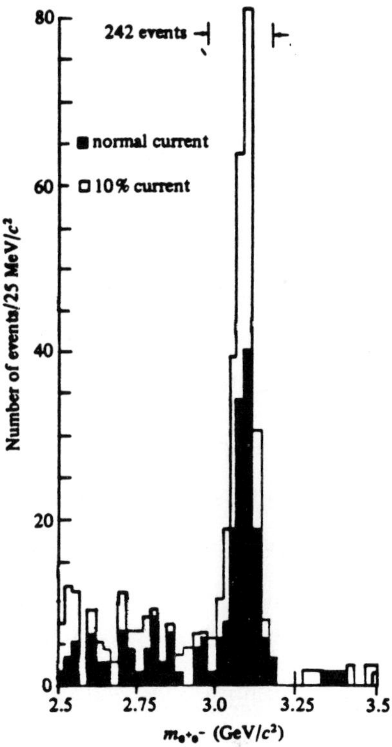

Figure 1.

The e^+e^- mass distribution in the reaction $p + Be \rightarrow e^+e^- + X$ from the MIT-BNL group.[1]

at $\sqrt{s} = 7.4$ GeV. With an extremely intense beam, $\sim 10^{12}$ protons/pulse, and a high resolution double arm spectrometer, they observed a narrow peak at $m(e^+e^-) = 3.1$ GeV/c^2 with a very small cross section, $\sigma \cong 10^{-34}$ cm^2. The e^+e^- mass distribution is shown in Fig. 1.

The SLAC-LBL group[2] at the e^+e^- storage, SPEAR, with the Mark I detector scanned the region around 3 GeV and observed a huge increase in multi-hadronic , e^+e^-, $\mu^+\mu^-$ and two body hadronic events at 3.105 GeV. The hadronic cross section as a function of $\sqrt{s}$ is shown in Fig. 2. The width of this resonance was less than the machine resolution, $\Gamma \leq 3$ MeV.

These results were immediately confirmed by groups[3] at the ADONE storage ring. All three results appeared in the same remarkable December 2, 1974 publication of Physical Review Letters.

1.2 e^+e^- DETECTOR

The principle e^+e^- detectors used for high statistics J/ψ physics are the Mark I,[4] MARK II,[5] Crystal Ball,[6] DM2,[7] and the MARK III[8] detectors. The total number of produced J/ψ events taken by each detector is listed in table 1.1.

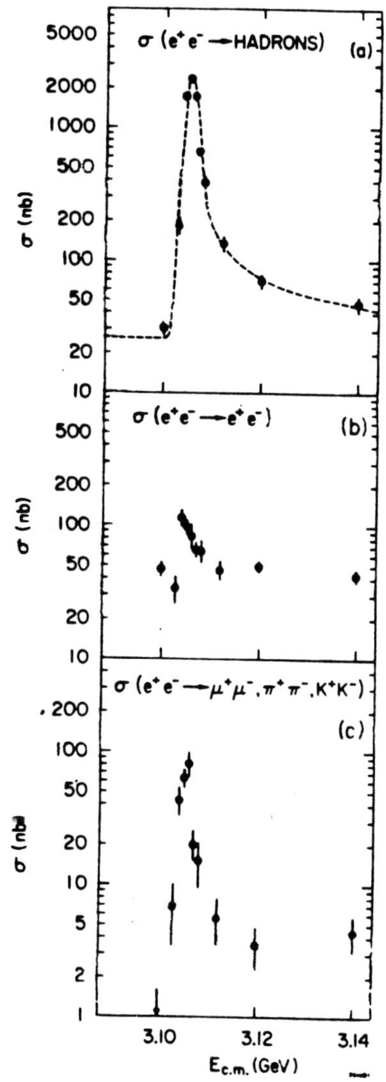

Figure 2.

The hadronic cross section in the reaction $e^+e^- \to$ hadrons (a), e^+e^- pairs (b), and muon, pion and kaon pairs (c), as a function of $\sqrt{s}$ from the SLAC-LBL group.[2]

Table 1.1 J/ψ Events Samples

# Produced J/ψ's	Run Date	Group
$\sim 150,000$	1974-1975	Mark I
1.3×10^6	Fall 78, Spring 79	MARK II
2.2×10^6	Fall 78, Spring 79, Fall 80	Crystal Ball
8.6×10^6	1982, Fall 83	DM2
5.8×10^6	Fall 82, Spring 83, Spring 85	MARK III

The Mark I detector, a general purpose e^+e^- detector, is the prototype for most e^+e^- experiments at PEP, PETRA, LEP, SLC and now BPC (Beijing). It featured a cylindrical spark chamber with 4 layers around the beam pipe inside a solenoidal magnet with 0.4 Tesla field. The drift chamber was enclosed by the 48 TOF counters and this was followed by the magnet coil and the shower counter (24 lead scintillator shower counters). The TOF resolution was about 500 ps. The shower counter was used to identify electrons.

The MARK II detector was a new detector built by the Mark I group and installed at SPEAR. It used a larger magnet with a 0.41 Tesla field and a large liquid argon calorimeter. The cylindical drift chamber contained 16 layers and was surrounded by 48 TOF counters with $\sim$ 300 ps resolution. The liquid argon calorimeter was installed behind the magnet coil and had eight modules covering $\sim$64% of the solid angle and an energy resolution of $0.14/\sqrt{E}$ GeV.

The Crystal Ball detector is a specialized neutral detector with 672 NaI crystals. It covers 94% of 4π and has an energy resolution of $\sigma_E/E = 0.026/E^{\frac{1}{4}}$. The beam pipe is surrounded by 3 layers of drift tubes with charge division readout. The crystals and the drift tubes enable identification of charged and neutral tracks, measurements of the direction of charge tracks and the measurement of the direction and energy of photons and electrons.

The DM2 experiment is at the ORSAY storage ring, DCI. The experiment was originally proposed to study resonance physics below the J/ψ. The detector has a 0.5 Tesla solenoidal magnet with proportional and drift chambers. The drift chamber is surrounded with 36 TOF counters and 36 water Cerenkov counters. The TOF resolution is 260 ps but the long bunch length of the DCI machine actually degrades the total TOF resolution to ~500 ps. The shower counter is a lead-scintillator proportional tube sandwich. The total coverage is 70% of 4π. The resolution is ~35% at 1 GeV.

The Mark III detector began running in 1982. It used the old Mark I magnet and replaced the original coil with a larger one to enable the insertion of a shower counter. The detector had a Beryllium beam pipe surrounded by a central trigger chamber with 400 micron resolution. This has been replaced in Summer 1987 with a high resolution straw chamber that should achieve 50 micron resolution. The main drift chamber has 24 axial sense wire planes and 0 stereo sense wire planes. Twelve of the axial planes have charge readout to enable energy loss measurements. Outside the drift chamber are 48 2" thick scintillation counters with phototube readout on each end. The TOF resolution is 170-200 picoseconds. Following the TOF counters is a 24 layer (12 radiation length) lead proportional tube shower counter with charge division readout. The resolution is $18\%/\sqrt{E(GeV)}$ and the chamber is efficient to detect photons with energies as low as 50 MeV/c^2. Beyond the shower counter is the magnet coil and flux return steel. Mounted outside the steel are two double layers of proportional tubes separated by 5" steel plates to provide muon detection. The table 1.2 summarizes relevant features.

The detectors have evolved with slightly better charged momentum resolution, but the important improvements are the low energy photon detection, larger solid angle coverage and better TOF resolution. In J/ψ physics virtually all reactions were exclusive topologies. They include reactions such as $J/\psi \rightarrow \gamma\omega\omega$ which has four charged tracks and 5 photons. Many reactions were possible to measure because of kinematic constraint fits in topologies with photons improved the mass resolution in the 10-20 MeV/c^2 range. This was the key to overcoming the poor energy resolution of the shower counters.

Future detectors could improve the performance in several areas. This includes TOF measurement on the endcaps and high resolution measurements near the beam to enable reconstruction of soft tracks and very forward tracks.

Table 1.2 e^+e^- Detectors for J/ψ Physics

	σ_p/p (1 GeV)	σ_{TOF}	Shower E @ 100% eff.	Shower Counter Coverage
MARK I	20 MeV	400 pc	—	—
MARK II	18 MeV	300 pc	~400 MeV	65%
DM2	35 MeV	550 pc	~100 MeV	70%
MARK III	21 MeV	190 pc	~100 MeV	94%
CRYSTAL BALL	26 MeV for E_γ	—	—	93%

2. Early J/ψ Measurements

The early measurements of the J/ψ properties include the width, the spin-parity, G-parity, isospin and the SU(3) classification. These measurements were done by the Mark I, DASP, and Frascati groups.

2.1 J/ψ WIDTH

The resonant amplitude for a resonance produced in e^+e^- production that decays into a final state f

$$e^+e^- \to J/\psi \to f$$

at a center mass energy $\sqrt{s}$ and spin J is,

$$\sigma_f = \pi \frac{2J+1}{S} \frac{\Gamma_e \Gamma_f}{(m-\sqrt{s})^2 + (\Gamma/2)^2}$$

where Γ_e and Γ_f are the partial widths into e^+e^- and f, respectively. Experimentally, the resolution of the storage ring (SPEAR has an energy spread of 2.6 MeV) is too poor to measure the width directly. Instead, the widths are obtained by measuring the integral of the rate with a scan which will include the gaussian beam resolution. Defining $\Sigma_f = \int \sigma_f \, d\sqrt{s}$, then

$$\Sigma_e = \frac{6\pi^2}{m^2} \frac{\Gamma_e^2}{\Gamma} , \quad \Sigma_\mu = \frac{6\pi^2}{m^2} \frac{\Gamma_e \Gamma_\mu}{\Gamma} , \quad \Sigma_h = \frac{6\pi^2}{m^2} \frac{\Gamma_e \Gamma_h}{\Gamma}$$

where Σ_e, Σ_μ, and Σ_h are the integrated e^+e^-, $\mu^+\mu^-$ and hadronic rates, respectively, over $\sqrt{s}$. With these quantities, the partial e^+e^-, $\mu^+\mu^-$, and hadronic widths can be derived. The measurements from the Mark I,[9] $\gamma\gamma$ (ADONE),[10] MEA (ADONE),[11] and DASP[12] groups and the Particle Data Group[13] summary are shown in table 2.1.

Table 2.1 J/ψ Widths

Γ_e (KeV)	Γ_μ (KeV)	Γ_h (KeV)	Γ (KeV)	Group
4.8 ± 0.6	4.8 ± 0.6	59 ± 14	69 ± 15	Mark I
4.6 ± 0.8	—	59 ± 24	68 ± 26	$\gamma\gamma$ (ADONE)
4.6 ± 1.0	5.0 ± 1.0	50 ± 25	60 ± 25	MEA(ADONE)
4.4 ± 0.6	4.4 ± 0.6	—	58 ± 13	DASP
4.60 ± 0.39	4.85 ± 0.51	57.3 ± 10.9	63.0 ± 8.6	PDG

2.2 J/ψ Spin-Parity

The J/ψ spin-parity was determined from a study of the interference between $e^+e^- \to J/\psi \to \mu^+\mu^-$ and $e^+e^- \to \gamma \to \mu^+\mu^-$. For the different J/ψ spin-parity assignments the coupling is,

$$J^{PC} \qquad \frac{d\sigma}{d\Omega}$$

$$0^{-+} \quad \frac{2}{3}\frac{\alpha^2}{S}\left[\frac{3}{8}(1+\cos^2\theta) + \frac{g^4}{e^4}\frac{s^2}{(m^2-s)^2+m^2\Gamma^2}\right]$$

$$1^- \quad \frac{\alpha^2}{4s}(1+\cos^2\theta)\left[\frac{1-2g^2}{e^2}\frac{(m^2-s)s}{(m^2-s)^2-m^2\Gamma^2} + \frac{g^4}{e^4}\frac{s^2}{(m^2-s)^2+m^2\Gamma^2}\right]$$

$$1^+ \quad \frac{\alpha^2}{4s}\left[(1+\cos^2\theta)(1+\frac{g^4}{e^4}\frac{s^2}{(m^2-s)^2+m^2\Gamma^2}) - 2\cos\theta\frac{g^2}{e^2}\frac{(m^2-s)s}{(m^2-s)^2+m^2\Gamma^2}\right]$$

where $g^2 = 12\pi\Gamma_e/m$. The 1^{--} case has $\sigma_{\mu^+\mu^-}$ interference and the 1^+ case has a forward-backward asymmetry $(F - B/F + B)$. These measurements were done by the Mark I[14] and the ADONE[15] groups. The $\mu^+\mu^-$ cross section had a dip slightly below the peak of the J/ψ and the forward backward asymmetry of the $\mu^+\mu^-$ pairs was consistent with spin 1.

2.3 G-Parity and Isospin

The Mark I group[15] measured the pion multiplicity,

$$J/\psi \to n\pi\text{'s}$$

on and off the J/ψ peak. The ratio of the number of events with pions on and off the peak was about a factor 5-7 larger for odd pion decays (3,5,7) relative to even pion decays (4,6). This demonstrated that the G-parity was odd.

The isospin of the J/ψ can be determined by a study of the three body pion decay. In this topology the dominant mode is $J/\psi \to \rho\pi$. If the J/ψ is an isoscalar then,

$$B(J/\psi \to \pi^+\rho^-) = B(J/\psi \to \pi^-\rho^+) = B(J/\psi \to \pi^0\rho^0)$$

The Mark I group measured,

$$\frac{B(\rho^0\pi^0)}{B(\rho^+\pi^-) + B(\rho^-\pi^+)} = 0.59 \pm .17$$

which is consistent with $I = 0$.

2.4 SU(3) Classification

The SU(3) classification of the J/ψ can be identified by comparing the decays into pairs of mesons. The pattern of decays determines if the J/ψ is an SU(3) singlet, the eighth component an octet or a mixture. These measurements were performed by the Mark I[16] group in a study of hadronic decays. If the J/ψ has odd charge conjugation and it is an SU(3) singlet it cannot decay into K^+K^- or $K^*\bar{K}^*$ but instead into $K\bar{K}^*$ or $K\bar{K}^{**}$. The relative branching ratios are,

$$\frac{B(J/\psi \to K^+K^-)}{B(J/\psi \to K\bar{K}^*)} \cong 10^{-2}$$

For $J/\psi \to$ vector + pseudoscalars, an SU(3) singlet should decay equally into $\pi^+\rho^-$, $\pi^0\rho^0$, $\pi^-\rho^+$, K^+K^{*-}, K^-K^{*+}, $K^0\bar{K}^{*0}$, and $\bar{K}^0K^{*0}$. The branching ratios ($\times 10^{-3}$) are

$$\frac{\pi^+\rho^- + \pi^0\rho^0 + \pi^-\rho^+}{3} : \frac{K^+K^{*-} + K^-K^{*+}}{2} : \frac{K^0\bar{K}^{*0} + \bar{K}^0K^{*0}}{2} = .43 : .32 : .27$$

For $J/\psi \to$ vector + tensor, an SU(3) singlet should decay equally into $\rho^0 A_2$, $K^{*0}\ \bar{K}^{**0}$, ωf and $\phi f'$. Including phase space corrections the ratio of rates are,

$$1.0 : 0.9 : 1.01 : 0.78 = 1.0 : 1.0 \pm .5 : 0.7 \pm .3 : 0.3 \pm .2$$

where the left side is the prediction and the right side the branching ratios. The combined pattern of two-body decays is consistent for the J/ψ to be identified as an SU(3) singlet.

3. Introduction to J/ψ Models

In this section simple theoretical models are discussed to provide guidelines for an experimentalist understanding of J/ψ decays.[19] In the first section the general features are explained including the OZI rule and QCD predictions. The next section covers hadronic decays and the use of SU(3) symmetry. The last two sections cover radiative decays into conventional $q\bar{q}$ mesons and two gluon states.

3.1 General Features

The J/ψ is a vector, $J^{PC} = 1^{--}$ ($^3S_1$) bound state of $c\bar{c}$. The puzzling aspect of this high mass state was it was very narrow width, $\Gamma \simeq 70$ KeV. The qualitative explanation was given by the Okubo, Zweig, Iizuka (OZI) rule.[20] The rule states that decays with disconnected quark lines are suppressed relative to decays with continuous quark lines. Another paraphrase of this rule is decays are suppressed if initial state quarks annihilate and final state quarks materialize.

Examples of OZI allowed decays are $\phi \to K^+K^-$ and $\psi'' \to D^+D^-$, $D^0\bar{D}^0$. The decay rates are listed in table 3.1.

Table 3.1 OZI allowed decays

Decay	Percentage
$\phi \to K^+K^-$	50%
$\phi \to K_SK_L$	35%
$\phi \to 3\pi$	15%
$\psi'' \to D^+D^-$	$\sim$ 50%
$\psi'' \to D^0\bar{D}^0$	$\sim$ 50%

The OZI violating ϕ decay into 3π is very suppressed because the decay into K^+K^- and K_SK_L is actually phase space suppressed and hence the matrix element for the OZI allowed relative to the suppressed decay is very large. The ψ'' is above $c\bar{c}$ threshold and decays into open charm whereas the J/ψ is below $c\bar{c}$ threshold and cannot decay into states with open charm. Hence, the initial charm quarks annihilate and the final state light quarks materialize. The pattern of J/ψ decays, in contrast to the above table, is strikingly different. Instead of a few large decays there are many small two body decays that are at most a few tenths of a percent.

The quantitative aspects of the OZI rule and the J/ψ decays are provided by Quantum Chromodynamics (QCD). The key features for this discussion are;

1) Quarks are color triplets

2) Quark interactions occur via an octet of 8 massless color gauge bosons called gluons.

3) Quark-gluon strong coupling constant, $\alpha_s(Q^2)$ depends on the momentum transfer, Q^2 and as Q^2 increases $\alpha_s(Q^2)$ decreases.

4) Gluons can self interact.

5) Hadrons are color singlets and gluons are flavorless.

The QCD predictions for the J/ψ width give quantitative evidence for the third feature or what is often called asymptotic freedom.[21] The leptonic width procedes via the annihilation diagram, and the rate is

$$\Gamma(J/\psi \to e^+e^-) = \frac{4\alpha^2 q^2}{m^2} \mid R(0) \mid^2$$

The hadronic decay procedes via the three gluon diagram,

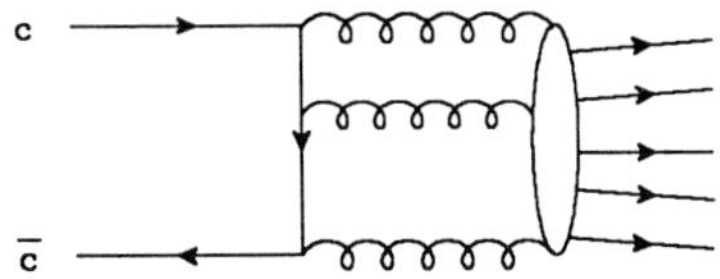

Diagram 1. Hadronic J/ψ Decays

and the rate is

$$\Gamma(J/\psi \to 3g) = \frac{40}{81\pi}(\pi^2 - 9)\frac{\alpha_s^3}{m^2} \mid R(0) \mid^2$$

The ratio becomes,

$$\frac{\Gamma(J/\psi \to \text{hadrons})}{\Gamma(J/\psi \to e^+e^-)} = \frac{5}{18\pi}(\pi^2 - 9)\frac{\alpha_s^2}{\alpha^2}(\frac{2/3}{q})^2$$

and since, $\Gamma(J/\psi \to e^+e^-) = 4.8$ and $\Gamma(J/\psi \to \text{hadrons}) = 69$ KeV we obtain, $\alpha_s(m = 3.1) \cong 0.2$. When this is applied to the ϕ we obtain $\alpha_s(m = 1.02) \cong 0.5$ and this is quantitative evidence for asymptotic freedom.

3.2 Hadronic Decays

The J/ψ, as explained in Section 2.4, is an SU(3) singlet. The two body decays into pairs of SU(3) octets can be predicted if SU(3) symmetry were exact. An example is the decay $J/\psi \to$ vector + tensor. The exact SU(3) symmetric decay would predict,

$$\rho A_2 : \omega f : \phi f' = 3 : 1 : 1$$

The measurements correcting for phase space and normalizing to the ρA_2 rate are

$$\rho A_2 : \omega f : \phi f' = 3 : 0.8 \pm 0.3 : 0.5 \pm .2$$

In the vector + pseudoscalar modes, the ratio of $\rho\pi$ to $K^{\pm}K^{\mp *}$ is expected to be 1.5 and the measurement is ~ 3. These results demonstrate that the J/ψ behaves approximately as an SU(3) singlet but there are corrections. These corrections[22] includes SU(3) breaking terms for the quark mass differences and the electromagnetic diagrams. The Rosner formulation[23] defined the wavefunction as a combination of non-strange, strange and non-$q\bar{q}$ components,

$$x| \frac{u\bar{u}+d\bar{d}}{2}\rangle + y| s\bar{s}\rangle + z| gg\rangle$$

where $| gg\rangle$ represents the non-$q\bar{q}$ part of the wavefunction and x, y and z are coefficients. This formulation provided a simple framework to connect vector radiative decays and the $\gamma\gamma$ widths of pseudoscalar and tensors. These methods are techniques of measuring the quark content of mesons with electromagnetic probes ($\gamma\gamma$ widths and radiative decays) and strong probes (hadronic J/ψ decays).

The early predictions were applied to understand the quark content of the η and η' in $J/\psi \rightarrow$ vector + pseudoscalar decays. The model predicted the non-$q\bar{q}$ content of the η' was 35%. An extension[24] of this model to include the double OZI (DOZI) violating terms,

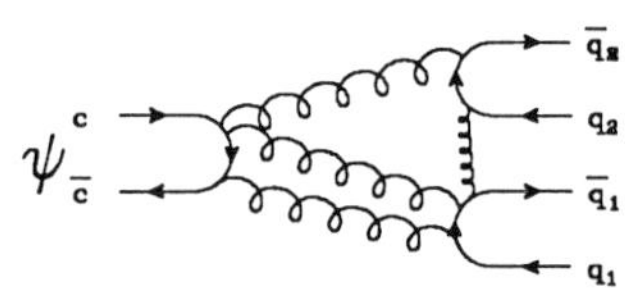

Diagram 2. Double OZI violating J/ψ Decays

produced a fit where the η' was wholly made up of $q\bar{q}$. The model had a $\sim 10\%$ DOZI violation and this yielded a η, η' mixing angle of $\theta = -19^o$ which is quite compatible with the mixing angle from $\gamma\gamma$ experiments on the η and η'.[25]

3.3 RADIATIVE J/ψ DECAYS TO CONVENTIONAL $q\bar{q}$ STATES

In this section the radiative decays to conventional $q\bar{q}$ states are discussed for the pseudoscalars and tensors. The basic diagrams[26] for these decays are,

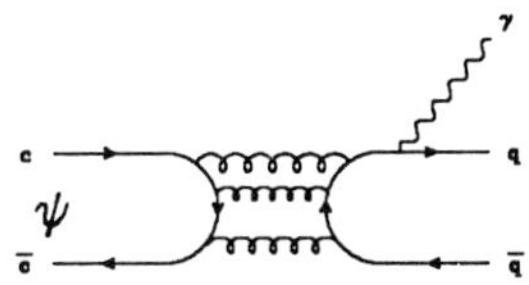

Diagram 3. Radiative diagram I.

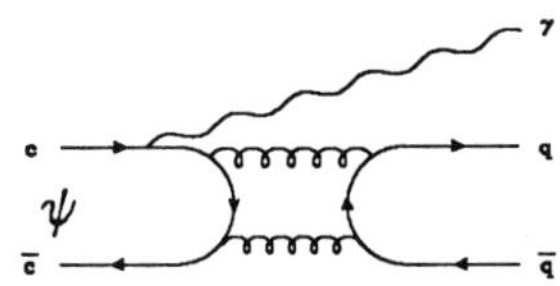

Diagram 4. Radiative diagram II.

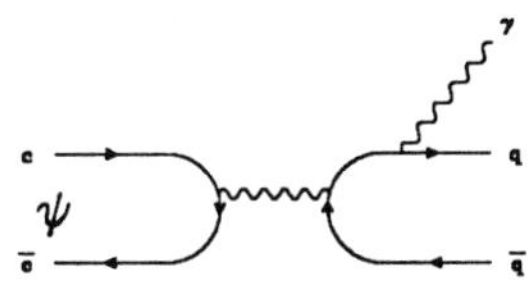

Diagram 5. Radiative diagram III.

In diagram I, the photon emission occurs after the OZI emission. The $q\bar{q}$ states produced in this way have the SU(3) quantum numbers of the photon. In diagram II, the photon emission occurs before the OZI emission and the $q\bar{q}$ state is an SU(3) singlet. In the last diagram the J/ψ decay occurs via a virtual photon and the $q\bar{q}$ state has the same quantum numbers as resonances seen in $\gamma\gamma$ production. The rates into $J/\psi \to \gamma + (A_2,\ f,\ f'),\ (\pi^0, \eta, \eta')$ are shown in table 3.3a.

Table 3.3a Tensor and Pseudoscalar Predictions

Diagram	A_2	f	f'	π^0	η	η'
I	9	1	2	3	$\cos^2\theta$	$\sin^2\theta$
II	0	4	2	0	$\sin^2\theta$	$\cos^2\theta$
III	9	25	2	3	$(\cos\theta - \sin\theta\ 2\sqrt{2})^2$	$(\sin\theta - \cos\theta\ 2\sqrt{2})^2$

where $\eta = \eta_8\ \cos\theta + \eta_1\ \sin\theta,\ \eta' = -\eta_8\ \sin\theta + \eta_1,\ \cos\theta$. From the $\gamma\gamma$ experiments[24] on the η and η', the angle is $\theta \cong -19.3°$. The experimental branching ratios are shown in table 3.3b.

Table 3.3b Tensor and Pseudoscalar Rates

Mode	Rate (%)
$J/\psi \to \gamma A_2$	not seen
$J/\psi \to \gamma\pi^0$	0.004 ± 0.001
$J/\psi \to \gamma f$	$0.136 \pm 0.008 \pm .022$
$J/\psi \to \gamma f'$	$0.06 \pm .014 \pm .012$
$J/\psi \to \gamma\eta$	$0.086 \pm .008$
$J/\psi \to \gamma\eta'$	$0.42 \pm .05$

The f, f' ratios, normalized by phase space, are

$$\frac{B(J/\psi \to \gamma f)/p_f^5}{B(J/\psi \to \gamma f')/p_{f'}^5} \cong 1.4$$

where $B(f' \to K\bar{K}) = 1.0$ is assumed. If the f and f' are ideally mixed we expect 2. The η, η' rates, normalized by phase space, are

$$\frac{B(J/\psi \to \gamma\eta)/P_\eta^3}{B(J/\psi \to \gamma\eta')/P_{\eta'}^3} \cong .17$$

This ratio predicts a mixing angle of 22° which agrees with $\gamma\gamma$ experiments. The $J/\psi \rightarrow \gamma\pi^0$ rate is very small and can be related via the vector dominance model to $J/\psi \rightarrow \rho\pi$. The overall pattern of decays agrees very well with diagram II. The simple picture of diagram II reproduces the data and it may be unnecessary to add gluonic or $c\bar{c}$ components to the η or the η'.[17,18]

3.4 Radiative Decays to Gluonic States

The existence of colorless, flavorless, quarkless isoscalar states made up of gluons was suggested soon after the J/ψ discovery.[28] The gluonium states may be produced in radiative J/ψ decays in the following diagram,

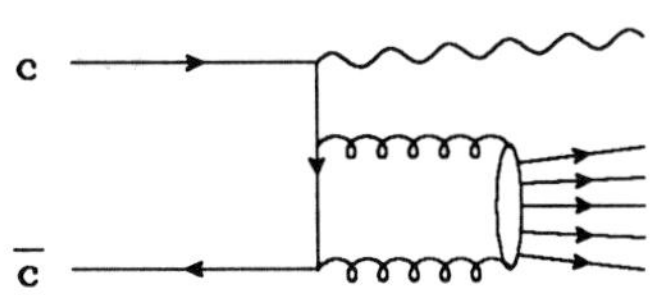

Diagram 6. Radiative J/ψ Decays

The rate, to order α_s, is

$$\frac{\Gamma(J/\psi \rightarrow \gamma gg)}{\Gamma(J/\psi \rightarrow ggg)} = \frac{36\,\alpha q^2}{5\,\alpha_s} = \frac{16\,\alpha}{5\,\alpha_s} \approx 10\%$$

A naive picture for two gluon resonances is to treat the gluons in analogy to the quark model.[29] Adding up the orbital and spin angular momentum and forming color singlets which are symmetric under interchange of gluon indices yields the spin-parity states listed in table 3.4.

Table 3.4 J^{PC} spin-parity of 2 gluon states

L	S=0	S=1	S=2
0	0^{++}	—	2^{++}
1	—	$0^{-+}, 1^{-+}, 2^{-+}$	—
2	2^{++}	—	$0^{++}, 1^{++}, 2^{++}, 3^{++}, 4^{++}$

The states should be even g-parity and if the gluons are on mass shell, spin 1 states should be suppressed. Neither odd pion states nor spin 1 states are seen in radiative J/ψ decays.

In addition, the gluonic states should have the following features,

1. Flavor Symmetry
2. Masses in the 1-2 GeV/c^2 Region
3. Suppressed in $\gamma\gamma$ Production
4. Widths, 10-30 MeV/c^2

Flavor symmetry follows from being formed from two flavorless gluons.[30] The decays of a state into pairs of pseudoscalars should follow,

$$\mathrm{K\bar{K}} : \eta\eta : \pi\pi = 3 : 0.5 : 6$$

Flavor symmetry may be altered from other considerations. It has been suggested that cavity perturbation in the MIT bag model will depend on the quark mass and could produce a larger coupling to heavier quark states. The coupling of gluons to quarks may enhance $s\bar{s}$ production relative to $u\bar{u} + d\bar{d}$.[31] This enhancement of strange meson production in glueball candidates, the θ and the ι, is observed. Other general arguments have been applied to glueball decays into two pseudoscalar.[32] It may be possible that flavor symmetry is changed by rescattering effects. These could cause large corrections as are seen the Cabbibo depressed decays $D^0 \to \pi^+\pi^-$ and $D^0 \to K^+K^-$.[27]

The bag model[33] provides qualititive estimates of glueball masses. In this model, gluons are confined to a spherical cavity. The modes of the gluon fields are transverse electric (TE) with parity $(-1)^{J+1}$ and transverse magnetic (TM) with parity $(-1)^J$. The boundary conditions lead to equations of constraint which enable mass predictions. The lowest lying $(TE)^2$ gluon states will have a mass near one GeV/c^2 with $J^{PC} = 0^{++}$ and 2^{++}. The lowest lying $(TM)^2$ gluon states will have a mass near 1.6 GeV/c^2 with $J^{PC} = 0^{++}$ and 2^{++}. The mixed (TM)(TE) gluon states will have a mass near 1.3 GeV/c^2 with $J^{PC} = 0^{-+}, 1^{-+}, 2^{-+}$.

In addition to the Bag Model, QCD lattice gauge theories are refining calculations and estimate a scalar glueball mass near 1 GeV. Unfortunately most of these predictions of the glueball mass spectrum have not had the success the charmonium model has had in predicting the mass spectrum of the χ states. One major problem is the lack of any evidence for scalar glueball candidates in J/ψ radiative decays.

The widths of glueball candidates have been expected to be narrow, 10-30 MeV. This follows from a loose interpretation of the OZI rule. The gluons in the glueball would annihilate and create a $q\bar{q}$ pair that would form the meson secondaries. This process is OZI suppressed and should cause the state to be narrow. This explanation is controversial and not widely accepted.[34]

The glueball states are expected to be suppressed in $\gamma\gamma$ production since they have no charge. The $\gamma\gamma$ production of the pseudoscalars, η and η', and the tensors, f and f', are well understood. The partial widths couple as, $\Gamma_{\gamma\gamma} \propto q^4$. A means of quantifying the glueball character of states has been suggested by Chanowitz.[35] The ratio, called stickyness, is defined for a state X as

$$S_X = \frac{\Gamma(J/\psi \to \gamma X)/PS(J/\psi \to \gamma X)}{\Gamma(X \to \gamma\gamma)/PS(X \to \gamma\gamma)}$$

The stickyness ratio will be large for gluonium states. We expect resonances with large gluonic components to be produced in radiative J/ψ decays and resonances with large $q\bar{q}$ components to be produced in $\gamma\gamma$ productions. This pattern is observed for the iota and theta mesons as will be discussed in later sections.

4. η_c Decays

The η_c ($J^{PC} = 0^{-+}$, $^1S_0$) is the lowest lying $c\bar{c}$ state. Since it is spin zero it will couple to two photons or two gluons. The two photon diagram,[36]

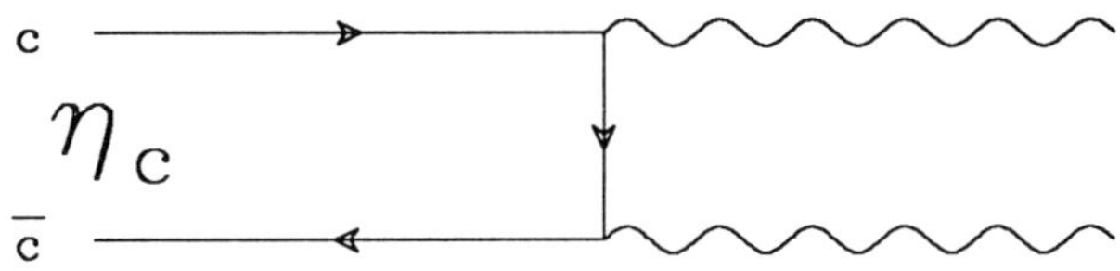

Diagram 7. Two gamma decay of the η_c

will have a partial rate,

$$\Gamma(\eta_c \to \gamma\gamma) = \frac{4\alpha^2}{m^2}\, q^4 \mid R(0) \mid^2 \;*\; 3$$

and the two gluon decay diagram,

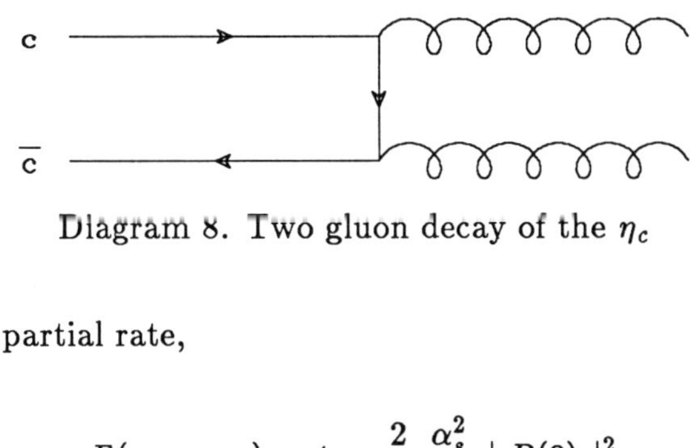

Diagram 8. Two gluon decay of the η_c

will have a partial rate,

$$\Gamma(\eta_c \to gg) = 4 \;*\; \frac{2}{3}\,\frac{\alpha_s^2}{m^2} \mid R(0) \mid^2$$

These partial widths can be compared to the J/ψ decays, yielding

$$\frac{\Gamma(\eta_c \to \text{hadrons})}{\Gamma(J/\psi \to \text{hadrons})} = \frac{\Gamma(\eta_c \to \gamma\gamma)}{\Gamma(J/\psi \to e^+e^-)} = 3\, q^2$$

which predicts $\Gamma(\eta_c \to \gamma\gamma) = 9$ KeV.

The transition rates for a quark spin flip will be an M1 transition for the radiative decay from a vector to a pseudoscalar,

$$\Gamma(^3S_1 \to ^1 S_0) = \frac{16}{3}\,(2J_f+1)\left(\frac{q}{2m_q}\right)^2 \alpha \mid m_{if} \mid^2 \omega^3.$$

For $m(J/\psi) = 3.095$ and $m(\eta_c) = 2.980$, the width is predicted to be

$$\Gamma(J/\psi \to \gamma\eta_c) = 2.4 \text{ KeV}$$

and yields a branching ratio, $B(J/\psi \to \gamma\eta_c) \cong 3\%$.

4.1 η_c HISTORY

The first claim for the observation of the η_c was from the DASP[37] group in the mode, $J/\psi \to \gamma\ \eta_c$, $\eta_c \to \gamma\gamma$. The mass was $2.82 \pm .014$ GeV/c^2 and the rate was $B(J/\psi \to \gamma\eta_c) \cdot B(\eta_c \to \gamma\gamma) = (1.4 \pm 0.4) \times 10^{-4}$. Theoretically, the $m(J/\psi) - m(\eta_c)$ mass difference and the product branching ratios were both too large. The Crystal Ball group[38] searched for this decay in $J/\psi \to \gamma\gamma\gamma$ and set a limit of $< 1.0 \times 10^{-4}$ at 90% C.L. Later the Crystal Ball and MARK II groups had evidence for the η_c mass near 2.98 GeV/c^2. The Crystal Ball group[39] looked into the inclusive modes $J/\psi \to \gamma + X$ and $\psi' \to \gamma + X$ and observed a state in both data sets at $m(\eta_c) = 2984 \pm 5$ Mev/c^2 with a width of $\Gamma(\eta_c) = 11.5^{+4.5}_{-4.0}$ MeV/c^2 and branching ratios of $B(J/\psi \to \gamma\eta_c) = (1.27 \pm 0.36)\%$ and $B(\psi' \to \gamma\eta_c) = (0.28 \pm 0.06)\%$. The MARK II group[40] looked into ψ' radiative into the exclusive states $p\bar{p}$, 4π, $\pi\pi KK$, $\pi\pi p\bar{p}$ and $K_S K^\pm \pi^\mp$. They observed a peak at 2980 ± 8 MeV/c^2 and a width < 40 MeV/c^2.

4.2 HADRONIC η_c MEASUREMENTS

The MARK III[41] and the DM2[42] groups have continued to study the η_c by measuring many η_c hadronic decay modes in radiative J/ψ decays. The η_C mass distributions for different hadronic decay modes from the MARK III group are shown in Fig. 3. The various decay modes are summarized in table 4.2. The MARK III masses have an average of 2982 MeV, the DM2 value is about 10 MeV lower. The most precise mass measurement is $2982^{+2.7}_{-2.3}$ MeV from the ISR $p\bar{p}$ gas jet experiment.[43]

The $\eta_c \to$ vector - vector decays disagree between the MARK III and DM2 values. For an SU(3) singlet η_c decay the vector decays should be

$$K^{*0}K^{*0} : \phi\phi : \rho\rho : \omega\omega = 0.5 : 1 : 1 : 1$$

The MARK III values of the ratios are

$$K^{*0}K^{*0} : \phi\phi : \rho\rho : \omega\omega = \ .22 \pm .13 \ : \ 1 \ : < 0.64 \ : < .34$$

and for the DM2 values the ratios are

$$\phi\phi : \rho\rho = 1 : 1.66 \pm 0.5$$

The DM2 value for the $\phi\phi$ channel is lower than the MARK III value and the DM2 observes a $\rho^0\rho^0$ signal whereas the MARK III group sets an upper limit. The MARK III η_c publication discussed the trend of larger rates with $s\bar{s}$ content. The two DM2 measurements are compatible with exact SU(3) flavor symmetry.

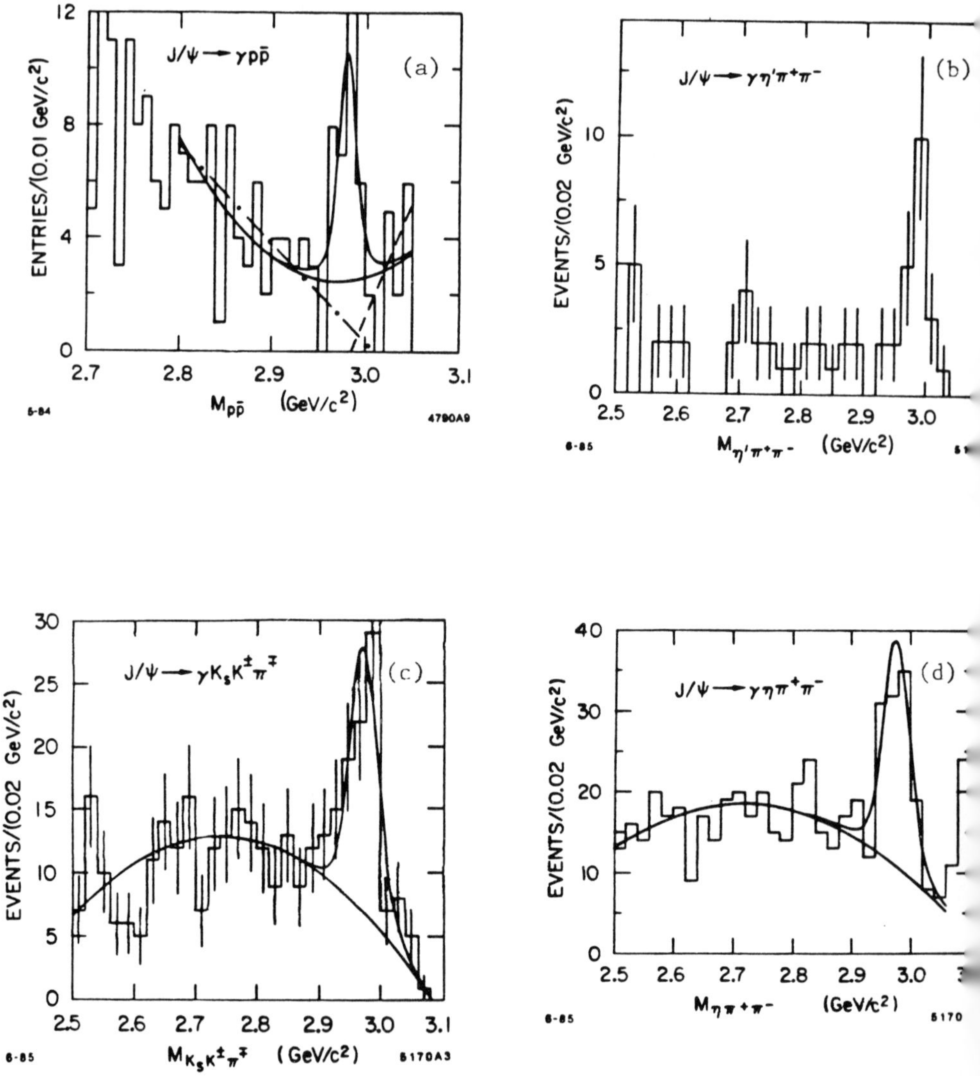

Figure 3.

The η_C mass distributions for $\eta_C \to p\bar{p}$ (a), $\eta_C \to \eta' \pi^+ \pi^-$ (b), $\eta_C \to K_S K^{\pm} \pi^{\mp}$ (c), and $\eta_C \to \eta \pi^+ \pi^-$ (d) from the MARK III group.[41]

4.3 η_c Spin-Parity Test

The η_c spin-parity was first measured by MARK III in the sequential decay, $J/\psi \to \gamma\eta_c$, $\eta_c \to \phi\phi$. In the decay sequence where both ϕ's decay into K^+K^-, the ϕ decay planes are orthogonal for odd parity and parallel for even parity.[44] This can be generalized for arbitrary spin.[45]

The most sensitive variable to test the spin is the azimuthal angle, χ, between the ϕ decay planes. When integrated over all other angles, the distribution has the form

$$\frac{dn}{d\chi} = 1 + \beta\cos(2\chi)$$

where β will different values depending on the spin, The χ angle distribution was fit for different spin-parity hypothesis and compared to $J^{PC} = 0^{-+}$. The angular distributions are shown in in Fig. 4. The MARK III[46] and DM2[47] results are shown in table 4.3.

Table 4.3 η_c spin-parity tests

J^P	$L_{\phi\phi}$	β	MARK III Likelihood	DM2 Likelihood
0^-	1	-1	1	1
0^+	0	$+.667$	1.8×10^8	1.2×10^{12}
1^-	1	0	2200	2.8×10^4
1^+	2	0	2200	2.8×10^4
2^+	0	$+.07$	5100	8.3×10^4
2^-	1	$-.4$	55	192
2^-	3	$-.6$	12	26
0^+	2	$+.333$	-	1.0×10^7

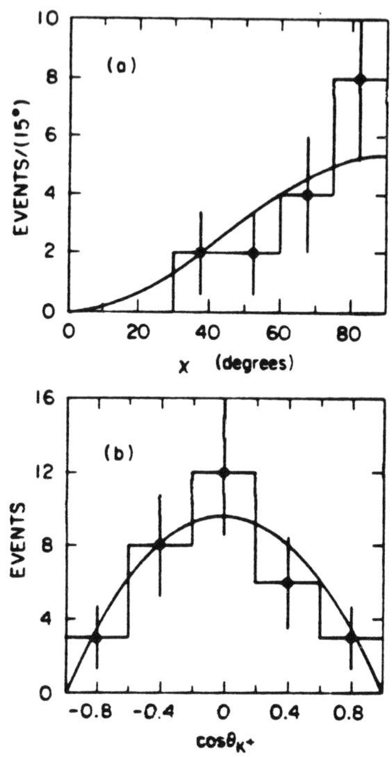

Figure 4.

The χ angle (a) and the K^+ angle (b) in the ϕ rest frame distributions with the pseudoscalar hypothesis curve from the MARK III group.[46]

All values relative to $J^P = 0^-$ are largely excluded. The test can be extended to include all the angles to the joint distribution

$$\frac{d^3n}{d\chi d\, \cos\theta_1\, d\, \cos\theta_2} = -\,\beta \sin^2\theta_1 \sin^2\theta_2 \sin^2\chi + \frac{1}{2}(1+\beta)$$
$$(\sin^2\theta_1\cos^2\theta_2 + \cos^2\theta_1\sin^2\theta_2 + \frac{1}{2}\sin 2\theta_1 \sin 2\theta_2 \cos\chi)$$

where θ_1 and θ_2 are the polar angles of the two K^+'s in their respective ϕ rest frames. The likelihood rejection of the $J^P = 2^-$ (F wave) case improves to 120 (MARK III) and 150 (DM2).

5. Review of $J/\psi \to$ Vector + Pseudoscalars

The hadronic decay $J/\psi \to$ Vector + Pseudoscalars has been extensive measured by the MARK III group[51] and recently by the DM2 group.[49] The Dalitz plot of the decay $J/\psi \to \rho\pi$ from MARK III group is shown in Fig. 5. The results are summarized in table 5.

Table 5. Vector-Pseudoscalar rates

Decay Mode	Final State	Branching Ratio $\times 10^{-3}$	
		MARK III	DM2
$\rho\pi$	$\pi^+\pi^-\pi^0$	$13.3 \pm 0.3 \pm 1.5$	12.7 ± 9
$K^{*+}K^- + K^{*-}K^+$	$K^\pm\pi^\mp K_S$ $K^+K^-\pi^0$	$5.0 \pm 0.23 \pm 0.6$ $6.0 \pm 0.3 \pm 0.7$	5.4 ± 4.4 —
$K^{*0}\overline{K^0} + \overline{K^{*0}}K^0$	$K^\pm\pi^\mp K_S$	$3.9 \pm 0.2 \pm 0.6$	$3.9 \pm .6$
$\omega\eta$	$\pi^0\pi^+\pi^-\eta$ $\pi^0\pi^+\pi^-\pi^+\pi^-\pi^0$	$1.9 \pm 0.2 \pm 0.3$ —	1.5 ± 2.5 1.2 ± 2.2
$\omega\eta'$	$\pi^0\pi^+\pi^-\gamma\rho^0$ $\pi^0\pi^+\pi^-\eta\pi^+\pi^-$	$0.39 \pm 0.11 \pm 0.06$ $0.43^{+.19}_{-.22} \pm 0.07$	$.40 \pm .11$ —
$\phi\eta$	K^+K^- + neutrals $K^+K^-\eta$ $K^+K^-\pi^0\pi^+\pi^-$	$0.69 \pm 0.07 \pm 0.08$ $0.64 \pm 0.15 \pm 0.08$ $0.61 \pm 0.14 \pm 0.08$	— $.58 \pm .16$ $.70 \pm .13$
$\phi\eta'$	$K^+K^-\gamma\rho^0$ $K^+K^-\eta\pi^+\pi^-$	$0.39 \pm 0.10 \pm 0.06$ $0.365 \pm 0.04 \pm 0.05$	$.45 \pm .08$ $.36 \pm .07$
$\omega\pi^0$	$\pi^0\pi^+\pi^-\pi^0$	$0.67 \pm 0.06 \pm 0.11$	$.51 \pm 0.8$
$\rho^0\eta$	$\pi^+\pi^-\eta$	$0.18 \pm 0.02 \pm 0.04$	$.19 \pm .05$
$\rho^0\eta'$	$\pi^+\pi^-\eta\pi^+\pi^-$	< 0.1	$.08 \pm .03$
$\phi\pi^0$	$K^+K^-\pi^0$	< 0.013	< 0.013

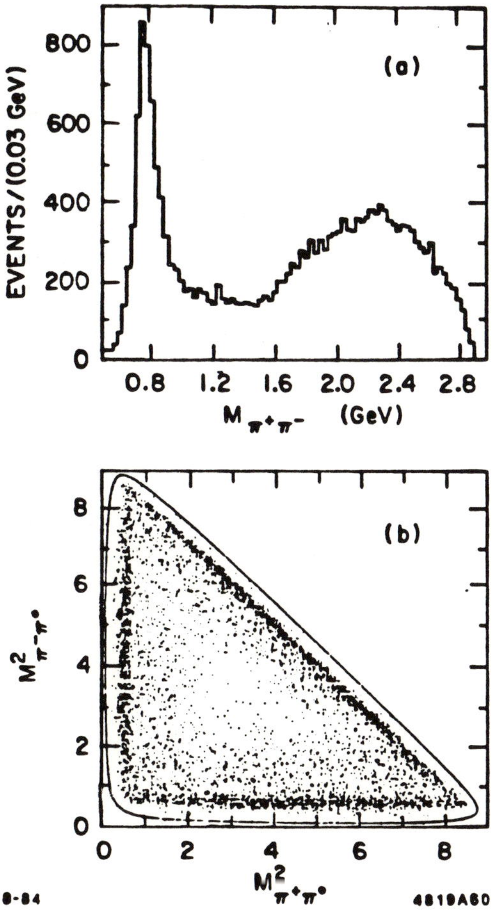

Figure 5.

The $\pi^+\pi^-$ mass distribution (a) and the Dalitz plot (b) of the decay $J/\psi \to \rho\pi$ from the MARK III group.[48]

These measurements enable tests of the quark content of mesons. The J/ψ hadronic decay should follow the decay of an SU(3) singlet with small electromagnetic and SU(3) corrections. There are purely electromagnetic decays which are observed ($J/\psi \to \rho\eta$ and $\omega\pi^0$). The MARK III publication included a model that estimated the quark content of the η and η' and found the $q\bar{q}$ content accounting for $\sim 65\%$ of the η'. The rest could be attributed to gluons. Later the model was refined[24] to include small OZI breaking terms (and another parameter) which now predicts the η' to be fully accountable in term of $q\bar{q}$ content.

The total decay rate of $J/\psi \to$ vector + pseudoscalars has been a puzzle. When compared to the ψ' decay rates for vector + pseudoscalars the J/ψ rates are much too large. The relative rates should be proportional to the leptonic rates,

$$\frac{B(\psi' \to hadrons)}{B(J/\psi \to hadrons)} \sim \frac{B(\psi' \to e^+e^-)}{B(J/\psi \to e^+e^-)} = 0.135 \pm 0.023$$

but the $\rho\pi$ and $K^*\bar{K}$ rates are a factor 20 too large for the J/ψ. Several predictions claim that this could be evidence for a vector glueball state near 3 GeV that mixes with the J/ψ causing an anomalous rate in certain vector+pseudoscalar modes.[52]

The hadronic decays are being extended to $J/\psi \to$ Vector + Tensor and Vector + Scalar. Since the quark content of the tensor nonet is well known, precise measurements in this sector will enable further tests of the models. In the decays into Vector + Scalar, the $J/\psi \to \phi S^*$ is quite evident whereas the decay $J/\psi \to \rho\delta$ is not observed.[50] If the δ isovector member of the scalar nonet, $J/\psi \to \rho\delta$ should be as large in the vector-scalar decay as $J/\psi \to \rho\pi$ is in the vector-pseudoscalar decays. This is additional evidence suggesting that the S^* and the δ are not the members of the scalar nonet.

6. Radiative Decays, $\gamma\pi^0$, $\gamma\eta$, $\gamma\eta'$ γf, $\gamma f'$

The J/ψ radiative decays to conventional pseudoscalars, π^0, η, η' and tensors f, f' are important tests of our understanding of the mechanisms for J/ψ radiative decays and tests of mixing.

6.1 $J/\psi \to \gamma\pi^0$

The decay, $J/\psi \to \gamma\pi^0$ is expected to occur via vector meson dominance through $J/\psi \to \rho^0\pi^0$. The measurements by the DASP,[53] DM2,[54] and Crystal Ball[55] groups are listed in table 6.1.

Table 6.1 $J/\psi \to \gamma\pi^0$ rates

$B(J/\psi \to \gamma\pi^0) \times 10^{-5}$	Group
7.3 ± 4.7	DASP
$3.7 \pm .7 \pm .6$	DM2
$3.6 \pm 1.1 \pm 0.7$	CB

The rate expected from VDM is

$$\Gamma(J/\psi \to \gamma\pi^0) = \frac{4\pi\alpha}{g_\rho^2}\Gamma(J/\psi \to \rho^0\pi^0) \approx 1\ eV$$

This predicts a rate of $\simeq 2 \times 10^{-5}$ which is in agreement with the above measurements.

6.2 $J/\psi \to \gamma'\eta,\ \gamma\eta'$

The radiative decays for the η and η' have been measured into modes, $\eta \to \gamma\gamma$, $\pi^0\pi^+\pi^-$ and $\eta' \to \gamma\gamma$, $\eta\pi^+\pi^-$, $\gamma\rho^0$. All measurements are in excellent agreement, The η results from the CNTR,[56] DASP,[57] Crystal Ball,[58] and DM2[54] groups are listed in table 6.2a.

Table 6.2a $J/\psi \to \gamma\eta$ rates

$B(J/\psi \to \gamma\eta) \times 10^{-3}$	Mode	Group
1.3 ± 0.4	3γ	CNTR
0.82 ± 0.10	3γ	DASP
$0.88 \pm 0.08 \pm 0.11$	3γ	CB
$.85 \pm .39 \pm .15$	$\gamma\gamma\gamma$	DM2

The η' results from the CNTR,[60] DASP,[61] MARK II ,[62] Crystal Ball,[63] DM2,[54] and MARK III[64] groups are listed in table 6.2b.

Table 6.2b $J/\psi \to \gamma\eta'$ rates

$B(J/\psi \to \gamma\eta') \times 10^{-3}$	Mode	Group
2.4 ± 0.7	$\gamma\gamma\rho$	CNTR
2.9 ± 1.1	$\gamma\gamma\gamma$	DASP
3.4 ± 0.7	$\gamma\gamma\rho$	MKII
$4.1 \pm 0.3 \pm 0.6$	$\gamma(\eta\pi\pi, \gamma\gamma, \gamma\rho)$	CB
$4.8 \pm .49 \pm .86$	$\gamma\gamma\gamma$	DM2
$4.7 \pm 0.2 \pm 0.7$	$\gamma\gamma\rho$	MK3

Assuming the decay procedes through an SU(3) flavor singlet coupling; the relative rates are

$$\frac{B(J/\psi \to \gamma\eta')/p^3_{\eta'}}{B(J/\psi \to \gamma\eta)/p^3_{\eta}} = \frac{1}{\tan^2\theta}$$

where θ is mixing angle. The measurements predict an angle of $\theta = -22°$. This is in very good agreement with the angle $\theta = -23°\pm3°\pm1°$ from the Gilman and Kauffman estimate.[25] This agree implies that the $\eta - \eta'$ mixing is quite conventional and not altered by additional gluonic coupling of the η or η'. Many models[65] were developed to explain what appeared to an anomalously large η' rate that was implied when a mixing angle of $\theta = -11^0$ is assumed as predicted by the quadratic mass mixing formula.

6.3 $J/\psi \to \gamma f_2(1285)$, $\gamma f_2'(1515)$

The radiative decays into the tensors, $f_2(1285)$ and $f_2'(1515)$ are summarized in the following tables. The f results from the PLUTO,[66] DASP,[67] Crystal Ball,[68] MARK III ,[69] and DM2[54] group are listed in table 6.3a.

Table 6.3a $J/\psi \to \gamma f$ rates

$B(J/\psi \to \gamma f(1270)) \times 10^{-3}$	Mode	Group
2.0 ± 0.7	$\gamma\pi^+\pi^-$	PLUTO
1.2 ± 0.6	$\gamma\pi^+\pi^-$	DASP
1.38 ± 0.39	$\gamma\pi^0\pi^0$	CB
$2.0 \pm .12 \pm .34$	$\gamma\pi^+\pi^-$	MK3
$1.5 \pm .36 \pm .23$	$\gamma\pi^+\pi^-$	DM2
$1.04 \pm .22 \pm .18$	$\gamma\pi^0\pi^0$	DM2

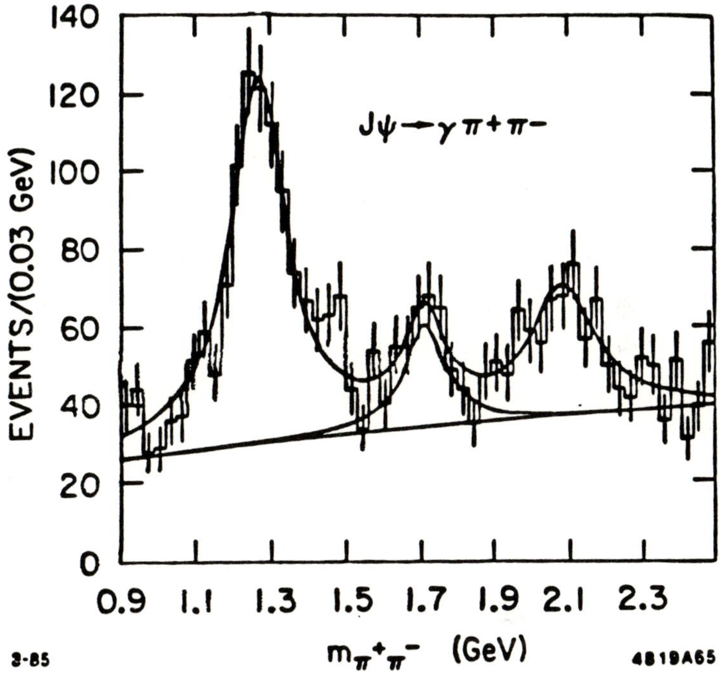

Figure 6.

The $\pi^+\pi^-$ mass distribution in the decay $J/\psi \to \gamma\pi^+\pi^-$ from the MARK III group.[69]

The f' results from the MARK II ,[70] MARK III ,[69] and DM2[54] groups are listed in table 6.3b.

Table 6.3b $J/\psi \to \gamma f'$ rates

$B(J/\psi \to \gamma f') \times 10^{-3}$	Mode	Group
1.8 ± 1.0	γK^+K^-	MK2
$3.0 \pm 0.7 \pm 0.6$	γK^+K^-	MK3
$1.5 \pm 0.3 \pm 0.3$	$\gamma K_S K_S$	DM2
$2.5 \pm 0.6 \pm 0.4$	γK^+K^-	DM2

The $\pi^+\pi^-$ mass distribution in the decay $J/\psi \to \gamma\pi^+\pi^-$ and the K^+K^- mass distribution in the decay $J/\psi \to \gamma K^+K^-$ are shown in Figs. 6 and 7.

The helicity amplitudes for the f have been measured by PLUTO,[71,66] Crystal Ball,[72] MARK III ,[69] and DM2[54] groups and the measurements which included a phase in the fit are listed in table 6.3c.

Table 6.3c f Helicity Amplitudes

x	y	ϕ	Group
0.6 ± 0.1	0.2 ± 0.2	near 0	MARK III
1.1 ± 0.1	0.2 ± 0.1	near 0	DM2

All the values are consistent with $x \cong 1$ and $y \cong 0$ where the helicity 2 amplitude is zero. This has been calculated by Korner *et al.*,[73] to lowest order QCD and x and y are predicted to have roughly equal values. Another model by Li and Shen[74] assume a gluonium hypothesis for the f and obtain a suppression of helicity. A conjecture

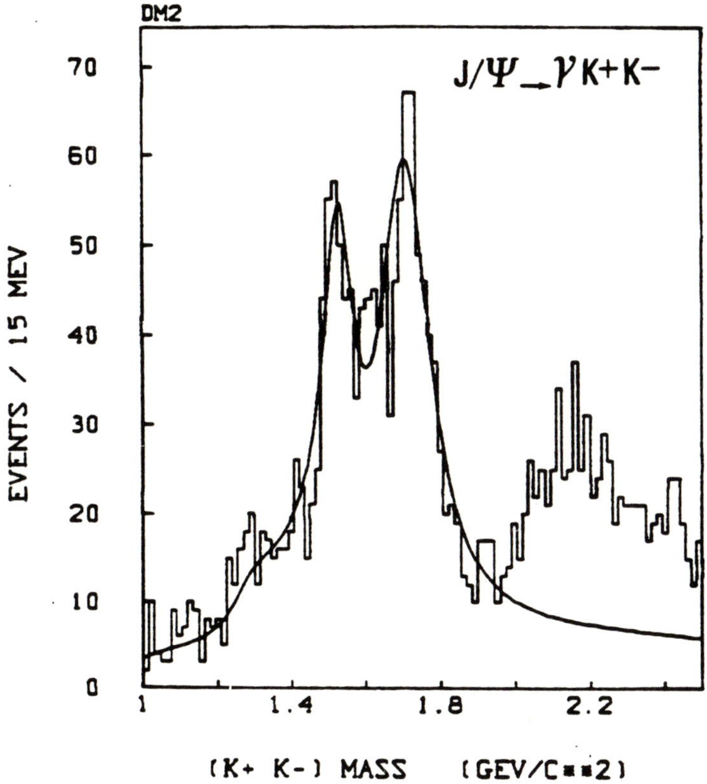

Figure 7.

The K^+K^- mass distribution in the decay $J/\psi \to \gamma K^+K^-$ from the DM2 group.[54]

by Close[75] suggests that the $L_Z \cong 0$ for the $q\bar{q}$ pair will suppress helicity 2.

The ratio of the f, f' rates is predicted to be,

$$\frac{B(J/\psi \to \gamma f')/P.S.\ (f')}{B(J/\psi \to \gamma f)/P.S.\ (f)} = \tan^2\theta$$

If we use the value $\theta = 35°$ from two photon measurements and we assume a form of the matrix elements[73] we obtain excellent agreement.

7. Review of the $\iota/\eta(1440)$ and $E/f_1(1420)$ Mesons

The $\iota/\eta(1440)$ and $E/f_1(1420)$ mesons* are of great interest in hadron spectroscopy. The $E/f_1(1420)$ meson was first observed in 1967 decaying into $K\bar{K}\pi$. There has been experimental controversy as to whether its spin is 0^{-+} or 1^{++}. The $\iota/\eta(1440)$ is a resonance seen in radiative J/ψ decays in $K\bar{K}\pi$ with a slightly higher mass and width than that of the $E/f_1(1420)$. In this section the experimental evidence for the iota is reviewed, the $E/f_1(1420)$ results are summarized and the interpretations are discussed.

The $\iota/\eta(1440)$ resonance was first observed by the Mark II group[76] in $J/\psi \rightarrow \gamma K_S K^{\pm}\pi^{\mp}$. It was identified as the $E/f_1(1420)$ since the mass and width were similar. Later the Crystal Ball group[77] performed an isobar analysis on the resonance in the mode $J/\psi \rightarrow \gamma\pi^0 K^+K^-$ and reported it was $J^{PC} = 0^{-+}$ with a dominant quasi-two body mode $J/\psi \rightarrow \gamma\iota$, $\iota \rightarrow \delta\pi$, $\delta \rightarrow K^+K^-$. The Crystal Ball group named this resonance the ι, emphasizing it was not the $E/f_1(1420)$, although the $E/f_1(1420)$ has been reported to be $J^{PC} = 0^{-+}$ in several experiments as will be discussed later. The parameters of the $\iota/\eta(1440)$ from the MARK II ,[76] Crystal Ball,[77] MARK III ,[64] and DM2.[54] are listed in table 7a. The $K\bar{K}\pi$ mass distribution from the decay $J/\psi \rightarrow \gamma K\bar{K}\pi$ from the DM2 group is shown in Fig. 8.

⋆ The iota , D, and E mesons have been renamed the $\eta(1440)$, $f_1(1285)$, and $f_1(1420)$, respectively, by the Particle Data Group. In this paper they will be call the $\iota/\eta(1440)$, $D/f_1(1285)$, and $E/f_1(1420)$.

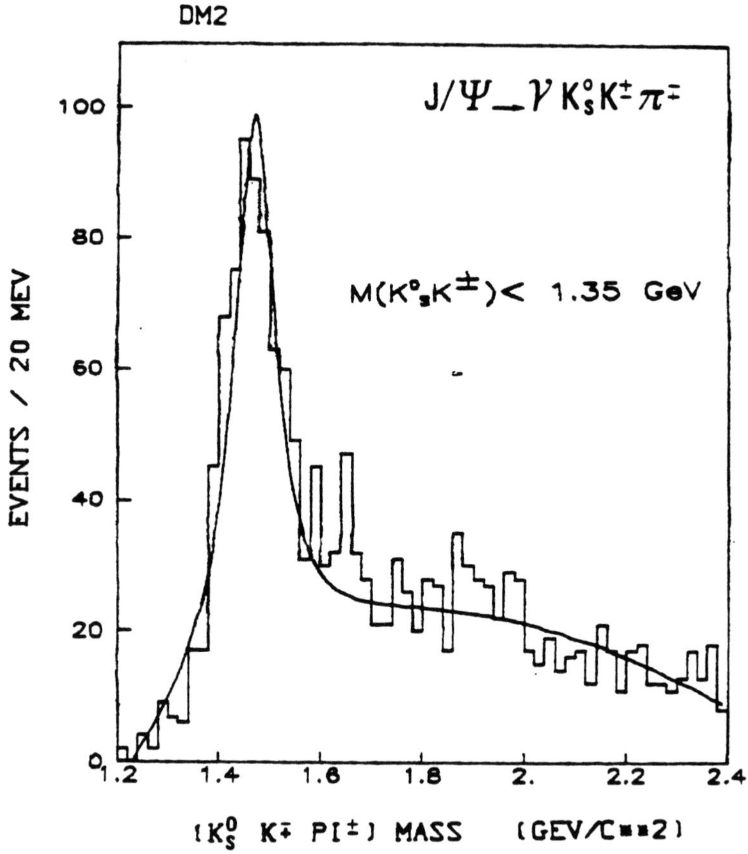

Figure 8.

The $K\bar{K}\pi$ mass distribution from the decay $J/\psi \rightarrow \gamma K\bar{K}\pi$ from the DM2 group.[54]

Table 7a. Iota parameters

Mass	Γ	$B(J/\psi \to \gamma\iota) \times 10^{-3}$	Mode	Group
$1440.0 \pm 10.0 \pm 15.0$	$50 \pm 30 \pm 20$	4.3 ± 1.7	$K_S K^{\pm}\pi^{\mp}$	MK2
$1440.0 \pm 20.0 \pm 15.0$	$55 \pm 20 \pm 30$	4.3 ± 1.2	$K^+K^-\pi^0$	CB
1461 ± 5	101 ± 10	$4.9 \pm 0.2 \pm 0.8$	$K^+K^-\pi^0$	MK3
$1456 \pm 5 \pm 6$	$95 \pm 10 \pm 15$	$5.0 \pm 0.3 \pm 0.8$	$K_S K^{\mp}\pi^{\pm}$	MK3
1451 ± 3	96.6 ± 10	$4.1 \pm .5 \pm 6$	$K^+K^-\pi^0$	DM2
$1460 \pm 3 \pm 8$	$100 \pm 12 \pm 15$	$4.1 \pm 0.6 \pm 0.9$	$K_S K^{\mp}\pi^{\pm}$	DM2

The ι has been spin-parity analyzed with the Jacob-Berman technique for decays into 3 pseudoscalars. This method is independent of the quasi-two body decays in the $K\bar{K}\pi$ system. The results from the both MARK III[64] and DM2[54] groups are $J^{PC} = 0^{-+}$.

The line shape of the ι as seen in $K^+K^-\pi^0$, $K_S K^{\pm}\pi^{\mp}$, and $K_S K_S \pi^0$ has been fit[78] and it does not fit a single Breit-Wigner curve decaying into 3-body phase space. The study concluded that the line shape could be fit by a curve with two Breit-Wigner resonances or a single Breit-Wigner resonance decaying into a coupled channel with K^*K and $\delta\pi$.

In $\gamma\gamma$ production the $\iota/\eta(1440)$ has not been seen, the upper limit is given as $\Gamma_{\iota(1440)\to\gamma\gamma} \cdot B(\iota/\eta(1440) \to K\bar{K}\pi) < 1.6$ KeV at 95% C.L.[79] This aspect plus the large radiative branching ratio provide strong evidence for the $\iota/\eta(1440)$ to be identified as a gluonium state. When stickiness ($S_X = \frac{\Gamma(J/\psi \to \gamma X)/K_\gamma^3}{\Gamma(X \to \gamma\gamma)/m_X^3}$) is compared between the $\iota/\eta(1440)$, the η' and the η, there is a dramatic increase for the $\iota/\eta(1440)$,

$$S_\eta \ : \ S_{\eta'} \ : \ S_{\iota/\eta(1440)} = 1 \ : \ 4 \ : > 65$$

The ι decay into $\iota \to \delta\pi$, $\delta \to \eta\pi$ has been investigated. The results from the Crystal Ball,[80] MARK III ,[59] and DM2 groups[42] are quite similiar. There is no evidence for $\iota \to \delta\pi$, $\delta \to \eta\pi$. In the spectrum there is evidence for $\delta \to \eta\pi$ but there is a structure at the $D/f_1(1285)$ region and below 1400 MeV. In fact there is a dip at the E/ι region at 1440 MeV. The mass of the upper structure is summarized in table 7b.

Table 7b. $\eta\pi\pi$ resonance parameters

Mode	Mass	Γ	BR(10^{-4})	Group
$\eta\pi^+\pi^-$, $\eta \to \gamma\gamma$	1391.5	52 ± 9	$4.1 \pm 3 \pm 1$	DM2
$\eta\pi^+\pi^-$, $\eta \to \gamma\gamma$	$1382 \pm .6$	69 ± 23	$5.2 \pm 1.8 \pm .5$	MARK III
$\eta\pi^+\pi^-$, $\eta \to 3\pi$	1400 ± 7	62 ± 16	$5.2 \pm 1.2 \pm 0.5$	MARK III

The radiative decay of the ι into $\gamma\rho^0$ has been investigated by the Crystal Ball,[81] DM2,[54] and MARK III groups.[64] The event statistics is limited but there is a structure near 1400 MeV. The results are listed in table 7c.

Table 7c. $\gamma\rho$ Resonance Paramters

Mode	Mass	Γ	BR(10^{-4})	Group
$\gamma\rho^0$	$1420 \pm 15 \pm 20$	$133 \pm 55 \pm 30$	$1.1 \pm .24 \pm .25$	MK3
$\gamma\rho^0$	1401 ± 18	174 ± 44	$0.9 \pm 0.2 \pm .14$	DM2
$\gamma\rho^0$	1390 ± 25	185^{+110}_{-80}	$1.9 \pm .5 \pm .4$	CB

The structure was fit with a single Breit-Wigner. It is possible the structure includes contributions from the $\eta(1275)$ and $\iota/\eta(1440)$ shifting the combined mass to $\sim$ 1440 MeV. There has been no definitive spin-parity determination, although the angular distributions are consistent with $J^{PC} = 0^{-+}$.

In hadronic production, the study of the $E/f_1(1420)$ meson has a long and complicated history.[82] The results are summarized in Table 7d.[83]

Table 7d. E Hadronic Production

Reaction	Mode	J^{PC}	Group
$p\bar{p} \to E\pi\pi, E \to K\bar{K}\pi$	$\delta\pi$, K$^*K$	0^{-+}	Ballion *et al.*, 1967
$\pi^- p \to En, E \to K\bar{K}\pi$	K$^*K$	1^{++}	Dionisi *et al.*, 1980
$(\pi^+/p)p \to (\pi^+/p)E, E \to K\bar{K}\pi$	K$^*K$	1^{++}	Armstrong *et al.*, 1984
$\pi^- p \to En, E \to \eta\pi\pi$	$\delta\pi$	0^{-+}	Ando *et al.*, 1985
$\pi^- p \to En, E \to K\bar{K}\pi$	$\delta\pi$, K$^*K$	0^{-+}	Chung *et al.*, 1985
$\gamma\gamma^* \to E, E \to K\bar{K}\pi$	K$^*K$(?)	1^{++}	TPC, MARK II , 1986

The hadronic production results indicate either two or more wrong results or two or more states ($J^{PC} = 0^{-+}$ and 1^{++}). The $\gamma\gamma^*$ results yield conclusive evidence that there is a 1^{++} meson at 1.42 GeV/c^2. If there is a 0^{-+} meson at 1.42 GeV/c^2, perhaps the iota is a compound resonance with one part being this pseudoscalar seen hadronic production. In addition this state has not been seen in the following modes[84,85,86] listed in table 7e.

Table 7e. Reactions not producing the E or Iota

Mode	Group
$\gamma\gamma \to \iota, \iota \to K\bar{K}\pi$	TPC(Aihara)
$\gamma\gamma \to E, E \to \eta\pi\pi$	CB(Cooper)
$K^- p \to E\Lambda, E \to K\bar{K}\pi$	LASS(Aston)
$\gamma\gamma^* \to E, E \to \eta\pi\pi$	MARK II (Gidal)

Unraveling the ι/E puzzle is an important issue in meson spectroscopy. Aside from the spin-parity question, the combined experimental data is not explained by the naive quark model. The quark content of the E is inconsistent in production and decay. The reasons to believe the E is mostly $u\bar{u} + d\bar{d}$ are: and not K^-p.

a) It is produced in π^-p production

b) It is produced in $\gamma\gamma^*$ production.

c) It appears to be produced in $J/\psi \to \omega E$ and not $J/\psi \to \phi E$.[87]

The reasons to believe the E is mostly $s\bar{s}$ are:

1) E decays into $K\bar{K}\pi$, mostly KK* and not $\delta\pi$.

2) If the D and E are the 1^{++} isoscalars, the heavier E should contain $s\bar{s}$ quarks.

In addition a resonance in $K\bar{K}\pi$, called the $D'(1530)$[88] has been claimed to the 1^{++} isoscalar partner of the $D/f_1(1285)$. If this is correct, the E becomes an extra 1^{++} meson that does not fit into the quark model nonets.

What is the 1^{++} $E/f_1(1420)$? A possible suggestion[89] is that it is a hybrid state $g(u\bar{u} + d\bar{d})$. This would allow $\gamma\gamma^*$ production and decay into $s\bar{s}$ final states. The exciting possibility is that the state is infact an exotic $J^{PC} = 1^{-+}$ hybrid state. Other models include non-ideal mixing[82] and a different nonet ($J^{pc} = 1^{+-}$) assignment.[90]

8. Review of the $\theta(1700)$ Resonance

The $\theta(1700)$ was first observed in the reaction $J/\psi \to \gamma\eta\eta$ by the Crystal Ball group.[77] They obtained spin-parity measurement of $J^{PC} = 2^{++}$. The measurement did not separate the f' signal. Later the Mark II group observed the θ in the mode $J/\psi \to \gamma K^+K^-$ and attempted to separate out the f' signal from the θ. The high statistics data from the MARK III[69] and the DM2 group[54] data provide clear evidence that the f' and the $\theta(1700)$ appear separated in $J/\psi \to \gamma K^+K^-$ and $\gamma K_S K_S$. The $J/\psi \to \gamma\eta\eta$ mode from the Crystal Ball group[91] has been reanalyzed with the f' and θ signals fitted in the mass distribution. The $\eta\eta$ mass distribution from the decay $J/\psi \to \gamma\eta\eta$ from the Crystal Ball group is shown in Fig. 9. The results are summarized in table 8a.

Table 8a. Theta parameters

MODE	MASS(MeV)	Γ(MeV)	BR(10^{-4})	GROUP
K^+K^-	1707 ± 10	166 ± 33	$4.6 \pm 0.7 \pm 0.7$	DM2
K^+K^-	1720 ± 7	132 ± 15	$4.8 \pm 0.6 \pm 0.9$	MarkIII
$\eta\eta$	$1655 \pm 33 \pm 5$	219^{+74}_{-54}	$2.6 \pm .8 \pm .7$	CB

The spin-parity of the θ has been determined by the MARK III and DM2 groups to be $J^{PC} = 2^{++}$. The helicity applitudes appear in roughly equal amounts which is very different from the f' which has very little helicity 2.

A search for the θ in $\gamma\gamma$ production has been done by the TPC[92] and TASSO[93] groups. Their upper limits are summarized in Table 8b.

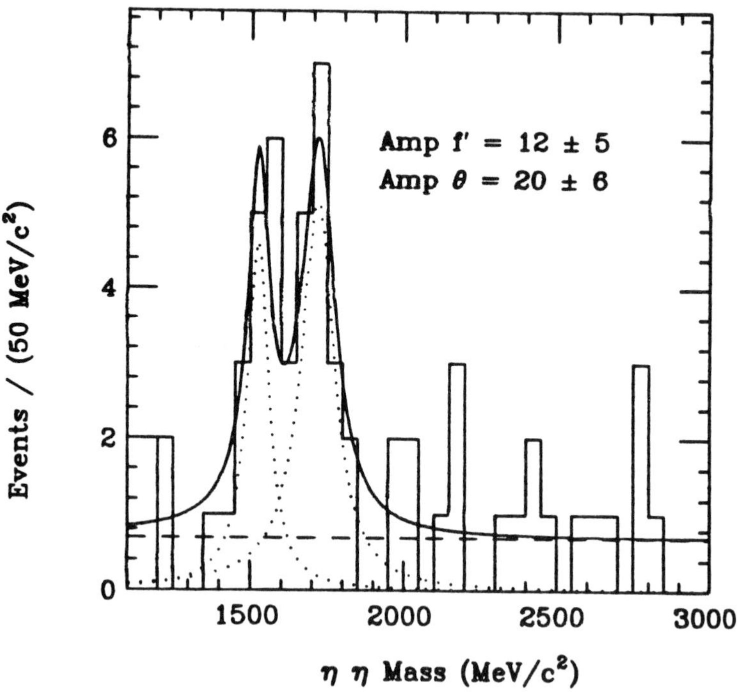

Figure 9.

The $\eta\eta$ mass distribution from the decay $J/\psi \to \gamma\eta\eta$ from the Crystal Ball group.[91]

Table 8b. $\gamma\gamma$ Productions Upper limits

MODE	$\Gamma_{\gamma\gamma}$(KeV)	GROUP
K^+K^-	<.28	TASSO
K^+K^-	<.10	TPC

Comparing the stickiness ratio, S_x, we observe

$$S_f \ : \ S_{f'} \ : \ S_\theta = 1 \ : \ 7 \ :> 14$$

It is evident that the θ is again quite different from the f and f'.

There is preliminary evidence that the θ decays into $\theta \rightarrow \pi^+\pi^-$. There is a peak in the $\pi^+\pi^-$ mass distribution with a mass and width consistent with the θ although the spin-parity analysis has not been done. If this is the θ the hadronic decay rates are

$$\mathrm{K\bar{K}} \ : \ \eta\eta \ : \ \pi\pi = 3 \ : \ 1 \ : \ 0.8$$

If the θ decayed as a pure singlet we expect

$$\mathrm{K\bar{K}} \ : \ \eta\eta \ : \ \pi\pi = 3 \ : \ 0.5 \ : \ 6$$

The $\pi^+\pi^-$ rate appears to be suppressed. It may be possible that the $s\bar{s}$ enhancement from Bag cavity perturbation is at work. It has been suggested that final state interactions cause this distortion.[27] This may be similiar to the problems that cause $B(D^0 \rightarrow \pi^+\pi^-) \neq B(D^0 \rightarrow K^+K^-)$ in Cabbibo suppressed decays.

A very striking result has been obtained by the LASS group.[94] They observe the f' in the reaction $K^-p \to \mathrm{K\bar{K}\Lambda}$ but not the θ. In this formation experiment $s\bar{s}$ resonances are produced. This indicates the θ is not is an $s\bar{s}$ resonance although it decays into $\mathrm{K\bar{K}}$ and $\eta\eta$.

There is evidence that the θ is produced in the reaction $J/\psi \to \omega\theta$ and $\phi\theta$. The DM2[49] and MARK III[95] groups observe a mass peak in the $\mathrm{K\bar{K}}$ mass distribution recoiling from the ω. In the $\phi\theta$ mode, the $\mathrm{K\bar{K}}$ mass recoiling against the ϕ appears as a shoulder above the f'.

The evidence for the θ to be identified as a glueball is fairly impressive. The key points are;

1. It is produced radiative J/ψ decays.

2. It is not produced in $\gamma\gamma$ nor $s\bar{s}$ production.

3. The helicity of the θ is very different from the f and f'.

4. It decays with approximate flavor symmetry.

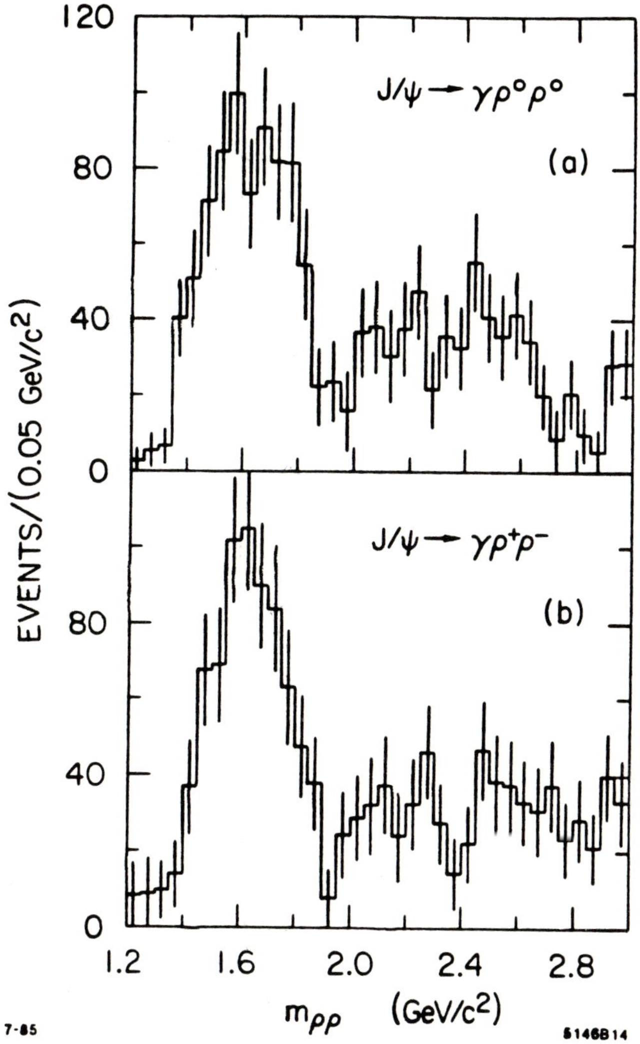

Figure 10.

The $\rho^0\rho^0$ (a) and the $\rho^+\rho^-$ (b) mass distributions in the decay $J/\psi \to \gamma\rho\rho$ from the MARK III group.[96]

9. Vector-Vector resonances in $J/\psi \rightarrow \gamma + (\rho\rho, \omega\omega, \phi\phi, \omega\phi)$

The radiative J/ψ decay into two vector meson combinations, ρ, ω, and ϕ, has been measured. The MARK II group[103] originally observed a large $\rho^0\rho^0$ signal near threshold at 1.7 GeV/c^2. This was thought to be associated with the θ resonance. Later the MARK III group[96] observed the same resonance in $\rho^0\rho^0$ and $\rho^+\rho^-$ and determined it spin parity to be largely pseudoscalar and not tensor. The DM2 group[54] has confirmed this result.

The search was extended to $\omega\omega$,[97,98] $\phi\omega$,[99] and $\phi\phi$[100,101] by the Mark III and DM2 groups. Both the $\omega\omega$ and $\phi\phi$ channels have a large pseudoscalar resonance near threshold. The spin-parity techniques used in these channels are similiar to the analysis performed on the $\eta_c \rightarrow \phi\phi$. The OZI violating decay into $\phi\omega$ has a small signal and a cluster of events near 2.23 GeV/c^2. The $\rho\rho$ mass distribution in the decay $J/\psi \rightarrow \gamma\rho\rho$ from the MARK III group is shown in Fig. 10 and the $\phi\phi$ mass distribution in the decay $J/\psi \rightarrow \gamma\phi\phi$ from the DM2 group is shown in Fig. 11. The results are summarized in table 9.

What is this vector-vector resonance near threshold? A pseudoscalar decay into two vectors is surprising because the vectors must be in a p-wave. Unless there is a VDM like connection between vector gluons and vector mesons this is unexpected. The glueball candidate $\phi\phi$ tensor, $g_T(2240)$, has not been identified in the 2 GeV region.[102] A possible solution[104] is this resonance is part of a single large pseudoscalar resonance that includes the ι. This would produce resonances in channels that would rise at threshold and fall when new channels are opening up in other modes.

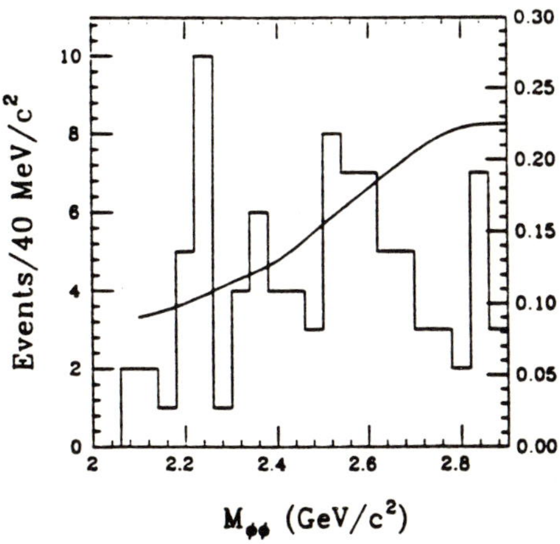

Figure 11.

The $\phi\phi$ mass distribution in the decay $J/\psi \to \gamma\phi\phi$ from the DM2 group.[98]

Table 9. Vector-Vector rates

MODE	MASS(GeV)	$BR(10^{-4})$	GROUP
$\rho^0\rho^0$	<2.0	$47 \pm 3 \pm 9$	MARK III
$\rho^+\rho^-$	<2.0	—	MARK III
$\omega\omega$	<2.0	$12.2 \pm .7 \pm 3.1$	MARK III
$\omega\omega$	<2.0	$14.1 \pm 2 \pm 4.2$	DM2
$\phi\omega$	1.7 - 3.1	$1.4 \pm 2.5 \pm .28$	MARK III
$\phi\phi\ (K^+K^-)$	<2.9	$3.1 \pm .3 \pm .6$	DM2
$\phi\phi\ (K^+K^-)$	2.1-2.4	$3.4 \pm .8$	MARK III
$\phi\phi\ (K_SK_L)$	2.1-2.4	$3.0 \pm .6$	MARK III

10. Review of $J/\psi \rightarrow \gamma\xi(2.2)$

In summer 1983 the MARK III group presented new results for a narrow state, the $\xi(2.2)$, seen in radiative J/ψ decays into $K\bar{K}$.[106] The signal, $J/\psi \rightarrow \gamma\xi$, $\xi \rightarrow K^+K^-$ had a branching ratio of $(3.8 \pm 1.3 \pm 0.9) \times 10^{-5}$ and a statistical significance of ~ 4.7 standard deviations based on 2.7 million J/ψ decays. Later the DM2[107] searched for this state in their first data sample and didn't see it. Subsequently, the Mark III group ran again at the J/ψ and confirmed the signal.[105] At the same time the DM2 group[54] finished analyzing their complete data sample (8.3 million J/ψ events) and set an upper limit for the state. The K^+K^- and K_SK_S mass distributions in the decay $J/\psi \rightarrow \gamma K\bar{K}$ from the MARK III group are shown in Fig. 12 and the K_SK_S mass distribution from the DM2 group is shown in Fig. 13. The final results are listed in table 10a.

Table 10a. $J/\psi \rightarrow \gamma\xi(2.2)$ rates

Decay	Mode	Mass	Width	BR$\times 10^{-5}$	Group
$J/\psi \rightarrow \gamma\xi$	γK^+K^-	$2.230 \pm .006 \pm .014$	$.026^{+.020}_{-.016} \pm .017$	$4.2^{+1.7}_{-1.4} \pm 0.8$	MARK III
$J/\psi \rightarrow \gamma\xi$	γK_SK_S	$2.232 \pm .007 \pm .007$	$.018^{+.020}_{-.015} \pm .010$	$3.1^{+1.6}_{-1.3} \pm 0.7$	MARK III
$J/\psi \rightarrow \gamma\xi$	γK^+K^-	—	0	< 1.2	DM2
$J/\psi \rightarrow \gamma\xi$	γK_SK_S	—	0	< 2.0	DM2

Other groups have searched for this state in different modes and reactions. The search includes production in Υ decays, $p\bar{p}$ annihilation, π^-p and K^-p reactions. In the later two there is evidence for a resonance near 2.2 GeV/c^2.

In Υ decays, the CLEO group[108] looked in the reaction, $\Upsilon(1s) \rightarrow \gamma\xi$, $\xi \rightarrow K^+K^-$ and K_SK_S. A limit of 2×10^{-4} and 9×10^{-5} has been

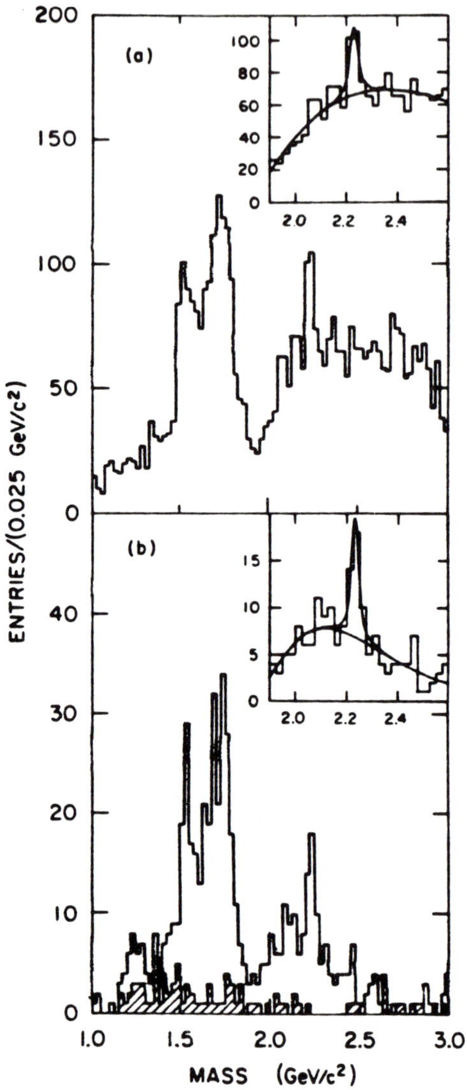

Figure 12.
The K^+K^- (a) and K_SK_S (b) mass distributions in the decay $J/\psi \to \gamma K\bar{K}$ from the MARK III group.[105]

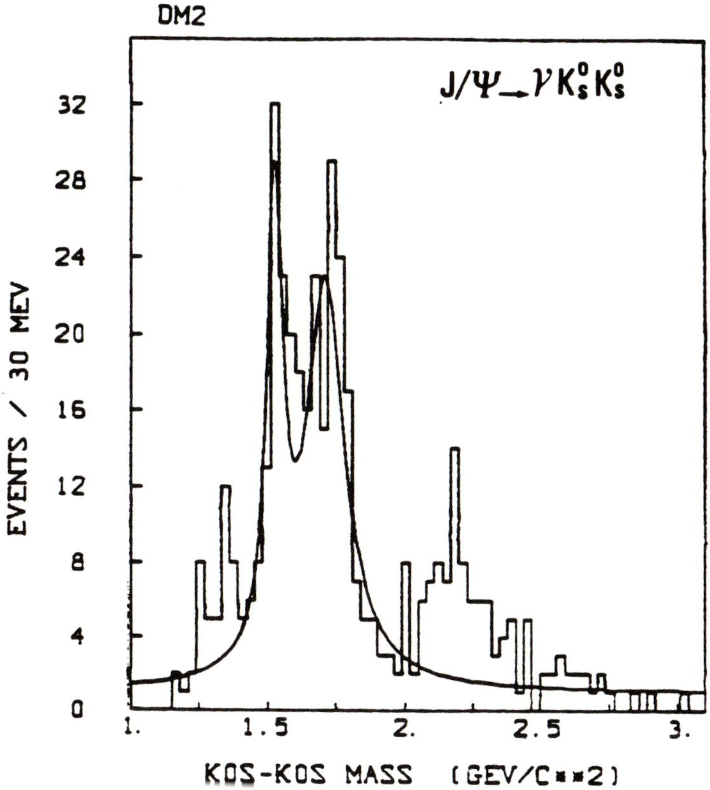

Figure 13.
The $K_S K_S$ mass distribution in the decay $J/\psi \to \gamma K_S K_S$ from the DM2 group.[54]

set at 90% C.L. In $p\bar{p}$ annihilation, the BNL[109] and LEAR(PSI70)[110] groups have set upper limits for $B(\xi \to p\bar{p}) \times B(\xi \to K^+K^-)$ for different spins and widths listed in table 10b.

Table 10b. $\xi(2.2) \to p\bar{p}$ upper limits

Spin(J^{PC})	Γ (MeV)	Limit (10^{-4})	Group
0^{++}	3.5	16	BNL
0^{++}	35	5.6	BNL
2^{++}	3.5	5.1	BNL
2^{++}	35	1.8	BNL
0^{++}	1	30	LEAR
0^{++}	3	18	LEAR
0^{++}	30	5.4	LEAR
2^{++}	1	7.7	LEAR
2^{++}	3	4.6	LEAR
2^{++}	30	1.3	LEAR

In $\pi^- p$ production, the GAMS group[111] reported a narrow structure at 2.22 GeV in the $\eta\eta'$ mode in the reaction $\pi p \to \eta\eta'\, n$, with incident pion beams at 38 and 100 GeV. The angular distributions are consistent with $J \geq 2$. Another unpublished result,[112] in the reaction $\pi^- p \to K_S K_S +$ neutrals, reported a peak at 2.2 GeV/c^2 with an incident pion momentum at 175 GeV/c. In $K^- p$ production, the LASS group[84] has evidence for a structure near 2.2 GeV in the reaction $K^- p \to K_S K_S \Lambda$. This structure again appears to have an angular distribution consistent with $J \geq 2$.

The theoretical models include identification as a high spin $s\bar{s}$ state,[113] a $\Lambda\bar{\Lambda}$ bound state,[114] a Higgs boson,[115] a glueball,[116] a hybrid state,[31] a technidilaton,[117] a neutral color scalar,[118] a Regge occurance,[119] a 4-quark state,[119] and a techipion.[120] The simpliest model, a high spin $s\bar{s}$ state, predicts an $\eta\eta'$ decay and production in $s\bar{s}$ formation experiments such as $K^-p \to K\bar{K}\Lambda$. The narrow width has been predicted by a such model[113] but disputed in another calculation.[114] Many models were developed to explain the narrow width. The conventional $s\bar{s}$ quark model prediction, predicts the $\eta\eta'$ decay and would be expected to appear in K^-p production. The narrow width in some models[113] is found to be wide by other calculations.[114]

11. Summary

The results from high statistics J/ψ experiments are providing new insights into meson spectroscopy. The main thrust has been the study of radiative J/ψ decays to understand the two gluon spectrum predicted by QCD. Other topics under investigation include the hadronic decays to measure the light quark content in mesons.

The J/ψ radiative spectrum appears to quantitatively follow the predictions for states formed from two qluons. The pseudoscalars and tensors decay at rates consistent with SU(3) symmetry. The are new resonances, the iota and the theta, which do not appear to seen in hadronic production and do not seem to fit into the conventional $q\bar{q}$ nonets. The large radiative J/ψ branching ratios and the small $\gamma\gamma$ upper limits support the glueball interpretation. The detailed theoretical predictions of the mass spectrum and the decays into specific mesons are not well understood.

The $E/f_1(1420)$ meson ($J^{PC} = 1^{++}$ case) has become even more puzzling with new measurements. The recent results indicate that the $E/f_1(1420)$ is being produced via non-strange quarks even though it decays into $K\bar{K}\pi$ and behaves as an $s\bar{s}$ state.

In other radiative decays, the $\xi(2.2)$ appears to be a high spin state now seen in $\pi^- p$ production in $\eta\eta'$ and in $K^- p$ production in $K\bar{K}$. There are pseudoscalar vector-vector resonances formed near threshold in $\rho\rho$, $\phi\phi$ and $\omega\omega$. These could be separate channels of a large single multi-channel resonance.

In hadronic decays, the $J/\psi \rightarrow$ Vector + Pseudoscalar branching ratios have been measured and now the $J/\psi \rightarrow$ Vector + Tensor and Vector + Scalar branching ratios are being determined. Models can fit the Vector + Pseudoscalar decays with conventional $\eta - \eta'$ mixing angles. The decay $J/\psi \to \delta\rho$ is missing suggesting that the δ is not the 0^{++} nonet isovector.

The results in the near future will continue to come from the analysis on the existing data from the DM2 and Mark III groups and possibly new results in gluonium resonances from LEAR. Any fundamental discoveries, such as the discovery of a scalar state in radiative decays or an exotic meson, will have to come from a very high statistics J/ψ study. The DM2 group has closed down. The Mark III group has not taken any new data (since early 1986) due to the SLC upgrade at SLAC and is not scheduled to run until late 1988 (?). Unless SPEAR luminosity could be significantly increased it is doubtful that the Mark III group would run at the J/ψ. The Beijing Spectrometer, BES, at the e^+e^- collider, BEPC, will have the opportunity to run at the J/ψ and could accomplish a high statistics study in the early 1990's.

References

1. J. J. Aubert *et al.*, Phys. Rev. Lett. **33**, 1404(1974).
2. J. Augustin *et al.*, Phys. Rev. Lett. **33**, 1406 (1974).
3. C. Bacci *et al.*, Phys. Rev. Lett. **33**, 1408 (1974).
4. J. Augustin *et al.*, Phys. Rev. Lett. **34**, 233 (1974).
5. G. Abrams *et al.*, Phys. Rev. Lett. **43**, 477 (1979).
   G. Abrams *et al.*, Phys. Rev. Lett. **43**, 481 (1979).
6. E. D. Bloom, *Proceedings of the 1979 International Symposium on Lepton and Photon Interactions at High Energies*, edited by T. Kirk and H. Abarbanel (Fermi National Accelerator Laboratory, Batavia, Ill., 1979), p. 92, and J. C. Tompkins, *Quantum Chromodynamics*, edited by A. Mosher (Stanford University, Stanford, California, 1980), p. 556.
7. J. Augustin *et al.*, Physica Scripta **23**, 623 (1981).
8. J. Roehrig *et al.*, Nucl. Instrum. Methods**226**, 319 (1984).
   D. Bernstein *et al.*, Nucl. Instrum. Methods**226**, 301 (1984).
   W. Toki *et al.*, Nucl. Instrum. Methods**219**, 479 (1984).
   R. Fabrizio *et al.*, Nucl. Instrum. Methods**227**, 220 (1984).
   J. Brown *et al.*, Nucl. Instrum. Methods**221**, 503 (1984).
9. A. Boyarski *et al.*, Phys. Rev. Lett. **34**, 1357 (1985).
10. R. Baldini *et al.*, Phys. Lett. **58B**, 471 (1975).
11. R. Esposito *et al.*, Nuovo. Cim. **14**, 73 (1975).
12. R. Brandelik *et al.*, Z. Phys. **C1**, 233 (1979).
13. M. Aguilar-Benitez *et al.*, Phys. Lett. **170B**,1(1986)
14. A. Boyarski *et al.*, Phys. Rev. Lett. **34**, 1357 (1975).

15. B. Jean-Marie *et al.*, Phys. Rev. Lett. **36**, 291 (1976).

16. F. Vanucci *et al.*, Phys. Rev. **D15**, 1814 (1977).

17. For η, η' mixing with $c\bar{c}$ states see H. Fritzsch and J. D. Jackson, Phys. Lett. **66B**,365(1977)

18. For η, η' mixing with gluonic components of the iota see E. Kawai, Phys. Lett. **124B**,262(1983) and W. Palmer, S. Pinsky and C. Bender, Phys. Rev. **D30**,1002(1984)

19. For an introduction to basic models see E. Leader and N. Predazzi, *An Introduction to Gauge Theories and the New Physics* (Cambridge University, London, 1982) and F. E. Close, *An Introduction to Quarks and Partons* (Academic Press, London, 1981). For a comprehensive review of meson spectroscopy see B. Diekmann, CERN-EP/86-112, for a summary of J/ψ decays see K. Konigsmann, Phys. Rep. **139**, 243 (1986), and U. Mallik,Stanford Linear Accelerator Center, Stanford University, Report No. SLAC-PUB-4238, Feb 1987, and for a recent review of hadron spectroscopy see S. Cooper, *Proceedings of the XXIII International Conference on High Energy Physics*, Edited by S. Loken, (World Scientific, Singapore, 1986).

20. S. Okubo, Phys. Lett. **5**, 163 (1963).

21. T. Appelquist and D. Politzer, Phys. Rev. Lett. **34**, 43 (1975).

22. S. Okubo, Phys. Rev. **D13**, 1994 (1976).
S. Rudaz, Phys. Rev. **D14**, 298 (1976).
H. Kowalski and T. Walsh, Phys. Rev. **D14**, 852 (1976).
H. Lipkin, Phys. Lett. **67B**, 65 (1977).
H. Haber, J. Perrier, Phys. Rev. **D32**, 296 (1985).

23. J. Rosner, Phys. Rev. **D27**, 1101 (1983).

24. A. Seiden, H. Sadrozinski, and H. Haber, Santa Cruz Institute of Particle Physics, Report No. SCIPP 87/73, June 1987.

25. F. Gilman and R. Kauffman, Stanford Linear Accelerator Center, Stanford University, Report No. SLAC-PUB-4301, April 1987, submitted to Physical Review D.

26. H. Lipkin and H. Rubinstein, Phys. Lett. **76B**,324 (1978)

27. J. Rosner, University of Chicago, Report No. EFI-87-26-Chicago, May 1987

28. P. Freund and Y. Nambu, Phys. Rev. Lett. **26**,1645 (1975) and H. Fritzsch and P. Minkowski, Nuovo. Cim.**30A**, 393 (1975)

29. D. Robson, Nucl. Phys. **B130**, 328 (1977)

30. H. Lipkin,Phys. Lett. **109B**, 326 (1982)

31. M. Chanowitz and S. Sharpe, Phys. Lett. **132B**, 413 (1983)

32. F. Close, Rutherford Appleton Laboratory, Report No. RAL-87-040, June 1987

33. R. Jaffe and K. Johnson, Phys. Lett. **60B**, 201 (1976).

34. A. Hendry and D. Lichtenberg, Fortschr. Phys.**33**,139(1985); M. Chanowitz, Lawrence Berkeley Laboratory, Report No. LBL-17930, June 1984

35. M. Chanowitz, *Proceedings of the Sixth International Workshop on Photon-Photon Collisions*, edited by R. L. Lander (World Scientific, Singapore, 1984), p. 95.

36. J. D. Jackson, *Proceedings of the SLAC Summer Institute*, 1976, edited by A. Mosher (Stanford University, 1976).
T. Appelquist, R. Barnett and K. Lane, Ann. Rev. Nucl. Part. Sci. **28**, 387 (1978).

37. W. Braunschweig *et al.*, Phys. Lett. **67B**, 1243 (1978).

38. R. Partridge *et al.*, Phys. Rev. Lett. **44**, 712 (1980).

39. R. Partridge *et al.*, Phys. Rev. Lett. **45**, 1150 (1980).
J. Gaiser *et al.*, Phys. Rev. **D43**, 711 (1986).

40. T. Himel *et al.*, Phys. Rev. Lett. **45**, 1146 (1980).

41. R. Baltrusaitus *et al.*, Phys. Rev. **D33**, 629 (1986).

42. Z. Ajaltouni *et al.*, Laboratoire de l'Accelerateur Lineaire (Orsay) Report No. LAL 86/29, October 1986.

43. C. Baglin *et al.*, Phys. Lett. **187B**, 191 (1987).

44. N. Chang and C. Nelson, Phys. Rev. Lett. **40**, 1617 (1978).

45. L. Trueman, Phys. Rev. **D18**, 3432 (1978).

46. R. Baltrusaitus *et al.*, Phys. Rev. Lett. **52**, 2126 (1984).

47. D. Bisello *et al.*, Phys. Lett. **179B**, 289 (1986).

48. K. Einsweiler, Ph.D. Thesis, Stanford Linear Accelerator Center, Stanford University, SLAC-Report-272, May 1984, unpublished.

49. G. Sklarz, DM2 Results presented at Hadron '87, KEK, April 1987.

50. U. Mallik, Stanford Linear Accelerator Center, Stanford University, Report No. SLAC-PUB-4238, Feb 1987.

51. R. M. Baltrusaitus *et al.*, Phys. Rev. **D32**,2883 (1985).

52. P. Freund and Y. Nambu, Phys. Rev. Lett. **34**, 1645 (1975),
W. Hou and A. Soni, Phys. Rev. Lett. **50**, 569 (1983),
S. Brodsky, G. Lepage and S.F. Tuan, Stanford Linear Acceler-

ator Center, Stanford University, Report No. SLAC-PUB-4276, May 1987, submitted to Physical Review Letters.

53. W. Braunschweig *et al.*, Phys. Lett. **67B**, 243 (1977).

54. J. Augustin *et al.*, , contributed paper to the International Symposium on Lepton and Photon Interactions at High Energies, Kyoto (Japan), August 19-24, 1985, Laboratoire de l'Accelerateur Lineaire (Orsay) Report No. LAL 85/27, October 1986

55. C. Edwards, Ph.D. thesis, California Institute of Technology, June 1985, unpublished.

56. W. Bartel *et al.*, Phys. Lett. **66B**, 489 (1977).

57. W. Brandelik *et al.*, Z. Phys. **C1**, 233 (1979).

58. E. Bloom and C. Peck, Ann. Rev. Nucl. Sci., **33**, 143 (1983).

59. R. M. Baltrusaitus *et al.*, contributed a paper to the 23rd International Conference on High Energy Physics, Berkeley, No. 3433, August 1986.

60. W. Bartel *et al.*, Phys. Lett. **64B**, 483 (1976).

61. W. Brandelik *et al.*, Z. Phys. **C1**, 233 (1979).

62. D. Scharre, Stanford Linear Accelerator Center, Stanford University, Report No. SLAC-PUB-2321, May 1979.

63. E. Bloom and C. Peck, Ann. Rev. Nucl. Sci. **33**, 143 (1983).

64. J. Richman, Ph.D. Thesis, California Institute of Technology, 1985, unpublished.

65. J. Jackson and P. Fritsch, Phys. Lett. **66B**, 365 (1977).
V. Novikov *et al.*, Nucl. Phys. **B165**, 55 (1980).

66. G. Alexander *et al.*, Phys. Lett. **72B**, 493 (1978).

67. W. Brandelik, Phys. Lett. **74B**, 292 (1978).

68. C. Edwards *et al.*, Phys. Rev. **D25**, 3065 (1982).

69. R. Baltrusaitus *et al.*, Phys. Rev. **D35**, 2077 (1987).

70. M. Franklin, Ph.D. Thesis, Stanford Linear Accelerator Center, Stanford University, SLAC-Report 254, 1983.

71. G. Alexander *et al.*, Phys. Lett. **76B**, 652 (1978).

72. C. Edwards *et al.*, Phys. Rev. **D25**, 3065 (1982).

73. J. Korner *et al.*, Nucl. Phys.**B229**, 115 (1983).

74. B. Li and Q. Shen, Phys. Lett. **126B**, 125 (1983).

75. F. Close, Phys. Rev. **D27**, 311 (1983).

76. D. Scharre *et al.*, Phys. Lett. **97B**, 329 (1980).

77. C. Edwards *et al.*, Phys. Rev. Lett. **49**, 259 (1982).

78. R. Partridge, XXII Recontre de Moriond, Les Arcs, France, March 1987.

79. H. Aihara *et al.*, Phys. Rev. **D57**, 51 (1986).

80. C. Edwards *et al.*, Phys. Rev. Lett. **51**, 850 (1083).

81. C. Edwards, Ph.D. Thesis, California Institute of Technology, 1985, unpublished.

82. S. Meshov, W. Palmer and S. Pinsky, XXII Recontre de Moriond, Los Arcs, France, March 1987; S. Chung, Proc. Aspen Winter Particle Physics Conference, Aspen, Colorado, January 1986.

83. P. Ballion *et al.*, Nuovo. Cim. **50A**, 393, (1967).
C. Dionisi *et al.*, Nucl. Phys. **B169**, 1 (1980).
T. Armstrong *et al.*, Phys. Lett. **146B**, 273 (1984).

A. Ando *et al.*, Phys. Rev. Lett. **57**, 1296 (1986).

S. Chung *et al.*, Phys. Rev. Lett. **55**, 779 (1985).

H. Aihara *et al.*, Phys. Rev. Lett. **57**, 2500 (1986).

G. Gidal *et al.*, Stanford Linear Accelerator Center, Stanford University, Report No. SLAC-PUB-4275, April 1987.

84. D. Aston *et al.*, Stanford Linear Accelerator Center, Stanford University, Report No. SLAC-PUB-4340, June 1987.

85. H. Aihara *et al.*, Phys. Rev. Lett. **57**, 51 (1986).
G. Gidal *et al.*, Stanford Linear Accelerator Center, Stanford University, Report No. SLAC-PUB-4274, April 1987.

86. S. Cooper, *Proceedings of the Second International Conference on Hadron Spectroscopy*, Edited by Y. Oyanagi, (National Laboratory for High Energy Physics, Japan, 1987)

87. J. Becker *et al.*, Phys. Rev. Lett. **59**, 186 (1987)

88. P. Gavillet *et al.*, Z. Phys. **C16**, 119 (1982).
D. Aston *et al.*, Stanford Linear Accelerator Center, Stanford University, Report No. SLAC-PUB-4340, June 1987.

89. M. Chanowitz, Phys. Lett. **187B**, 469 (1987).

90. A completely different explanation for $J^{PC} = 1^{+-}$ has been given by J. Iizuka, F. Masuda and T. Miura, University of Tsukuba, Report No. GK-U.TSUK/87-1, January 1987.

91. R. Lee, Ph.D. Thesis, Stanford Linear Accelerator Center, Stanford University, SLAC-Report-282, May 1985, unpublished.

92. H. Aihara *et al.*, Phys. Rev. Lett. **57**, 404 (1986).

93. M. Althoff *et al.*, Phys. Lett. **121B**, 216 (1983).

94. S. Suzuki, *Proceedings of the Second International Conference on Hadron Spectroscopy*, Edited by Y. Oyanagi, (National Laboratory for High Energy Physics, Japan, 1987)

95. L. Köpke, *Proceedings of the XXIII International Conference on High Energy Physics*, Edited by S. Loken, (World Scientific, Singapore, 1986).

96. R. M. Baltrusaitus *et al.*, Phys. Rev. **D33**, 1222 (1986)

97. R. M. Baltrusaitus *et al.*, Phys. Rev. Lett. **55**, 1723 (1985).

98. D. Bisello *et al.*, Phys. Lett. **192B**, 239 (1987).

99. R. M. Baltrusaitus *et al.*, contributed a paper to the 23rd International Conference on High Energy Physics, Berkeley, No. 346833, August 1986.

100. D. Bisello *et al.*, Phys. Lett. **179B**, 294(1986).

101. W. Wisiniewski, *Proceedings of the Second International Conference on Hadron Spectroscopy*, Edited by Y. Oyanagi, (National Laboratory for High Energy Physics, Japan, 1987)

102. A. Etkin *et al.*, Phys. Rev. Lett. **49**, 1620 (1982).

103. D. L. Burke *et al.*, Phys. Rev. Lett. **49**, 632 (1982).

104. N. Wermes,Stanford Linear Accelerator Center, Stanford University, Report No. SLAC-PUB-3730, July 1985,
N. Achasov and G. Shestakov,Phys. Lett. **156B**, 434 (1985).

105. R. M. Baltrusaitus *et al.*, Phys. Rev. Lett. **56**, 107 (1985).

106. K. Einsweiler, Stanford Linear Accelerator Center, Stanford University, Report No. SLAC-PUB-3702, September 1983.

107. J. Augustin *et al.*, contributed paper to the International Symposium on Lepton and Photon Interactions at High Energies,

Leipzig, August 1984, Laboratoire de l'Accelerateur Lineaire (Orsay) Report No. LAL 84/30, October 1986

108. S. Behrends *et al.*, Phys. Lett. **137B**, 277 (1984).

109. G. Christenson *et al.*, Brookhaven National Laboratory, Report No. BNL-38385-MC, July 1985.

110. G. Bardin *et al.*, CERN, Report No. CERN-EP/87-21, February 1987, submitted to Physics Letters B.

111. D. Alde *et al.*, Phys. Lett. **177B**, 120 (1986).

112. R. Delzenero, Ph.D. Thesis, University of Illinois at Chicago Circle, 1982, unpublished.

113. S. Godgrey, R. Kokoski and N. Isgur, Phys. Lett. **141B**, 439 (1984).
D. Kulshreshtha, Kaiserslautern, preprint, October 1984.

114. S. Ono, Phys. Rev. D35, 944 (1987).

115. H. Haber, G. Kane, Phys. Lett. **135B**, 196 (1984).
R. Wiley, Phys. Rev. Lett. **52**, 585 (1984).
R. Barnett, G. Senjanovic, and D. Wyler, Phys. Rev. D30, 1529 (1984).

116. B. Ward, Phys. Rev. D31, 2849 (1985).

117. K. Yamawaki, M. Bando, and K. Matumoto, Phys. Rev. Lett. **56**, 1335 (1986).

118. M. Shatz, Phys. Lett. **138B**, 209 (1984).

119. S. Pakvasa, M. Suzuki, and S. F. Tuan, Phys. Lett. **145B**, 135 (1984).
S. Pakvasa, M. Suzuki, and S. F. Tuan, Phys. Rev. D31, 2378 (1985).

120. H. Haber, Stanford Linear Accelerator Center, Stanford University, Report No. SLAC-PUB-3193, August 1983.

MESON SPECTROSCOPY VIEWED FROM J/ψ DECAY

Michael S Chanowitz
Lawrence Berkeley Laboratory
University of California
Berkeley, California 94720, USA

Abstract

The theoretical and experimental status of the search for gluonic states is reviewed. Progress clearly requires much higher statistics studies of J/ψ decay, as will be possible at the Beijing Electron Positron Collider.

1. INTRODUCTION

QCD and Existence of Gluonic States

Everyone is convinced that QCD is the correct theory of hadrons, chiefly because we can compute the short distance properties and find that they agree with what is observed in high energy experiments. But we still know very little about the properties of QCD at long distances, of order 1 fermi = 10^{-23} cm or larger. This is the physics that controls the spectrum of hadrons made from the light quarks – u, d, and s. Usually we do not consider a theory to be "solved" until we have understood the spectrum. I am not arguing against the popular conclusion that QCD is correct, but I want to suggest that we cannot say we understand the theory until we at least understand the spectrum qualitatively and understand at least some of the simpler bound states at a quantitative level. This is an ambitious goal that can be accomplished in the next ten to twenty years. It will result from more powerful theoretical calculations that will be possible with the next generations of computers and by future experiments that will go far beyond what has already been accomplished to clarify the complexity of the hadron spectrum. With high luminosity studies of J/ψ decay, BEPC can make critical contributions to this process.

I stressed above that we do not even understand the hadron spectrum in a qualitative way. The principal qualitative novelty of QCD is that it is a nonabelian gauge theory. As such the gauge bosons – gluons – are themselves color–charged and confined. This creates the possiblity that there will be states in the spectrum that contain gluonic constituents. Some, made only of gluons in first approximation, we call glueballs. Others, mixed states of quarks and gluons, I call meiktons from the ancient Greek word for a "mixed object". At this meeting we could call them "huěn hé". The most important qualitative question we can ask about the hadron spectrum is whether glueballs or $\bar{q}qg$ states exist.

We must be humble about proposing the answer to this question because we still do not have a deep understanding of the spectrum of ordinary hadrons made from u, d, and s. The nonrelativistic model seems to describe the spectrum qualitatively, but we know these hadrons are actually relativistic systems. In fact it is not even obvious that the static properties of hadrons must be dominated by the valence quarks.

The $\Pi^+ = |u\bar{d}\rangle$ has other components in its wave functions, such as $|\bar{u}dq\bar{q}\rangle$, $|u\bar{d}q\bar{q}q\bar{q}\rangle$. If those higher components were important in leading order, the hadrons would not have been in simple SU(3) flavor representations and quarks would not have been deduced from the spectrum as they were in 1964.[2]

The existence of valence quarks suggests that the concept of valence gluons might also be valid. Then the lightest glueballs would contain two confined valence gluons and there would also be hadrons made of a color octet $\bar{q}q$ and one valence gluon. This is what we expect in the bag model,[2] a relativistic model of hadrons that gives a good description of the ground state hadrons but not yet of the excited states.

I will discuss the question of the existence of gluonic states in more detail in section 3. Here I want to present an argument due to Bjorken[3] that glueballs would be required to exist because of confinement and the short–distance properties of QCD if all quarks were very heavy. Let us imagine that u, d, and s do not exist and that there is only one very heavy quark flavor, Q. Suppose $M_Q = 40$ GeV. Then there will be a $\bar{Q}Q$ vector meson bound state, $\Psi(\bar{Q}Q)$,

$$M_{\Psi(\bar{Q}Q)} < 2M_Q = 80 \text{ GeV} \qquad (1.1)$$

$\Psi(\bar{Q}Q)$ could be formed in e^+e^- annihilations

$$e^+e^- \rightarrow \Psi(\bar{Q}Q). \qquad (1.2)$$

Since Q is very heavy, Ψ is small enough that we can use the known, perturbative short distance properties to predict that Ψ decays to three gluons,

$$\Psi(\bar{Q}Q) \rightarrow ggg \qquad (1.3)$$

But in this imaginary experiment, what hadrons would appear in the detectors? Pairs of $\overline{Q}Q$ mesons would be too heavy. Since gluons are confined, the only possibility is to form color–singlet glueballs. If we next imagine introducing u, d, and s quarks, we expect the glueball states to remain in the spectrum. But since they might decay to hadrons made of u, d, s, we must understand the dynamics to know whether they would be narrow enough to appear as well–defined resonances.

Difficulties and General Strategy

The study of gluonic states is clearly not easy. Experiments are very difficult and theoretical understanding is poor. Serious experimental study began in 1980, when the iota was discovered in radiative J/ψ decay by the Mark II group[4]. Few definite conclusions have been reached in this period, though progress made during the last two years makes it seem probable that iota is indeed a glueball[5]. But many questions remain. Iota is poorly understood and even less is known about other possible glueballs. It is not known experimentally whether meiktons exist, though there are interesting experimental clues discussed in the following lectures. Because the subject is so very difficult, many fundamental discoveries remain to be made. Third generation experiments such as BES at BEPC are essential to make these discoveries.

BES is in a strategic position to contribute because it can study the most favored process for glueball production, radiative J/ψ decay:

$$\psi \rightarrow \gamma G \qquad (1.4)$$

For a very heavy $\overline{Q}Q$ system the inclusive radiative decay width can be calculated[6], since the bound state radius is small enough to be in the short distance domain of perturbation theory. In lowest order

$$\Gamma(\Psi \rightarrow \gamma x) = \Gamma(\Psi \rightarrow \gamma gg). \qquad (1.5)$$

In perturbation theory the dominant partial waves are $J^{PC}(gg) = 0^{++}, 0^{-+}, 2^{++}$. Since the two gluons are in a color singlet, we expect 0^{++}, 0^{-+}, and 2^{++} glueballs to be copiously produced.

The ψ is not heavy enough for eq. (1.5) to be precisely correct but it is valid as a rough approximation. The lowest order prediction is[6]

$$\frac{\Gamma(\psi \rightarrow \gamma X)}{\Gamma(\psi \rightarrow \text{hadrons})_{\text{strong}}} =$$

$$\frac{\Gamma(\psi \to \gamma g g)}{\Gamma(\psi \to g g g)} = \frac{16\alpha}{5\alpha_s} = 0.12 \qquad (1.6)$$

The ψ total width is

$$\Gamma_4(\mathrm{TOT}) = 63 \pm 9 \text{ keV} \qquad (1.7)$$

Of this total

$$\Gamma(\psi \to e^+e^- + \mu^+\mu^-) = 9.4 \pm 0.6 \text{ keV} \qquad (1.8)$$

from which we can estimate the electromagnetic contribution to hadron decay at

$$\Gamma(\psi \to \gamma^* \to \text{hadrons}) = R_{3.1} \cdot \Gamma(\psi \to e^+e^-) \qquad (1.9)$$
$$\tilde{=} 12 \text{ KeV}$$

The measured value[9] of the radiative decay width is

$$\Gamma(\psi \to \gamma x) \tilde{=} 4 \text{ keV} \qquad (1.10)$$

(obtained by using the theoretical spectrum to extrapolate the measurement at $X<0.4$ to small X).
Subtracting eqs. (1.8–1.10) from (1.7) we get

$$\Gamma(\psi \to \text{hadrons})_{\text{strong}} \tilde{=} 38 \text{ keV} \qquad (1.11)$$

Then

$$\frac{\Gamma(\psi \to \gamma X)}{\Gamma(\psi \to \text{hadron})_{\text{strong}}} = 0.11 \qquad (1.12)$$

in agreement with 0.12 from eq. (1.6). (The uncertainties are greater than the difference between (1.6) and (1.12).) Though the partial width $\Gamma(\psi \to \gamma X)$ agrees roughly with eq.(1.6), we should not expect the photon energy spectrum to be consistent with perturbation theory because the ψ radiative decays are dominated by the resonance region, $M_X = 1$–2.5 GeV . The prediction for the inclusive partial width $\Gamma(\gamma \to \gamma X)$ is understood in the "duality" sense familiar from deep inelastic scattering, where the sum over resonances agrees with the inclusive parton model prediction[21].

The fact that the resonance region dominates is good news. It means that radiative ψ decays are the perfect channel to explore resonances that couple to gluons. We therefore expect prominent decays to 0^{++}, 0^{-+}, and 2^{++} glueballs that are kinematically allowed. Heavier quarkonia cannot compete with J/ψ because the $\Psi_{\overline{Q}Q}$ production cross sections are much smaller and because a smaller fraction of the radiative decays are in the resonance region.

Stickiness is a useful quantity for identifying glueball candidates. It compares the fourth power of their electric charge. It should therefore be largest for glueballs, since gluons have bigger color charges than quarks and have no electric charges (naively $G \to \gamma\gamma$ proceeds by a $\overline{q}q$ loop – however see section 3). The stickiness S_X of a state X is defined as

$$S_X = \frac{\Gamma(\psi \to \gamma X)}{\Gamma(X \to \gamma\gamma)} \cdot \frac{\mathrm{LIPS}\ (X \to \gamma\gamma)}{\mathrm{LIPS}\ (\psi \to \gamma X)} \qquad (1.13)$$

where LIPS means Lorentz invariant phase space. It is only meaningful to consider the relative stickiness of states with the same J^{PC}. The stickiness of $\iota(1460)$ is very suggestive. Because of the improved upper limit on $\iota \to \gamma\gamma$ it is now

$$S_\iota : S\eta : S\eta = (>65) : 4 : 1 \qquad (1.14)$$

What we do <u>not</u> know about gluonic states is much greater than what we do know. Therefore what we know is very precious and must be used to the greatest possible advantage. What we know best is that gluonic states must be additional states in the spectrum beyond what is predicted in the $\overline{q}q$ spectrum of the nonrelativistic quark model. To use this obvious statement it is necessary to understand the spectrum of ordinary mesons very precisely. This is not easy because the $1–2\frac{1}{2}$ GeV region is filled with many $\overline{q}q$ nonets, requiring careful partial wave analysis that can only be done with accurate high statistics data.

A year ago it seemed that only three nonets were completely and unambiguously filled: π,ρ, and A_2/a_2. (For familiar states I often use the old particle names but in some cases I will use the new system of the Particle Data Group[8], sometimes giving both old and new names as in A_2/a_2.) We can be encouraged that in just the last year, two and perhaps four more nonets have been completed. As discussed in section 2, the $J^{PC} = 1^{-+}$ B/b_1 nonet and the $J^{PC} = 3^{--}$ $\rho_3(1690)$ nonet are unambiguously filled, and there are at least enough states to fill the $J^{PC} = 0^{-+}$ $\pi(1300)$ and $J^{PC} = 1^{++}$ A_1/a_1 nonets. Partial wave analysis of BES data will be of tremendous value, both in the radiative decay channel and for the direct hadronic decays.

Though ψ decay may be the single most valuable channel in searching for gluonic states, data from all experimental channels must eventually be combined to understand the total picture. As will be evident in the following lectures, there have been many important contributions from photon–photon scattering and from hadron scattering experiments with π, K, p, and $\bar{p}$ beams. High statistics experiments with low energy p beams at LEAR will provide interesting data in the next few years. PEP, Tristan, SLC and LEP will provide additional photon–photon scattering data. Hadron scattering experiments will continue at the CERN SPS, the KEK proton synchrotron, at Fermilab and Brookhaven.

Theoretical Overview

The theoretical perspective will be discussed more fully n section 3. Now I will just make a few brief remarks. The problem is that we still do not understand QCD at long distances, and that is also why the subject is still interesting. We should be encouraged that this theoretical ignorance is not a permanent condition. Serious quantitative calculations can be expected within the next ten to twenty years that can be tested by and provide guidance for the experimental studies.

For the present the only theoretical guidance comes from models and approximations that are at best very crude. These models can provide suggestions of what to explore and what might be found, but the dominant role must be played by experiment. That is why we need BEPC!

Since gluons are massless it is natural to use a relativistic model to represent their bound states. The bag model is a simple relativistic model of confinement that naturally predicts gluonic states containing valence gluons. In lowest order the glueball ground states are $J^{PC} = 0^{++}$ and 2^{++} two gluon bound states, while $J^{PC} = 0^{-+}$ and 2^{-+} are the first excited states[22]. Including first order corrections from cavity perturbation theory[22] the expected pattern from lowest to highest mass becomes

$$0^{++} < 0^{-+} < 2^{++} < 2^{-+} \qquad (1.14)$$

in the range from ~1 to 2.5 GeV. Iota (1460) could well be the 0^{-+} state and $\theta(1730)$ might be the 2^{++}. Then the 0^{++} state is expected between 1 and 1.3 GeV, for which there is no experimental support. The O^{++} mass might be affected by the structure of the vacuum[22], i.e. "vacuum mixing", which would tend to increase the mass.

The bag model also predicts $\bar{q}qg$ states. The ground state[2] consists of four nonets, in the order

$$0^{-+} < 1^{-+} < 1^{--} < 2^{-+} \qquad (1.15)$$

in the fit to $\iota(1460)$ and $\theta(1730)$ the I=1 members of these nonets range from 1200 to 1800 MeV. The 1^{-+} nonet is especially interesting since it has <u>exotic</u> J^{PC} that cannot occur in the nonrelativistic $\bar{q}q$ spectrum. Excited states[23] are expected in the order

$$0^{++} < 1^{++} < 1^{+-} < 2^{++} \qquad (1.16)$$

In the fit to ι and θ the I=1 members fall between 1800 and 2230 MeV.

$\bar{q}qg$ states are also expected in the flux–tube model[24], though a few hundred MeV heavier than the bag model estimates quoted above. I am not certain that these flux–tube excitations should be identified with the valence gluon states of the bag model. They may instead correspond to collective excitations, that would correspond to string modes in a string model or cavity excitations in the bag. This is an example of a qualitative theoretical question that is really not understood.

QCD sum rules have also been used to estimate the masses of gluonic states. The I=1 exotic $J^{PC} - 1^{-+}$ $\bar{q}qg$ is found to lie between 1.3 and 1.7 GeV[25]. Glueball masses have also been studied. One recent calculation[26] predicts 0^{++}glueball masses below 1 GeV, possibly very narrow. A light scalar glueball might have avoided detection in $\pi\pi$ scattering data if it were very narrow[27] but I do not think it would have escaped detection in $\psi \to \gamma\pi\pi$. An estimate for the 2^{++} glueball is consistent with the $\theta(1730)$ mass[28]. A 0^{-+} glueball is estimated at 1700 MeV, with an uncertainty that includes the iota mass[18].

For the future the most serious approach is certainly the lattice calculations. Since I am no expert and since there has just been a workshop in Beijing with many experts, I will only make a few comments. The present calculations typically give results for glueball masses in the same 1–2 GeV range as the models mentioned above. A recent review of lattice QCD was given by Iwasaki at the KEK Hadron '87 conference[29]. Present computations cannot reach large enough lattice sizes with enough statistics to be quantitatively reliable. Mixing with the $\bar{q}q$ sector is not yet included. But eventually future generations of supercomputers and special–purpose dedicated computers will provide reliable results.

Further discussion of the theoretical expectations are given in section 2.

Phenomenological Overview

Here I want to give a very quick impression of the status of the phenomenology. More complete discussions are given in sections 4 and 5.

In the last two years more evidence has accumulated in favor of the glueball interpretation of iota, $\iota(1460)$. The upper limit on $\iota \rightarrow \gamma\gamma$ has become very strong. Two pseudoscalar mesons have been observed, $\eta(1280)$ and $\eta(1400)$, which could fill the radially excited pseudoscalar nonet, leaving no room for iota as a $\bar{q}q$ state. However there is still much that is not understood. Iota itself is not well described by a single Brit–Wigner[4]. The evidence for $\eta(1400)$ is not completely clear, and it is even possible the data will require more pseudoscalars. I have already mentioned that bag model estimates place a meikton nonet in this mass region. As discussed in section 4, high statistics studies in radiative and hadronic ψ decay are crucial to clarify what could be an interesting and complicated system.

There are many other possible gluonic states to consider. They include the following:

$\theta(1730)$ is a 2^{++} resonance seen in radiative J/ψ decay. It cannot be understood as a $\bar{q}q$ meson. It might be a glueball, though the evidence is less clear than for iota. The failure to observe it in Kp scattering suggests that it has large missing decay modes, which would in turn make the glueball assignment very strong. High statistics studies of radiative J/ψ decay are needed to search for the missing modes.

C(1480) is a $J^{PC} = 1^{--}$ $\Phi\Pi$ resonance discovered in pion scattering[20]. $C \rightarrow \phi\pi$ is OIZ forbidden for any $\bar{q}q$ meson but could result from decay of a $\bar{q}qg$ state. If it is a $\bar{q}qg$ state it should be produced in hadronic J/ψ decay, such as $\psi \rightarrow C\Pi$.

$\xi(2230)$ has recently been observed in Kp scattering[22]. This and the observation that J>0 contradicts the Higgs interpretation considered previously. Other possibilities are that it is a 4^{++} $\bar{s}s$ state[22] or a $(\bar{u}u + \bar{d}d)g$ 2^{++} state[23]. As discussed in the 1984 BEPC workshop[33], $\lesssim$ 25,000,000 J/ψ decays are probably

needed to do a J^P determination of ξ.
Several different sets of data indicate interesting physics in the J = 1 channel in the E region, around1400 MeV. Tagged two photon scattering studies[24,25] find a narrow $\overline{K}K\Pi$ resonance at the E mass that must have $J^C = 1^+$. Confirmation of the $J^{PC} = 1^{++}$ D'(1530) in Kp scattering means there are too many $J^C = 1^+$ states for the $\overline{q}q$ spectrum. A signal in $\psi \rightarrow \omega\overline{K}K\pi$,is seen[4,26] with mass and width consistent with E. It is not seen in $\psi \rightarrow \phi\overline{K}K\Pi$, contrary to what is expected if E were an $\overline{s}s$ state as previously supposed. One hypothesis is that the resonance seen in these experiments at 1420 is a $J^{PC} = 1^{-+}$ $\overline{q}qg$ state, a J^{PC} exotic that cannot be a $\overline{q}q$ state in the nonrelativistic spectrum[27].
High statistics studies of $\psi \rightarrow \omega KK\pi$ could test this hypothesis. It is consistent with a preliminary report[26] that a $J^{PC} = 1^{-+}$ $\eta\pi$ resonance has been observed in the 1400 MeV region in πp scattering.
Other developments will be reviewed in less detail, including the 2^{++} $g_T\Phi\Phi$ structures[29], the O^{++} G(1590) observed[31] decaying to $\eta\eta$ and $\eta\eta'$ and three 0^{++} resonances found near $\overline{K}K$ threshold in a recent partial wave analysis[32].
The above topics are discussed in more detail in sections 4 and 5.

<u>Prospects for BEPC</u>

The study of gluonic states is just beginning. Though there have been many interesting developments in the last two years, most of the work remains to be done. Previous experiments have often lacked the statistical power to reach definitive conclusions. J/ψ decay is the most favorable channel for the study of gluonic states. The largest data samples are the ˜6,000,000 events of the Mark III and ˜8,000,000 of DM 2. They define the interesting statistical level for future experiments as 50 to 100 million events. At that level it should be possible to answer many of the questions discussed in these lectures and to find new phenomena that nobody today has anticipated. The work will rely heavily on the techniques of partial wave analysis.

This is clearly a difficult and ambitious program. The greatest difficulties are probably not in producing a sufficient number of J/ψ's but in detecting and recording the decays and in the off–line analysis.

The theoretical peak cross section for $e^+e^- \rightarrow \psi$ is

$$J_\psi = \frac{12\Pi}{\mu_\psi^2} B(\psi \to e^+e^-)$$
$$= (1.05 \pm .14).10^{-28}\ \mathrm{cm}^2 \qquad (1.17)$$

Smearing with the energy spread of the beam (done incorrectly by me at the 1984 workshop), the observed cross section is

$$\bar{\sigma}_\psi = \frac{\Gamma_\psi(\mathrm{TOT})}{\sqrt{2}\ \Delta\ E_{\mathrm{beam}}} \sigma_\psi$$
$$= (7.4 \pm 1.0).10^{-31}\ \mathrm{cm}^2 \qquad (1.18)$$

Here I assume $\Delta E_{\mathrm{beam}}/E_{\mathrm{beam}} = 4.1.10^{-4}$ at $E_{\mathrm{beam}} = M_\psi/2$, the RMS value extracted from the BEPC design parameters of 1984. Below the optimization energy the luminosity scales like E^2, $1.7 \cdot 10^{-32}$ so a peak luminosity of $1.7\ 10^{32}\ \mathrm{cm}^{-2}\ \mathrm{sec}^{-2}$ at $E_{\mathrm{beam}} = 2.8$ GeV corresponds to a luminosity at the J/ψ of

$$\mathscr{L}_\psi = 5.10^{31}\ \mathrm{cm}^{-2}\ \mathrm{sec}^{-2} \qquad (1.19)$$

Assuming an average luminosity of $\bar{\mathscr{L}}_\psi = \frac{1}{4} \mathscr{L}_\psi$ to take account of fill time and time to accelerate the beam, we then find for a run of 10^7 sec $\tilde{=} \frac{1}{3}$ year by the clock that

$$N_\psi = \bar{\sigma}_\psi \bar{\mathscr{L}}_\psi T \tilde{=} 9.10^7 \qquad (1.20)$$

50,000,000 J/ψ's is therefore a realistic possiblity if performance near the design parameters can eventually be achieved.

The problem of detecting, recording, and analyzing this volume of events was examined at the 1984 workshop and will be reexamined at this workshop. It is critical to determine if the planned computing capability will be sufficient. Here we can profit from the recent experience of high statistics fixed target experiments.

For example the LASS group has begun to report[32,33] many important results, several to be quoted in these lectures, which are based on a data sample of 140,000,000 recorded Kp events. Since the average multiplicity of their events is ~5, similar to J/ψ decay, the computing requirement is similar to what is needed for 100,000,000 J/ψ decays. LASS did the analysis using the SLAC mainframe to distribute events to a parallel microprocessor farm, while collaborators in Nagoya used a mainframe for the analysis.

BES might profit from the experience at LASS both in off–line computing and also perhaps in acquiring the essential techniques of partial wave analysis. The analysis of the LASS data is continuing and will provide more interesting results in the next few years.

In his lectures Mike Witherell will describe parallel processing systems being used by other fixed target experiments for high statistics studies that were inconceivable a few years ago.

2. "ORDINARY" MESONS

The spectrum of the "ordinary" mesons is itself interesting physics, that must be understood in detail to find and identify gluonic states.

For example, the success of the simple $\bar{q}q$ classification is a puzzle, since the light mesons are <u>not</u> nonrelativistic systems. We are lucky that spectroscopic properties are dominated by the valence quarks. If the $\overline{qq}qq$, $\overline{qqq}qqq$ wave function components were important spectroscopically, then there would not be simple $SU(3)_{flavor}$ representations and it would have been much harder to discover quarks and QCD. The spectroscopic dominance of the valence quarks is one reason to hope that a simple–minded view of gluonic states in terms of valence gluons may be qualitatively correct.

If our theoretical understanding were more advanced, it would be less important to understand the $\bar{q}q$ spectrum in such detail. But at present the only sure property of a gluonic state is that it is not (though it may mix with) a $\bar{q}q$ state. We need to understand the $\bar{q}q$ spectrum very precisely in order to identify the particles which do not belong to it.

<u>Overview of the spectrum</u>

The presently known mesons have J^{PC} quantum numbers that correspond to the hydrogen–like spectrum of the nonrelativistic quark model. The total angular momentum J is the sum

$$J = L + S \tag{2.1}$$

of orbital angular momentum and spin where

$$S = s_q + s_{\bar{q}} \tag{2.2}$$

S = 0 for the antisymmetric spin singlet

$$S = 0\colon \frac{1}{\sqrt{2}}(\uparrow\downarrow - \downarrow\uparrow), \tag{2.3}$$

and

$$S = 1\colon \uparrow\uparrow, \frac{1}{\sqrt{2}}(\uparrow\downarrow + \downarrow\uparrow), \downarrow\downarrow \tag{2.4}$$

for the symmetric triplet. Since q and $\bar{q}$ have opposite phases under P and C the parity and change conjugation of the $\bar{q}q$ state are

$$P = (-1)^{L+1} \qquad (2.5)$$

$$C = (-1)^{L+1}(-1)^{S+1} = (-1)^{L+S} \qquad (2.6)$$

Therefore the ground state with L = 0 has J = S, or $J^{PC} = 0^{-+}$ for the spin singlet and $J^{PC} = 1^{--}$ for the triplet. The orbitally excited states with L = 1 are then the singlet with J = L = 1 or $J^{PC} = 1^{+-}$ and the triplet with J = 1+1 = 0,1,2 and $J^{PC} = (0,1,2)^{++}$. The L = Z states are the singlet with $J^{PC} = 2^{-+}$, and the triplet with $J^{PC} = (1,2,3)^{--}$. Notice that this spectrum contains no state with $J^{PC} = 1^{-+}$. This is an example of exotic quantum numbers. If we find such a state, it must represent some form of new physics, perhaps a meikton.

There are also radial excitations corresponding to large values of the principal quantum numbers N = 1,2,.... The N = 2 states duplicate the N = 1 states at a larger mass. For instance, the N = 2, L = S = 0, $J^{PC} = 0^{-+}$ nonet is very important for our understanding of the glueball candidate iota.

Though the J^{PC} values of the observed mesons correspond to this nonrelativistic spectrum, the light mesons are certainly <u>not</u> nonrelativistic systems. The nonrelativistic model assumes v << c and an instantaneous potential. But for a typical meson radius

$$r \sim 0.8 \text{ fm} = 8 \cdot 10^{-14} \text{cm} \qquad (2.7)$$

the uncertainty principle tells us that

$$p \sim \frac{1}{r} \sim 250 \text{ MeV} \qquad (2.8)$$

which implies

$$v \sim 0(c) \qquad (2.9)$$

for $m_q \lesssim 300$ MeV. Since the quarks move at nearly the speed of light, the approximation of an instantaneous potential must eventually fail. There may be collective modes, for instance string modes in a string model or cavity excitations in a model based on the bag, that are not found in the nonrelativistic model. It is a controversial and deep dynamical question whether these states are identical to gluonic states made from valence gluons. My intuition, based on the bag model, is that they are not identical, but many theorists (especially the lattice theorists who start from the coarse–grained, strong coupling limit of the lattice) would disagree. Experiment and/or much more powerful theory will eventually decide.

Table 2.1 is a nonauthoritative, noncomprehensive, personal view of the status of the spectrum. As the dey show, the four sides of a box denote the $I = 1$, $I = \frac{1}{2}$, and two $I = 0$ members of a nonet. Solid lines are established, dashed lines require confirmation and crooked lines have ambiguous assignment (e.g., excited $J^P = 1^-$ could be radially or orbitally excited). One year ago my chart had only three complete nonets –π, ρ, a_2 — astonishingly meager for thirty years of hard work, the lesson being that mesons spectroscopy is very difficult. It is encouraging that one year later we can add two and perhaps four more nonets. The b_1 and ρ_3 nonets have been completed, and the a_1 and π'(radial) nonets seem to have at least enough (if not too many!) states.

Notice that I am mixing the old and new meson name systems, just to be sure you will be as confused as I am. For example, the $J^{PC} = 2^{++}$ $I = 1$ resonance that used to be A (1320) is by the new system called $a_2(1320)$. Sometimes I will write this as $A_2/a_2(1320)$. The 1987 Particle Data Group book[8] gives both old and new names to help us through the transition.

Groundstates: mixing and the 0IZ rule

Now we discuss the relationship between "ideal" mixing and the 0IZ (Okubo–Iizuka–Zweig) rule. For three flavors $\bar{u}$,d,s, there are nine $\bar{q}q$ combinations, a flavor "nonet" corresponding to one SU(3) octet. The octet contains two strange isodoublets, $(\bar{u}s, d\bar{s})$ and $(\bar{d}s, \bar{u}s)$, the isotriplet $(ud, \frac{1}{\sqrt{2}}(u\bar{u}–d\bar{d}), d\bar{u})$ and an isoscalar

$$X_8 = \frac{1}{\sqrt{6}} (u\bar{u} + d\bar{d} - 2s\bar{s}) \qquad (2.10)$$

The ninth member is the SU(3) singlet,

$$X_1 = \frac{1}{\sqrt{3}} (u\bar{u} + d\bar{d} + s\bar{s}) \qquad (2.11)$$

However, there is no reason to expect the $I = 0$ physical eigenstates to correspond to X_1 and X_8, since nothing forbids them to mix. I want to show that their flavor content is related to the OIZ rule. When the OIZ rule works well, then instead of X_1 and X_8 we expect <u>ideal</u> mixing for the eigenstates,

$$X_0 = \frac{1}{\sqrt{2}} (\bar{u}u + \bar{d}d) \qquad (2.12)$$

$$X_0' = \bar{s}s \qquad (2.13)$$

When there are large 0IZ violating forces in the flavor–singlet channel (as we would expect in QCD since gluons are flavor singlet) then the eigenstates will be nonideal and approximate X_1 and X_8. In general the mixing may be intermediate between these extremes.

To understand these ideas we begin with a very simple model that I would like to call the Idiot's Quark Model.
We suppose there are no forces at all! Then besides the kinetic energy terms which are flavor symmetric and which I do not bother to write, the free Hamiltonian contains just the quark mass terms

$$\mathscr{H}_{Free} = m(\bar{u}u + \bar{d}d) + m_s \bar{s}s \qquad (2.14)$$

Since there are no interactions, we have a trivial 0IZ rule

$$\frac{\bar{u}u + \bar{d}d}{\sqrt{2}} \mid \mathscr{H}_{Free} \mid \bar{s}s > = 0 \qquad (2.15)$$

Then the eigenstates are just determined by the quark masses and of course we have ideal mixing with X_0 and X_0'. The masses of the $\bar{q}q$ pairs are just the sum of their constituent masses, so of course

$$m_{I=1} = 2\,m$$

$$m_{I=\frac{1}{2}} = m + m_s$$

$$m_{X_0} = 2\,m$$

$$m_{X_0} = 2\,m_s \qquad (2.15)$$

From eq. (2.15) we can deduce two relations,

$$m_{I=1} = m_{X_0} \qquad (2.16)$$

and

$$2\,m_{I=\frac{1}{2}} = m_{X_0} + m_{X_0'} \qquad (2.17)$$

This may all seem very boring but I will now explain why it is interesting. The last two relations are valid not just for a free theory but for the interacting theory providing the 0IZ rule and SU(3) symmetry are approximately valid. When those two conditions are satisfied we expected ideal mixing as in (2.12) and (2.13) and we expect the relations (2.16) and (2.17) to be correct to a good approximation!

To show this we now introduce the interaction Hamiltonian $\mathscr{H}'$. Then the total Hamiltonian is $\mathscr{H} = \mathscr{H}_{\text{free}} + \mathscr{H}'$. If the 0IZ rule is valid then

$$< \bar{s}s \,|\, \mathscr{H}' \,|\, \frac{\bar{u}u \;+\; \bar{d}d}{\sqrt{2}} > \tilde{=} 0 \qquad (2.17)$$

as for the idiot quark model (2.15). Then the ideal mixtures X_0 and X_0', diagonalize the Hamiltonian and we have ideal mixing, (2.12) and (2.13).

The next step is to prove a little "theorem".

"Theorem": $SU(3)_{\text{Flavor}}$ symmetry and the 0IZ rule imply equal singlet and octet interaction energies (so–called "nonet symmetry" for the interaction energies).

<u>Proof</u>: From (2.10), (2.11), (2.12), and (2.13),

$$X_0 = \sqrt{1/3}\, X_8 + \sqrt{2/3}\, X_1 \qquad (2.18)$$

$$X_0' = -\sqrt{2/3}\, X_8 + \sqrt{1/3}\, X_1 \qquad (2.19)$$

Then the 0IZ rule implies

$$0 = < X_0 \,|\, \mathscr{H}' \,|\, X_0 >$$

$$= \frac{\sqrt{2}}{3}\left(\mathscr{H}'_{11} - \mathscr{H}'_{88} + \frac{1}{\sqrt{2}} \mathscr{H}'_{18} - \sqrt{2}\, \mathscr{H}'_{81}\right) \quad (2.20)$$

where $\mathscr{H}'_{11} < X_1 \mid \mathscr{H}' \mid X_1 >$ and so forth. From flavor SU(3) symmetry of the interactions,

$$\mathscr{H}'_{18} = \mathscr{H}'_{81} = 0 \tag{2.21}$$

so that (2.20) implies

$$\mathscr{H}'_{11} = \mathscr{H}'_{88} \tag{2.22}$$

This is just the result we wanted to prove.

Because of (2.22) the interactions do not violate the idiot relations (2.16) and (2.17). So we have shown that SU(3) symmetry and the 0IZ rule imply ideal mixing, (2.12) and (2.13), and the mass relations

$$m_{I=1} = m_{X_0} \tag{2.23}$$

$$2\, m_{I=1/2} = m_{X_0} + m_{X'_0} \tag{2.24}$$

Now consider some examples. The ideal nonet is the ground state $J^{PC} = 1^{--}$ ρ nonet, consisting of $\rho(770)$, $K^*(892)$, $\omega(783)$, and $\phi(1020)$. In fact, it was this nonet that led comrades 0,I, and Z to formulate the OIZ rule. Ideal mixing

$$\omega \sim \frac{\bar{u}u + \bar{d}d}{\sqrt{2}}$$

$$\phi \sim \bar{s}s$$

and the OIZ rule are indicated by the suppression of $\phi \to \rho\,\pi$ compared to $\phi \to \overline{K}K$ by two orders of magnitude in the square of the amplitudes (i.e. after removing the effect of phase space). The mass relations hold to about 2%:

$$700 \sim 783$$

and

$$2(892) \simeq 783 + 1020$$

The highest spin complete nonet, $J^{PC} = 3^{--}$, consists of $\rho_3(1690)$, $K_3^*(1780)$, $\omega_3(1670)$, and $\phi_3(1850)$. This is an $L = 2$ spin triplet state in the nonrelativistic classification. ϕ_3 is produced in K scattering and decays to $\overline{K}K$ and K^*K while ω_3 is produced in π scattering and decays to 3π and 5π, both suggesting ideal mixing and the 0IZ rule. The mass relations are valid to 1%!

The clearest example of nonideal mixing is provided by the π, K, η, η'nonet. The mass relations are terrible: $140 \neq 549$ and $2(494) \neq 549 + 958$. Instead the SU(3) mass formula for the octet component

$$m^2_{\eta_8} = \frac{4}{3} m^2_K - \frac{1}{3} m^2_\pi = (565)^2 \simeq m^2_\eta \quad (2.25)$$

gives a very good approximation to the η mass. Eq. (2.25) follows from SU(3) and the assumption that the SU(3) breaking Hamiltonian is in the octet representation, as is true in QCD where SU(3) breaking is due to quark masses (the Gell Mann–Nishijima formula). I use the quadratic formula because current algebra implies it for near–Goldstone bosons associated with spontaneous breeding of chiral symmetry. Eq. (2.25) shows the mixing is nearer to singlet–octet than to ideal:

$$\eta \simeq \eta_8 = \frac{1}{\sqrt{6}}(\bar{u}u + \bar{d}d - 2\bar{s}s) \qquad (2.26)$$

$$\eta' \simeq \eta_1 = \frac{1}{\sqrt{3}}(\bar{u}u + \bar{d}d + \bar{s}s) \qquad (2.27)$$

Singlet–octet mixing with the singlet much heavier than the octet, $m_1^2 >> m_8^2$, can be understood if the singlet interaction energy is much bigger than the octet,

$$\mathcal{H}'_{11} >> \mathcal{H}'_{88} \qquad (2.28)$$

From eq. (2.20) we see that this requires a large 0IZ violating force. In QCD a large flavor singlet 0IZ violating force could arise from multigluon exchange. We expect it to be stronger in the $J = 0$ but not $J = 1$ states can couple to two real gluons (so that the two gluon mixing amplitude cannot have an imaginary part in the J = 1 channel and is suppressed). The problem of understanding $\eta - \eta'$ mixing is sometimes referred to as the "U(1) problem". Plausible qualitative explanations have been given but it is still not understood quantitatively.

Let's consider a simple model that contains the essential physics.[34] We neglect the u and d masses (the chiral SU(2) limit), so that $m_u = m_d = m_\pi = 0$, and we assume

$$\mathscr{H}'_{11} >> \mathscr{H}'_{88} \tilde{=} 0$$

Then in the 8–1 basis the mass matrix (squared) is

$$M = \begin{bmatrix} \frac{4}{3}\, m_K^2 & -\frac{2\sqrt{2}}{3}\, m_K^2 \\ \frac{-2\sqrt{2}}{3}\, m_K^2 & \frac{2}{3}\, m_K^2 + \mathscr{H}'_{11} \end{bmatrix} \qquad (2.29)$$

Diagonalizing, the eigenvalues are

$$m_\eta^2, m_{\eta'}^2 =$$

$$\frac{1}{2}\left[\mathscr{H}'_{11} + 2\, m_K^2 \pm \sqrt{9\ \mathscr{H}'^{\,2}_{11} - 4\ \mathscr{H}'_{11}\ m_K^2 + 4\ m_K^4} \right.$$

(2.30)

Summing the two roots we determine the free parameter

$$\mathscr{H}'_{11} = m_\eta^2 + m_{\eta'}^2 - 2\, m_K^2 = (0.85\ \mathrm{GeV})^2 \qquad (2.31)$$

From

$$\tan 2\Theta = -\frac{4\ \sqrt{2}\ m_K^2}{3\ \mathscr{H}'_{11} - 2\ m_K^2} \qquad (2.32)$$

we then find

$$\Theta = -19^\circ \qquad (2.33)$$

$$m\eta = 490\ \mathrm{MeV} \qquad (2.34)$$

$$m\eta' = 990 \text{ MeV} \qquad (2.35)$$

These three results came from one free parameter. As expected, (2.31) shows that $m\eta'$ is dominated by the large singlet OIZ violating interaction.

The usual quark model angle $\Theta = -11^\circ$ is obtained by assuming two free parameters, M_{11} and M_{18}, and fitting to $m\eta$ and $m\eta'$. A recent analysis[35] of many experiments favors $\Theta \cong -20^\circ$. For example, if $\psi \to \gamma + \eta/\eta'$ is proportional to the η_1 components of η and η', as we would naively expect from SU(3) symmetry, then[36]

$$\frac{\Gamma(\psi \longrightarrow \gamma\eta)}{\Gamma(\psi \longrightarrow \gamma\eta)} = \left[\frac{K\eta}{K\eta'}\right]^3 \tan^2\Theta \qquad (2.36)$$

Experimentally, the left side is 4.8 ± 0.2 which gives $\Theta \cong -22^\circ$

The important lesson of this analysis is that when ideal mixing fails it is dangerous to assume "nonet symmetry". The analysis[37] of hadronic ψ decays to pseudoscalar plus vector meson which suggested a glueball component in η' could more conservatively have been interpreted as a violation of "nonet symmetry". That analysis was abandoned because it predicted a too large $\gamma\gamma$ width for iota interpreted as the mixed glueball. A more recent analysis[38] that includes D0ZI (double 0ZI suppressed) interactions is equivalent to including "nonet symmetry" breaking terms.

Before leaving this subject I wnat to comment on a related misconception. It is sometimes said that the large η' mass can be understood if the η' mixes with a heavy glueball. It is true that if a rabbit eats an elephant the rabbit will get fatter. But if a rabbit and an elephant mix quantum–mechanically, then the rabbit will get lighter and the elephant heavier. The "levels repel" according to the old rule for two body mixing. Mixing with a heavier glueball does not help explain the γ' mass.

A Tour of Some Excited States

The 2^{++} nonet, along with the ground state 0^{-+} and 1^{--}, is one of the best established $\bar{q}q$ nonets, consisting of $A_2(1320)$, $K_2^*(1430)$, f(1275) and f'(1525). It is approximately ideal since the mass relations are good to a few percent and the f' is prominently produced in K scattering and decays chiefly to $\bar{K}K$ as expected for an $\bar{s}s$ state. With the conventional analysis we define the mass matrix

$$M = \begin{bmatrix} m_8 & \delta \\ \delta & m_1 \end{bmatrix}, \qquad (2.37)$$

δ and m_1 are free parameters while m_8 is fixed by the GellMann–Nishijima formula

$$m_8 = \tfrac{4}{3} m_K - \tfrac{1}{3} m_{A_2} \qquad (2.38)$$

Diagonalizing and fitting the parameters to the f and f' masses, we find the angle Θ in the ideal basis

$$F = \cos\Theta \frac{\bar{u}u + \bar{d}d}{\sqrt{2}} + \sin\Theta\, \bar{s}s$$

$$F' = -\sin\Theta \frac{\bar{u}u + \bar{d}d}{\sqrt{2}} + \cos\Theta\, \bar{s}s \qquad (2.39)$$

to be $\Theta = -6^\circ$, very nearly ideal. Our understanding of the A_2 nonet helps us to recognize that Θ (1730) must be new physics, as discussed in section 4.

The 0^{++} nonet wins the prize for the most confusing nonet. There is an established isovector δ/a_0 (980) which decays to ηΠ and $\bar{K}K$, even though it is just below $\bar{K}K$ threshold. This is very puzzling since a $\bar{q}q$ isovector does not contain $\bar{s}s$ quarks. The nearly degenerate $I = 0$ S^*/f_0 (975) decays to ΠΠ and also to $\bar{K}K$! Two other established isoscalars are ϵ/f (1300) (decays predominantly to ΠΠ) and G/f_0 (1590) which decays most strongly to ηη' and ηη. Two other possible states are f_0 (1240) and f_0 (1730).

The $\bar{K}/K$ (1350), seen as a Kπ resonance, is an established 0^{++}

isodoublet.

It is hard to make a nonet of these. $m_S = m_\delta$ suggests ideal mixing but the strong couplins to $\overline{K}K$ are then not understandable. It may be that δ and S^* are crytoexotic four quark states[39],

$$S^* \stackrel{?}{=} (\bar{u}u + \bar{d}d)\,\bar{s}s$$
$$\delta^+ \stackrel{?}{=} u\,\bar{d}\,s\,\bar{s}$$

which could explain the strong $\overline{K}K$ coupling and their degeneracy. The f_0 (1300), f_0 (1730) and K_0^* (1350) could be 6/9 of the $L = 1$ $\bar{q}q$ nonet, leaving the isovector still to be found. The $J = 0$ states between 1 and 2 GeV may be the most difficult to discover, since low partial waves have flat angular distributions that are most difficult to see above the background. Therefore high statistics data of very high quality is required. This may explain why there is still no clear indication of a scalar glueball in radiative J/ψ decay.

The radially excited pseudoscalars are crucial to interpretation of ι (1460). There may be a complete nonet consisting of Π (1300), K (1460), η (1275), and η (1400). Π (1300) is seen by five experiments. It is very broad though there is some disagreement about the width, with estimates ranging from 200 to 500 MeV – another example of the difficulty of measuring low partial waves. K(1460) is seen by two high statistics experiments, LASS and ACCMOR, as a $K\Pi\Pi$ resonance with $\Gamma \sim 250$ MeV. The isoscalars will be discussed with ι (1460) in section 4. There is still some question about the existence and properties of η (1400) which must be resolved in order to understand iota.

The $J^{PC} = 1^{+\pm}$ nonets are of great interest in connection with the indications of new physics in the
$J = 1$ channel, to be discussed in section 5. Both are only recently added to my list of complete nonets because of new results obtained by LASS from their 140,000,000 event sample of $K\rho$ scattering data[32]. LASS has discovered the $\bar{s}s$ isoscalar H'/h_1 (1400) to complete the B/b_1 nonet, consisting also of b_1 (1235), Q_b, and h_1 (1190). The h_1 (1400) is also evident in BNL data reported at Hadron '87[40] though the analysis is not completed. LASS[32] also confirms the D'/f_1 (1530) which is an excellent candidate for the $\bar{s}s$ state needed to complete the 1^{++} nonet consisting in addition of A_1/a_1 (1270), Q_a and D/f_1 (1285).

However there are still major questions about the A_1 nonet. As reviewed by Gilman[41], data from $\iota \to \pi\pi\pi\nu$ suggest a lower A_1 mass, though some experiments with ~1200 MeV approach the larger hadronic values ranging from 1240 to 1280 MeV. The uncertainty about the mass could be related to the large width, estimated between 300 and 500 MeV. The second major controversy concerns E(1420), previously regarded as the $\bar{s}s$ member of the nonet. There is controversy as to whether a 1^{++} E exists at all (see section 5), and there is no evidence for 1^{++} E production in $K\rho$ scattering. D(1285), E(1420), and D'(1530) would be too many $I = 0$, $J^{PC} = 1^{++}$ states for the $\bar{q}q$ spectrum.

I did not quote masses for Q_a and Q_b in the discussion above because they are not directly observed. The observed mass eigenstates, Q_1/K_1 (1280) and Q_2/K_1 (1400) are not $\mathcal{G}$ parity eigenstates ($\mathcal{G}$ being the SU(3) generalization of the SU(2) G–parity, G = (–1) C) and do not correspond to Q_a and Q_b. Expressed in the $\mathcal{G}$ parity basis the mass matrix is not diagonal,

$$M = \begin{bmatrix} m(Q_a) & \epsilon \\ \epsilon & m(Q_b) \end{bmatrix} \qquad (2.40)$$

Q_1 (1280) and Q_2 (1400) are the eigenstates that diagonalize (2.40). We can define a mixing angle ϕ:

$$Q_1 = \cos\phi\, Q_a + \sin\phi\, Q_b$$

$$Q_2 = -\sin\phi\, Q_a + \cos\phi\, Q_b \qquad (2.41)$$

Because $\Gamma(Q_1 \to \rho K) >> (Q_1 \to K^* \pi)$, while $\Gamma(Q_2 \to K^* \pi) >> \Gamma(Q_2 \to \rho K)$, LASS[42] and ACCMOR[243] found by using SU(3) symmetry that $\phi = 45^\circ$, which implies

$$m(Q_a) \sim m(Q_b) \sim \tfrac{1}{2}(m(Q_1) + m(Q_2)) \sim 1340 \text{ MeV} \qquad (2.42)$$

This satisfies the mass relation for an ideal 1^{++} nonet quite well if $E(1420) = \bar{s}s$ but not for D' (1530). The analysis should be reconsidered in light of the recent data.

3. PROPERTIES OF GLUONIC STATES

In this section I summarize some of the theoretical ideas about gluonic states. I will discuss their production in radiative and hadronic J/ψ decay, the spectrum according to the bag model, and some features of their decays. Nothing I say should be accepted uncritically, since as I have stressed repeatedly the state of the theory is very poor. The slogan for this section should be: ALL THEORISTS GUILTY UNTIL PROVEN INNOCENT.

Consider the naive perturbative estimates for radiative J/ψ decay to glueballs $G = |gg>$, huĕn–hé $\gamma = (M_{\bar{q}qg})|qqg>$, and ordinary mesons $M = |\bar{q}q>$. In lowest order $\psi \to \gamma X$ proceeds by $\psi \to \gamma gg$, and counting powers of α_s we find

$$\Gamma(\psi \to \gamma G) \sim 0(\alpha\alpha_s^2) \tag{3.1}$$

$$\Gamma(\psi \to \gamma M(\bar{q}qg) \sim 0(\alpha\alpha_s^3) \tag{3.2}$$

$$\Gamma(\psi \to \gamma M) \sim 0(\alpha\alpha_5^4) \tag{3.3}$$

implying a hierarchy of production rates

$$\Gamma(\psi \to \gamma G) > \Gamma(\psi \to \gamma\, M_{(\bar{q}qg)} > \Gamma(\psi \to \gamma M) \tag{3.4}$$

Equation (3.4) explains why iota was immediately regarded as a possible glueball: confused with the obscure 1^{++} E, it appeared in $\psi \to \gamma$"E" with the largest branching ratio of any radiative J/ψ decay. Since there is no reason to expect a 0^{-+} glueball to be produced in $\psi \to \gamma G$ much more copiously than a 2^{++} glueball, $\Theta(1700)$ would be a more convincing glueball candidate if it were observed with a larger branching ratio $B(\psi \to \gamma\Theta)$ (as the LASS data indirectly suggest). In the presently observed modes it occurs at a rate that is only $\frac{1}{4}$ of that observed so far for iota.

Next consider hadronic J/ψ decays, $\psi \to M + (G$ or $M(\bar{q}qg)$ or $M')$ where M is a meson of known flavor content, eg, ω or ϕ. Such decays have been studied systematically by the Mark III[4] and DM2. Direct hadronic J/ψ decay occurs in lowest order by $\Gamma\ (\psi \to ggg) \sim 0(\alpha_s^3)$. From the lowest order diagrams beginning from the three gluon intermediate state we obtain

$$\Gamma(\psi \rightarrow M(\bar{q}qg) \sim 0\,(\alpha_s^5) \quad (3.5)$$

$$\Gamma(\psi \rightarrow MG) \sim 0\,(\alpha_s^6) \quad (3.6)$$

$$\Gamma(\psi \rightarrow MM') \sim 0(\alpha_s^6) \quad (3.7)$$

The estimates (3.5) and (3.7) are reduced by α_s^2 for flavor configurations that are DOZI[38] (double OZI suppressed, such as $\psi \rightarrow \phi f$ or $\psi \rightarrow \omega f'$. Provided the M final state is not DOZI suppressed, the hierarchy is

$$\Gamma \sim \rightarrow M_{(\bar{q}qg)} > \Gamma(\psi \rightarrow MG) \sim \Gamma((\psi \rightarrow MM') \quad (3.8)$$

Eq. (3.8) suggests that $\bar{q}qg$ states may be favored in hadronic J/ψ decay.

Spectrum of Gluonic States

Next we briefly review the spectrum of gluonic states expected in the bag model. The lowest gluon cavity mode is g_{TE} or "transverse electric" and has $J^{PC} = 1^{--}$ (like the b_1 nonet). The higher energy transverse magnetic mode g_{TM} has more familiar vector quantum numbers $J^{PC} = 1^{++}$. The ground state glueballs contain $g_{TE}\,g_{TM}$ with $J^{PC} = 0^{++}, 2^{++}$ (since J = 1 is forbidden by Bose statistics). Excited glueball states made of $g_{TE}\,g_{TM}$ have $J^{PC} = 0^{-+}$ and 2^{-+}. (The 1^{-+} state that also naively arises by combining $1^{+-} \oplus 1^{--}$ is "spurious", being interpreted as the L = 1 motion of the 0^{++} ground state with respect to the artificially fixed frame of the bag. However, as discussed in section 2, such states might actually exist with larger masses as the result of cavity mode excitations). The degeneracy of the $0^{++} - 2^{++}$ and $0^{-+} - 2^{-+}$ pairs is lifted by the order αs gluon exchange corrections. Using cavity perturbation theory as systematically developed by Lee[45], three groups[46 47 48] have found the mass hierarchy

$$0^{++} < 0^{-+} < 2^{++} < 2^{-+} \tag{3.9}$$

In the calculation of ref. 46, which includes gluon self energies as parameters, the 0^{++} and 2^{-+} glueballs are estimated at 1–1.3 GeV and 2.3 GeV respectively if $\iota(1460)$ and $\Theta(1730)$ are assumed to be the 0^{-+} and 2^{++} glueballs.

The $\bar{q}qg_{TE}$ ground state consists of four nonets, the spin singlet $0^{-+} \oplus 1^{+-} = 1^{--}$ and the triplet $1^{--} + 1^{+-} = (0,1,2)^{-+}$. The 1^{-+} $\bar{q}qg_{TE}$ nonet is exotic and cannot be confused with ordinary $\bar{q}q$ states. The excited $\bar{q}qg_{TM}$ states include the singlet $0^{-+} = 1^{--} = 1^{+-}$ and the triplet $1^{--} + 1^{--} = (0,1,2)^{++}$, the same quantum numbers as the L = 1 $\bar{q}q$ states. The gluon exchange corrections imply[2]

$$0^{-+} < 1^{-+} < 1^{--} < 2^{--} \tag{3.10}$$

for the $\bar{q}qg_{TE}$ nonets and[23]

$$0^{++} < 1^{++} < 1^{+-} < 2^{++} \tag{3.11}$$

for $\bar{q}qg_{TM}$. The mass estimates based on refs. 46 and 13 respectively are shown in tables 3.1 and 3.2, again assuming $\iota(1460)$ and $\Theta(1730)$ are the 0^{-+} and 2^{++} glueballs.

Notice in table 3.1 the 0^{-+} nonet is in the same mass region that we expect the radially excited 0^{-+} $\bar{q}q$ nonet. Perhaps the $\pi(1300)$ nonet discussed in sections 2

	0^{-+}	1^{-+}	1^{--}	2^{-+}
ρ	1200	1410	1640	1790
ω	1300	1510	1650	1890
K	1410	1590	1800	1940
ϕ	1630	1800	1980	2100

Table 3.1: bag model estimates of $\bar{q}qg_{TE}$ masses (MeV) from ref.46, with $C_{TE}/C_{TM} = \frac{1}{2}$ in order to fit ι and Θ

	0^{++}	1^{++}	1^{+-}	2^{++}
ρ	1800	1940	2130	2230
ω	1900	2040	2130	2320
K	1980	2110	2260	2350
ϕ	2220	2310	2400	2510

Table 3.2: Bag model estimates of $\bar{q}qg_{TM}$ masses (MeV) from ref. 13 with $C_{TE}/C_{TM} = \frac{1}{2}$ in order to fit ι and Θ .

and 4 are $\bar{q}qg$ states or mixtures of $\bar{q}qg$ and $\bar{q}q$. This hypothesis is supported by the strong upper limits on $\eta(1280) \to \gamma\gamma$ and $\eta(1400) \to \gamma\gamma$ discussed in section 4, which are more easily understood for $\bar{q}qg$ states than for $\bar{q}q$. Further discussion of this possiblility is given is section 4.

Also in table 3.1 we see that the exotic 1^{-+} ρ and ω are in the mass region corresponding to the new physics in the $J^C = 1^+$ channel discussed in section 5. Notice also that the $\bar{q}qg$ states in table 3.1 are the third possible source of 1^{--} excitations around 1.5 – 2 GeV, along with the N = 1, L = 2 and N = 2, L = 0 $\bar{q}q$ excitations discussed in section 2. Nobody promised that life would be easy! It is only relatively easy in the exotic channels like 1^{-+} which cannot be confused with "ordinary" physics.

Decays of glueballs and meiktons

Glueballs are SU(3) flavor singlets but it is dangerous to use this as a criterion for identifying them because large SU(3) breaking may effect their decays. For example, in lowest order perturbation theory

$$M(gg \to \bar{q}q)_{J=0} \;\alpha\; Mq \tag{3.12}$$

so that decays to $\bar{s}s$ are much larger than $\bar{u}u + \bar{d}d$ for the 0^{++} and 0^{-+} glueballs. Eq.(3.12) is a familiar consequence of "helicity conservation" – the same reason that $\Gamma(\pi \to \mu\nu) >> \Gamma(\pi \to e\nu)$. (The decay $gg \to \bar{q}q\bar{q}q$ occurs in the same order in α_s but with smaller phase space). This might explain why $\iota \to \overline{K}K\pi$ is dominant.

A second possible source of large SU(3) breaking emerges in cavity perturbation theory[13]: the $g_{TM}ss$ coupling is much larger than $g_{TM}(\bar{u}u + \bar{d}d)$. Since $\iota = g_{TM}g_{TE}$ in the bag model even the $gg \to \bar{q}q\bar{q}q$ decays may be dominated by $(\bar{s}s)\gamma\ (\bar{u}u + \bar{d}d)_8$ ("8" = color octet) that hadronize to $\overline{K}K\pi$.

Kinematic or form factor efforts can also cause large SU(3) breaking. For instance, Θ decays to $\overline{K}K$ and $\pi\pi$ in the ratio 4:1, a big SU(3) violation. As argued elsewhere[49] this could be due to kinematics: in a model with $\Theta \to \bar{u}u + \bar{d}d + \bar{s}s$ that is SU(3) symmetric at the quark level, the bigger phase space for multibody decays open to $\bar{u}u + \bar{d}d$ would cause a smaller fraction of $\bar{u}u + \bar{d}d$ to hadronize to $\pi\pi$ than $\bar{s}s$ to $\overline{K}K$. This is equivalent to the observation that the pion form factor may fall off more rapidly than the K[51].

Another source of SU(3) violation is the gluon "discoloration" mechanism[52] by which a second gg pair is created by vacuum polarization. The four gluon system may then couple pair–wise to η or η', which appear in radiative J/ψ decay to have large gg couplings. This might explain $\Theta \to \eta\eta$ and suggests a search for $\Theta\eta\eta'$ and more generally a study of $\psi \to \gamma\eta\eta'$ or $\gamma\eta'\eta'$.

The safest conclusion of this discussion is that there are no reliable predictions about the SU(3) symmetry properties of glueball decays.

A seemingly safe prediction, discussed in section 1, is that glueballs should have smaller $\gamma\gamma$ couplings than $\bar{q}q$ mesons. In perturbation theory, $M(gg \to \gamma\gamma) \sim \alpha\alpha_s$ proceeds by a $\bar{q}q$ loop while $M(\bar{q}q \to \gamma\gamma) \sim \alpha$ goes directly. However for light J = 0 glueballs this seemingly obvious result may be wrong![52]. Because of the chiral and trace anomalies light G = 0 glueballs could couple to $\gamma\gamma$ in order α with no power of α_s!. This strange result is reviewed in refs 53 and

54.

A second related conclusion[55 53 54] is that the large coupling of the $\eta'(958)$ to gg indicated by the large value of $\Gamma(\psi \to \gamma\eta')$ can be understood qualitatively by the chiral anomaly and approximate chiral symmetry, with no need to assume a big glueball component in η'. For simplicity neglect the $\eta - \eta'$ mixing and assume that $\eta' \cong \eta$. Consider the chiral SU(2) limit, $m_u = m_d = m_s = 0$. Then the QCD chiral anomaly for the singlet axial current is

$$\delta A_1 = \frac{\sqrt{3}\ \alpha_s}{2\sqrt{2}\cdot\pi} G\tilde{G} \tag{3.13}$$

where $G\tilde{G} = \epsilon_{\mu\nu\alpha\beta} G^{\mu\nu} G^{\alpha\rho}$ and $G^{\mu\nu}$ is the gluon field strength tensor. The singlet PCAC constant F_1 is defined by

$$< 0 \mid A_1^{\mu} \mid \eta' > = ik^{\mu} F_1 \tag{3.14}$$

Taking the divergence and using (3.13) we find

$$< 0 \mid \frac{\sqrt{3}\ \alpha_s}{2\ \sqrt{2}\pi} G\tilde{G} \mid \eta' > = m\eta'^2\ F_1. \tag{3.15}$$

Eq (3.15) tells us that the η'gg coupling is $0(\alpha_g^{-1})$ times a typical coupling of a meson to a quark bilinear current, without the suppression factor we would have expected from a $\bar{q}q$ loop! Together with the remarks in section 2 about $\eta - \eta'$ - glueball mixing and the η' mass, I conclude that there is no compelling reason to assume that η' has a large glueball component.

Finally I want to make two comments about huěn–hé or meikton decay modes. The first concerns the flavor structure of the final state for $\bar{l}l'$g states, where $\bar{l}, l'$ can be u or d. Consider decays that proceed by $g \to \bar{s}s$. The first step is

$$(\bar{l}l')_8 g \to (\bar{l}l')_8\ (\bar{s}s)_8 \tag{3.16}$$

where the subscript 8 denotes color octet. We can then form a final state of two hadrons either by rearrangement

$$(\bar{l}s)_1\ (\bar{s}l')_1 \tag{3.17}$$

or by soft gluon exchange

$$(\bar{l}l')_1\ (\bar{s}s)_1 \tag{3.18}$$

with the subscript 1 denoting *color singlet.* Eq. (3.17) should usually dominate but (3.18) might occur an appreciable fraction of the time. If (3.18) does occur, it may be the signature of a meikton decay, since such final states are OIZ forbidden decays of all $\bar{q}q$ mesons. The decay C(1480) $\rightarrow \phi\pi$ and the possible decay $\xi(2230) \rightarrow \phi\omega$ are examples discussed below.

The second comment[27] concerns two body decays of the ground state huen he $\bar{q}qg_{TE}$'s. The first step of the decay

$$(\bar{q}q)_8^{l=0}\ g_{TE} \rightarrow (\bar{q}q)_8^{l=0}\ (\bar{q}'q')_8^{l=1} \tag{3.19}$$

leads to two other color octet $\bar{q}q$ pairs, one in the $J^{PC} = 0^{-+}$ or 1^{--} state of an $l = 0$ $\bar{q}q$ pair, the other with the $J^{PC} = 1^{++}$ of the TE gluon. In the bag model one member of the 1^{++} $\bar{q}q$ pair is in an excited cavity mode $j^P = \frac{1}{2}^-$ or $\frac{3}{2}^-$ (j – j coupling). After rearrangement to make color singlets, as in eq. (3.17), the naive expectation is to have one ground state meson $J^{PC} = 0^{-+}$ of 1^{--}, and one excited meson incorporating the excited quark, $J^{PC} = 1^{+-}$ or $(0, 1, 2)^{++}$ depending on the particular initial state in question. This is also the expectation of the flux tube model[24]. However, the dynamics of bag fission is not at all understood, and another possibility is that the excited quark "loses" its angular momentum to the orbital angular momentum of one of the two newly formed bags, resulting in two ground state mesons which are in a p-wave with respect to one another. For example, the $I = 1$, $J^{PC} = 1^{-+}$ exotic huen–he/meikton could decay to $(\pi b_1)^{l=0}$ by the first mechanism or to $(\pi\eta)^{l=1}$ by the second. Depending on the huen–he/meikton masses, phase space might sometimes favor or even require the second mechanism. The bottom line is this: *all models are guilty until proven innocent.* Until our understanding is much greater than it is now, we should search in the kinematically allowed channels. The two body $l = 1$ decays are experimentally easier than the $l = 0$ decays which are quasi two body.

4. CANDIDATE GLUONIC STATES

In this section I will discuss the status of the search for gluonic states based on what is now known experimentally. I will indicate which experimental studies will be needed for future progress. Many of these studies will require high statistics measurements of J/ψ decays.

The particular states discussed are $\iota(1460)$, $\theta(1730)$, C(1480), $\xi(2230)$ and, briefly, several other possible resonances that do not appear strongly in J/ψ decay. Of these, iota is best established as a glueball: the only remaining uncertainty concerns $\eta(1400)$. If $\eta(1400)$ were confirmed by a J^{PC} analysis of $\psi \rightarrow \gamma\eta\pi\pi$, the last doubts about iota as a glueball would be removed. I will also discuss the possibility that the data indicates 0^{-+} $\bar{q}qg$ states in the iota mass region.

The most interesting new result about $\theta(1730)$ is that LASS did not observe it in their 140,000,000 event Kp scattering data. As I will explain, this probably means that θ has many unobserved decay modes, that $\Gamma(\psi \rightarrow \gamma\theta)$ is much bigger than the present lower limit, and that θ is indeed the 2^{++} glueball! To prove this we must find the missing θ decay modes. Again high statistics analysis of radiative J/ψ decay is essential.

C(1480) is a $J^{PC} = 1^{--}$ $\phi\pi$ resonance, an example of the $\bar{q}qg$ "signature" decay modes discussed in section 3 which are OIZ forbidden for $\bar{q}q$ mesons. It may then occur in hadronic J/ψ decay. Similarly there is an unconfirmed possiblility that $\xi(2230)$ decays to $\phi\omega$, another huen he "signature" mode. Again more statistics are needed to measure the spin of ξ and to confirm or deny the $\phi\omega$ decay, in order to decide between the 2^{++} $(\bar{u}u + \bar{d}d)g_{TM}$ hypothesis[23] or the 4^{++} $\bar{s}s$ hypothesis[22].

Iota (1460) and 0^{-+} $\bar{q}qg$

First a quick summary of the properties of iota. Measured in $\psi \rightarrow \pi K^+K^-\pi^o$ by the Mark III the mass and width are

$$M = 1461 \pm 5 \pm 5$$

$$\Gamma = 101 \pm 10 \pm 10 \qquad (4.1)$$

Spin–parity[4] is $J^{PC} = 0^{-+}$ and the production rate in radiative J/ψ decay is

$$B(\psi \to \gamma\iota)\, B(\iota \to \overline{K}K\pi) = (5.0 \pm 0.3 \pm 0.8)\cdot 10^{-3} \qquad (4.2)$$

A coupled–channel analysis suggests (but should not be regarded as establishing) that ι may also decay to $\rho\rho$ and $\omega\omega$ with

$$B(\psi \to \gamma_\iota)\, B(\iota \to \rho\rho) = (1.5 \pm 0.2) \cdot 10^{-3} \qquad (4.3)$$

$$B(\psi \to \gamma_\iota)\, B(\iota \to \omega\omega) = (0.3 \pm 0.1)\ 10^{-3} \qquad (4.4)$$

Iota may have other undiscovered decays, an interesting possibility being $\eta'\pi\pi$.

From (4.2) – (4.4) we can conclude that $B(\psi \to \gamma\iota)$ is <u>very big</u>, $\gtrsim (5\text{–}7) \cdot 10^{-3}$ depending on whether $\rho\rho$ and $\omega\omega$ are included. It is the most prominent state seen in $\psi \to \gamma X$ the second being η' with $B(\psi \to \gamma\eta') = (4.2 \pm 0.5)\ 10^{-3}$. It is also a large fraction of all $\psi \to \gamma X$. Comparing with eq. (1.10),

$$\frac{B(\psi \to \gamma_\iota)}{B(\psi \to \gamma X)_{\text{inclusive}}} \gtrsim 7\text{–}10\ \% \qquad (4.5)$$

The upper limit on $\iota \to \gamma\gamma$ is much improved since 1984. Today the best upper limit is[56]

$$\Gamma(\iota \to \gamma\gamma)\, B(\iota \to \overline{K}K\pi) < 1.6 \text{ keV } (95\% \text{ CL}) \qquad (4.6)$$

As noted in section 2, this improved limit contradicted $\gamma - \gamma' - \iota$ mixing models[37] that assumed "nonet symmetry" and predicted much bigger values for $\iota \to \gamma\gamma$. The discussion of the unsuppressed $\gamma\gamma$ coupling of a light 0^{-+} glueball only applies if it is the lightest SU(3) singlet pseudoscalar particle and therefore does not apply to iota since η' is lighter. (In fact, applied to the η' the analysis mentioned in section 3 leads[53 54] to the usual prediction for $\eta' \to \gamma\gamma$. The improved limit (4.6) is encouraging for the glueball interpretation.

Because of the improved $\iota \to \gamma\gamma$ upper limit, the lower limit on the stickiness of iota is also dramatically improved. Normalizing to $S\eta = 1$ we now find

$$S_\iota : S\eta' : S\eta : = (>65) : 4 : 1 \qquad (4.7)$$

a very striking result. Together with the evidence that there are enough pseudoscalars to fill a radially excited $\bar{q}q$ nonet, I would assess the confidence level for the glueball interpretaion at about 95%. (This is a preliminary value since I have not yet run the full–scale Monte Carlo).

Because of eq. (4.7) it is very unlikely that iota is a $\bar{q}q$ meson. We could be even more confident if the radially excited 0^{-+} nonet were completely filled. The $\pi(1300)$ and K(1460) were discussed in section 2. Here we discuss the two isoscalars. One, $\eta(1275)$, seems well established,[57] seen at Argonne and KEK in $\pi p \to \eta\pi\pi n$ and also at BNL[41] in $\pi p \to \bar{K}K\pi n$, in both cases in the $\delta\pi$ channel.

The second isoscalar is less well established. KEK[57] sees it in their latest data sample at $M = 1390 \pm 10$ with $\Gamma = 45 \pm 16$, previously reported in PRL by the same collaboration as 1420 ± 3 and 31 ± 7. BNL[41] observes a $\bar{K}K\pi$ 0^{-+} state at 1421 ± 3 and $\Gamma \sim 60$, in better agreement with the older KEK result. Comparing with eq. (4.1) we see that neither the KEK nor BNL states correspond well to iota. Therefore if either the KEK or BNL states are confirmed, there are too many $I = 0, J^{PC} = 0^{-+}$ states for just a $\bar{q}q$ nonet, as expected if iota is a glueball. If both the KEK and BNL states are confirmed <u>and</u> if they are different, then there are too many states for even a glueball plus a $\bar{q}q$ nonet. In that case we begin to consider the possibility of a second nonet in the region, consisting of $\bar{q}qg$ states as in table 3.1!

In fact a $\bar{q}qg$ interpretation of $\eta(1275)$ and $\eta(1400)$ is suggested by the strong upper limit[58]
$\gamma\gamma \to \eta(1275)/\eta(1400) \to \eta\pi\pi$ of less that 0.3 keV for either state. At least one of the $\bar{q}q$ isoscalars would be expected to appear at a larger rate. This result could be explained if $\eta(1275)$ or $\eta(1400)$ are $\bar{q}qg$ states which therefore have somewhat suppressed $\gamma\gamma$ couplings. The $I = 0$ $\bar{q}q$ states could be broader and at larger masses

($\gtrsim$ 1450 MeV), for which the $\gamma\gamma \to \eta\pi\pi$ upper limits are much weaker.

A similar remark applies to the $J^P = 0^-$ $\rho\rho$ and $\omega\omega$ structures seen by the Mark III in radiative ψ decay[4 59]. If they are not attributed to iota then they represent additional 0^{-+} isoscalars. If $\eta(1275)$ and $\eta(1400)$ are in a $\bar{q}qg$ nonet, then the $\rho\rho$ and $\omega\omega$ resonance could belong to the radially excited $\bar{q}q$ nonet. This would remove the conflict with the $\gamma\gamma \to \eta\pi\pi$ upper limits.

The $\psi \to \gamma\bar{K}K\pi$ signal for iota is not well described by a single Breit–Wigner[4]. The Mark III group has considered two explanations: a coupled–channel fit assuming $\delta\pi$ and K^*K or a two resonance fit in which the largest signal goes to the heavier resonance. In the second case, the heavier resonance would be dominantly a glueball and the lighter could be the $\eta(1400)$ seen at KEK and/or BNL. A decisive conclusion may require more data.

It is very important to establish the existence of $\eta(1400)$. In $\psi \to \gamma\eta\pi\pi$ there are enhancements at 1280 and 1400 which could be due to $\eta(1275)$ and $\eta(1400)$. It is essential to measure the spin–parity. The one at 1280 has been interpreted by the DM2 group as D(1285), but without measuring J^P the conclusion is uncertain.

To summarize, the very large stickiness of iota and the growing evidence for (at least) a complete $\bar{q}q$ nonet imply that iota is a glueball at the 95% confidence level. To raise the confidence level to 99.9% we need to establish the existence of $\eta(1400)$. This could be done in $\psi \to \gamma\eta\pi\pi$ by determining the spin–parity of the $\eta\pi\pi$ enhancement seen at 1400 MeV. Further pseudoscalars could also emerge in the complex structure of the iota peak and in the $\rho\rho$ and $\omega\omega$ channels. If in addition to $\eta(1275)$, $\eta(1400)$, and $\iota(1460)$ a fourth $I = 0$, $J^{PC} = 0^{-+}$ state were found, we would suspect the presence of two nonets ($\bar{q}q$ and $\bar{q}qg$) and a glueball!

Theta (1730): missing decays?

Since θ has been observed by three different groups in three decay modes, with spin $J = 2$ determined in two of the modes, there is no question it is a genuine resonance. All this evidence comes from radiative J/ψ decay. There is no firm evidence of θ production from any other source. There are bumps at the right mass in hadronic J/ψ decay with no spin determinations. As discussed below, the most interesting new result is that θ is not observed in Kp scattering.

θ has been observed decaying to $\eta\eta$, $\bar{K}K$ and $\pi\pi$, with

J = 2 determined in the first two. The $\overline{K}K$ channel is most prominent and gives[26]

$$M,\Gamma = 1720 \pm 10 \pm 10,\ 130 \pm 20 \qquad (4.8)$$

consistent with the more difficult measurement in the $\eta\eta$ channel

$$M,\Gamma = 1670 \pm 50,\ 160 \pm 80 \qquad (4.9)$$

The observed branching ratios are

$$\Gamma(\psi \rightarrow \gamma\Theta)B(\Theta \rightarrow \overline{K}K) = (9.6 \pm 1.2 \pm 1.8) \cdot 10^{-4} \qquad (4.10a)$$
$$\Gamma(\psi \rightarrow \gamma\Theta)B(\Theta \rightarrow \eta\eta) = (3.8 \pm 1.6) \cdot 10^{-4} \qquad (4.10b)$$
$$\Gamma(\psi \rightarrow \gamma\Theta)B(\Theta \rightarrow \pi\pi) = (2.4 \pm 1.6 \pm 1.5) \cdot 10^{-4} \qquad (4.10c)$$

Adding all three channels,

$$B(\psi \rightarrow \gamma\Theta) \gtrsim 1.6 \cdot 10^{-3} \qquad (4.10d)$$

The right side of (4.10) is comparable to $B(\psi \rightarrow \gamma f)$ and an order of magnitude bigger than $B(\psi \rightarrow \gamma f')$.

The upper limit on $\Theta \rightarrow \gamma\gamma$ has become rather tight, with the best limit from the TPC[62]

$$\Gamma(\Theta \rightarrow \gamma\gamma)B(\Theta \rightarrow \overline{K}K) < 0.2 \text{ keV (95\% CL)} \qquad (4.11)$$

The right hand side is an order of magnitude smaller than $\Gamma(f \rightarrow \gamma\gamma) \simeq 2\frac{1}{2}$ keV and a factor two bigger than $\Gamma(f' \rightarrow \gamma\gamma) \simeq 0.1$ keV. Assuming the s–wave amplitudes dominate over the d–wave we can compute the stickiness ratio

$$S_\Theta : Sf' : Sf = (>20) : 3 : 1 \qquad (4.12)$$

This is impressive though less striking than eq. (4.7) for iota.

A possible Θ signal is seen by WA76[62] in central production, $\pi^+ p \rightarrow \pi^+(K^+ K^-)p$, with $M = 1742 \pm 10$ and $\Gamma = 127 \pm 30$. However, the signal is rather small and the evidence for J = 2 is not definitive. No Θ enhancement is evident in the Y_4^0

moment.

Though the high statistics LASS experiment[32 33] sees f'/f_2 (1525) very clearly in $K^-p \to \overline{K}K\Lambda$ no Θ signal is observed. This is an extremely interesting result. The observation of f' establishes K exchange as expected, with the production mechanism

$$KK \to f' \to \overline{K}K$$

where one of the initial K's is the virtual exchange meson. On the other hand Θ is quite broad, $\Gamma \sim 130$ MeV, and $\frac{2}{3}$ of the *known* decay modes are to $\overline{K}K$. Therefore *if* the known decays ($\overline{K}K, \eta\eta$, and $\pi\pi$) are actually predominant, then we would have $\Gamma(\Theta \to \overline{K}K) \sim 90$ MeV indicating a large $g\Theta\overline{K}K$ coupling, and we would expect a large cross section for $K\rho \to \Theta\Lambda \to \overline{K}K\Lambda$ contrary to what is observed.

To be crudely quantitative, the $f' \to KK$ partial width satisfies

$$\Gamma(f' \to \overline{K}K) \lesssim \Gamma\ \mathrm{TOTAL}\ (f') = 70\ \mathrm{MeV}$$

Taking account of d–wave phase space we would conclude that if $\Gamma(\Theta \to \overline{K}K) \sim 90$ MeV then $g\Theta\overline{K}K \sim gf'KK$ and that σ $(\overline{K}Kp \to f'\Lambda \to K)$ and $\sigma(Kp \to \Theta\Lambda \to \overline{K}K\Lambda)$ should be roughly comparable. Instead there is a strong f' signal, of order 1000 events in K^+K^- and K_sK_s each, and no Θ signal at all! Apparently we should conclude that $\Gamma(\Theta \to \overline{K}K) << 90$ MeV or that

$$B(\Theta \to \overline{K}K) << 1 \tag{4.13}$$

which means from eq. (4.10) that

$$B(\psi \to \gamma\Theta) >> 1 \cdot 10^{-3} \tag{4.14}$$

This is a very exciting possibility. It means that Θ, like $\iota(1460)$, may have the large production rate expected for a 2^{++} glueball in radiative J/ψ decay! To verify this hypothesis we must find the missing decays of Θ in $\psi \to \gamma\Theta$. As discussed below, this will probably require the high statistics studies that could be done at BEPC.

To be more precise about eqs. (4.13) and (4.14) we await a

quantitative upper limit from LASS for $Kp \rightarrow \Theta\Lambda \rightarrow \overline{K}K\Lambda$, which could be used in a coupled-channel analysis to estimate $B(\Theta \rightarrow \overline{K}K)$. An earlier analysis, not including the latest LASS data, has been done by Longacre[61]. He considered the channels $\psi \rightarrow \gamma + KK/\eta\eta/\pi\pi$ and $\pi\pi \rightarrow \overline{K}K/\eta\eta/\pi\pi$, with the results $B(\Theta \rightarrow \overline{K}K) - 0.38^{+.09}_{-.19}$ and $B(\Theta \rightarrow \pi\pi) < 0.04$. When the LASS data is included it is likely that $B(\Theta \rightarrow \overline{K}K)$ will be at most 20%. This would imply $B(\psi \rightarrow \gamma\Theta) \gtrsim 5 \cdot 10^{-3}$ and that at least 70% of all decays are still unobserved!

It seems clear that θ is not a $\overline{q}q$ state. If it were it would be a radial excitation of f and f'. The low mass would suggest predominant $\overline{u}u + \overline{d}d$ content, but then the prominence of the $\overline{K}K$ mode and the tight $\gamma\gamma$ limit, eq. (4.11) are difficult to understand. If on the other hand we assume Θ is predominantly $\overline{s}s$, there are many problems:

- $m_\Theta - m_{f'}$ is too small.
- $\Gamma(\psi \rightarrow \gamma\Theta)/\Gamma(\psi \rightarrow \gamma f')$ is too large.
- $\sigma(K\rho \rightarrow \Theta\Lambda)/\sigma(Kp \rightarrow f'\Lambda)$ is too small.
- There should be a lighter $\overline{u}u + \overline{d}d$ partner.

In fact the BNL$\pi\pi \rightarrow \overline{K}K$ analysis[61] suggest a broad 2^{++} $\overline{u}u + \overline{d}d$ excitation at 1858 MeV. We would then expect the $\overline{s}s$ partner at ~ 2100 MeV. It is therefore very probable that Θ is some kind of new physics.

Both the Mark III and DM2 have observed $\overline{K}K$ enhancements[64] in hadronic J/ψ decay that have been attributed to Θ. In $\psi \rightarrow \omega K^+K^-$ the Mark III observes a K^+K^- enchancement at $1731 \pm 10 \pm 10$ (or 1750 ± 17 from DM2) with $\Gamma = 110^{+45}_{-35} \pm 15$ with a branching ratio from the ψ of $(4.5^{+1.2}_{-1.1} \pm 1.0) \cdot 10^{-4}$. In $\psi \rightarrow \phi K^+K^-$ the signal is at a significantly lower mass, $1671^{+15}_{-17} \pm 10$ (or 1686 ± 14 for DM2) with $\Gamma = 126^{+60}_{-40} \pm 15$ and a branching ratio of $(3.4^{+1.0}_{-0.8} \pm 0.9) \cdot 10^{-4}$. The low mass of the signal seen with ϕ suggests that it may be something other than Θ. The spin of these two enhancements is not yet measured. Since there are many mesons in this mass region, we should not conclude that either is Θ until the spin is established.

Before the Kp scattering data was presented by LASS it seemed that Θ must be something new perhaps a glueball or perhaps a meikton. But because $B(\psi \rightarrow \gamma\Theta)$ might not be very big, eq. (4.10d), it was not clear that a glueball interpretation should be preferred. If eqs. (4.13) and (4.14) are the correct interpretation of the LASS data, then $B(\psi \rightarrow \gamma\Theta)$ is very, very big and the glueball interpretation is very strong. To prove this it is essential to find the missing Θ decay modes.

Some possible decay channels are

$$(\eta\eta')_{L=2}$$
$$(a_2\pi)_{L=1,3} \rightarrow \rho\pi\pi, \eta\pi\pi$$
$$(a_1\pi)_{L=1,3} \rightarrow \rho\pi\pi$$
$$(f_2\eta)_{L=1,3} \rightarrow \pi\pi\eta$$
$$(K_2K)_{L=1,3} \rightarrow \overline{K}K\pi$$

The nominal Q– value for $\Theta \rightarrow f_2\eta$ is negative (1730–1270–549<0), but since Θ and f_2 are both rather broad the decay may still occur. All the above decays will be experimentally very challenging. To discover them will require partial wave analysis of high quality, high statistics data from radiative J/ψ decay. The $\eta\eta'$ mode will also require good photon and/or π^0 detection. Two other possible decay modes are already known to be small: the Mark III has obtained the upper limits[59].

$$B(\psi \rightarrow \gamma\Theta)B(\Theta \rightarrow \rho\rho) < 5\cdot 10^{-4}$$
$$B(\psi \rightarrow \gamma\Theta)B(\Theta \rightarrow \omega\omega) < 2.4\cdot 10^{-4}$$

C(1480) = a $\phi\pi$ resonance

An isovector $\phi\pi$ resonance is reported[64] by the Lepton–F collaboration at Serpukhov in $\pi^- p \rightarrow \phi\pi^0 n$. The mass is 1480 ±40 MeV and the width is 130± 60 MeV. $J^{PC} = 1^{--}$ is established by first determining that the angular distribution is consistent with pion exchange. This means that C couples to $\pi\pi$, implying natural $J^{PC} = 0^{++}, 1^{--}, 2^{++}, 3^{--}$,.... Since it also couples to $\phi\pi$ it must have negative charge conjugation, reducing the possibilities to

$J^{PC} = 1^{--}, 3^{--}$....Finally a fit to the angular distribution establishes $J^{PC} = 1^{--}$. An upper limit

$$\Gamma(C \rightarrow \omega\pi^0) < 2\,\Gamma(C \rightarrow \phi\pi^0) \qquad (4.15)$$

is also reported.

The decay $C \rightarrow \phi\pi$ is a nice example of the "signature" meikton decays discussed in section 3 which can occur by soft gluon exchange eq. (3.) but are 0IZ suppressed decays for any $\bar{q}q$ meson. It is then natural to speculate that C(1480) could be the I=1 $\bar{q}qg_{TE}$ state, expected in table 3.1 at ~1640 MeV. It would then decay to $\phi\pi$ by the sequence

$$C^+ \overset{?}{=} (u\bar{d})_8^{0^{-+}} g_{TE} \rightarrow (u\bar{d})_8^{0^{-+}} (s\bar{s})_8^{1^{+-}} \qquad (4.16)$$

$$\xrightarrow[\text{gluon exchange}]{} \phi\pi^+$$

Some other possible decay modes[46] are

$$\begin{aligned}
&(h\pi)_{L=0} \longrightarrow \rho\pi\pi \\
&(a_1\pi)_{L=0} \longrightarrow \rho\pi\pi \\
&(\omega\pi)_{L=1} \longrightarrow \pi\pi\pi\pi \\
&(\rho\eta)_{L=1} \longrightarrow \pi\pi\eta \\
&(K^*K)_{L=1} \longrightarrow \bar{K}K\pi \\
&(K^+K^-)_{L=1} \\
&(K_SK_L)_{L=1} \\
&(\pi\pi)_{L=1} \\
&(\rho\rho)_{L=1}
\end{aligned} \qquad (4.17)$$

The first two L=0 decays are of the type (see section 3)in which the angularly excited quark from $g_{TE} \longrightarrow \bar{q}q$ keeps its angular excitation in the final state h or a_1, while the remaining L=1 decays are of the type in which the quark "loses" its angular excitation to the relative orbital angular momentum of the two newly formed bags. The first two modes may be difficult to study experimentally because of the large widths of h and a_1.

Since there is no reason to expect $g_{TE} \rightarrow \bar{u}u + \bar{d}d$ to be suppressed relative to $g_{TE} \rightarrow \bar{s}s$, the same mechanism that causes $C \rightarrow \omega\pi$ should also cause $C \rightarrow \omega\pi$ at least at a comparable level. If $C = \bar{q}qg_{TE}$ we therefore expect

$$\Gamma(C \rightarrow \omega\pi) \gtrsim \Gamma(C \rightarrow \phi\pi) \qquad (4.18)$$

which is not yet incompatible with the experimental bound (4.15). In contrast, if C were a four quark state, eg., $C^+ = u\bar{d}s\bar{s}$, then by the OIZ rule we would expect

$$\Gamma(C \rightarrow \omega\pi) \ll \Gamma(C \rightarrow \phi\pi) \qquad (4.19)$$

It is therefore interesting to push the study of $C \rightarrow \omega\pi$ as far as possible. I do not understand how a $\bar{q}q\bar{s}s$ hypothesis can explain the 130 MeV width of C(1480), since such a state will "fall apart" into two color singlet hadrons with a very large width,[39] perhaps so large that it would not be recognizable as a resonance.

<u>$\xi(2230)$: 4^{++} $\bar{s}s$ or 2^{++} Meikton</u> ?

$\xi(2230)$ has been an object of much interest since its discovery by the Mark III in radiative J/ψ decay.[4] It is seen in $\psi \rightarrow \gamma K^+K^-$ with $M = 2230 \pm 16 \pm 14$ and $\Gamma = 26^{+20}_{-16} \pm 17$ with

$$B(\psi \rightarrow \gamma\xi)B(\xi \rightarrow K^+K^-) = (4.2^{+1.7}_{-1.4}\ 1.8)\cdot 10^{-5} \qquad (4.20)$$

and in the more background–free K_SK_S channel at $2232 \pm 7 \pm 7$ with $\Gamma = 18^{+23}_{-15} \pm 10$ with

$$B(\psi \rightarrow \gamma\xi)B(\xi \rightarrow K_SK_S) = (3.1^{+1.6}_{-1.3} \pm 1.7)\cdot 10^{-5} \quad . \qquad (4.21)$$

Since the width is consistent with zero, there was originally speculation that ξ might be a Higgs boson in a two doublet model. This is now ruled out, both by a report from Mark III that $J > 0$ and because ξ is produced in Kp scattering (where a Higgs boson would have too small a coupling to be seen) also with $J > 0$. However the spin of ξ is still not known.

The rather small rate $B(\psi \rightarrow \gamma\xi \rightarrow \gamma\overline{K}K)$ does not set off the glueball alarm. DM2 does not observe the ξ but the disagreement is not serious since their upper limit[65] is consistent with the Mark III if $\Gamma\xi \geq 30$ MeV or for any width if $B(\psi \rightarrow \gamma\xi \rightarrow K^+K^-) \sim 2\cdot10^{-5}$, the latter being a 1σ fluctuation on the Mark III measurement of $(4.2^{+1.7}_{-1.4} \pm 1.8)\cdot10^{-5}$.

The $\xi(2230)$ has been seen by LASS[22] in $K^-p \rightarrow (K^+K^- + K_SK_S)\Lambda$ with $J \geq 2$ clearly established and a suggestion of $J \geq 4$ from an enhancement at the ξ mass in the Y^0_8 moment. If $J = 4$ were confirmed, it would verify a suggestion[22] that ξ is the 4^{++} $\bar{s}s$ state, partner to the $f_4(2030)$.

GAMS reports[66 57] an $\eta\eta'$ enhancement in $\pi^-p \rightarrow \eta'\eta n$ with $M = 2220 \pm 10$, $\Gamma < 60$, and $J \geq 2$, which could be due to the ξ. If it is ξ and if it is identified with a similar signal seen by MIS–IHEP in $\pi p \rightarrow K_SK_Sn$, then

$B(\eta'\eta) > 2B(\overline{K}K)$, which would be inconsistant with an $\bar{s}s$ interpretation.

Sharpe and I suggested that ξ might be a $(\bar{u}u + \bar{d}d)\,g_{TM}$ meikton which would decay to $\overline{K}K$ because of the enhanced $g_{TM}\,\bar{s}s$ coupling mentioned above[23]. Under this hypothesis it could have $J^{PC} = 0^{++}$ or 2^{++}. We suggested a search for $\xi \rightarrow \phi\omega$ since by the gluon exchange mechanism the decay would proceed by

$$\begin{aligned}(\bar{u}u + \bar{d}d)^3_8\,S_1\,g_{TM} &\rightarrow (\bar{u}u + \bar{d}d)^3_8\,S_1(\bar{s}s)^3_8\,S_1\\ &\sim \omega_8\phi_8\\ &\xrightarrow[\text{gluon exch.}]{} \omega\phi\end{aligned} \tag{4.22}$$

The second line emphasizes that after $g_{TM} \rightarrow \bar{s}s$ the four quarks are essentially an ω and a ϕ in color octets, which can become $\omega\phi$ by soft gluon exchange. The Mark III reports[4] a 90% upper limit $B(\phi \rightarrow \gamma\xi)B(\xi \rightarrow \phi\omega)$ $< 6\cdot10^{-5}$ based on the data shown in ref. 4. It is intriguing that 6 of the ~ 50 events on the plot fall in the ξ bin, the largest bin on the plot. If these six events are attributed to ξ, they correspond to a branching ratio of $B(\phi \rightarrow \gamma\xi)B(\xi \rightarrow \omega\phi) \sim 3\cdot10^{-5}$, which is as large as the signals in K^+K^- or K_SK_S. Clearly we want to see more statistics. If ξ is the 2^{++} meikton then its dominant decay would be to $K^*\bar{K}^*$. Under the $\bar{q}qg_{TM}$ hypothesis $\xi(2230)$ is the state at 2320 MeV in table 3.2.

It is crucial to measure the spin of the ξ. Gao estimated at the 1984 BEPC workshop that at least 25,000,000 J/ψ decays will be needed. This estimate is supported by the fact that with 6,000,000 decays the Mark III group has been unable to determine the ξ spin.

Other gluonic candidates, not seen in J/ψ decay

Finally I will discuss very briefly some other objects that have been mentioned as gluonic candidates. These include the 2^{++} channel at 2–2.3 GeV, the 0^{++} channel around 1 GeV, and two $\eta\eta$ resonances at 1590 and 1750. A common theme is that none is observed in radiative J/ψ decay, with stringent upper limits in each case. Since perturbation theory[7] suggests that the 0^{++} and 2^{++} digluon channels in $\psi \rightarrow \gamma gg$ are as important as the 0^{-+} channel in which the large iota signal is observed, it would be surprising if any of these objects were glueballs.

G(1590) is distinguished principally by its pattern of decays, in the ratios[66]

$$\eta\eta' : \eta\eta : \bar{K}K : \pi\pi \simeq 3 : 1 : (<1) : (<0.3) \qquad (4.23)$$

Assuming SU(3) symmetry this pattern cannot be explained by any $\bar{q}q$ assignment. The large couplings of η and η' to the two gluon channel that is suggested by radiative decay ψ data implies a mechanism[67] discussed in section 3 by which $G \rightarrow \eta\eta$ or $\eta\eta'$ could dominate if G were a glueball. But the failure to observe $G \rightarrow \eta\eta$ in the Θ region, probably $B(\psi \rightarrow \gamma G) < 10^{-4}$, weighs against the glueball interpretation. A more recent hypothesis[68] is that G could be an octet component of a 0^{++} meikton nonet, which would be naturally suppressed in radiative ψ decay and would still have large $\eta\eta'$ and $\eta\eta$ decays. A similar limit must apply to a second recently announced $\eta\eta$ resonance, X(1750), seen in a high statistics study of $\pi p \rightarrow \eta\eta n \rightarrow \gamma\gamma\gamma\gamma n$ at Serpukhov by the GAMS collaboration[64]. The mass is 1755 ± 8 and the width is bounded by $\Gamma < 50$ MeV. $J^{PC} = 0^{++}$ is perferred but 2^{++} is possible. The X mass is in the Θ region but the width is significantly smaller.

A partial wave analysis[32] of the 0^{++} $\pi\pi$ and $\overline{K}K$ channels near 1 GeV suggests the possibility of three isoscalar resonances. I am not competent to judge the model–independence of the conclusions or whether uncertainties in the data might effect the qualitative conclusions. One of these resonances, $S_1(991)$, has approximately flavor independent couplings to $\pi\pi$ and $\overline{K}K$, and is therefore suggested as a possible glueball. This motivation is counter to the discussion in section 3 that J = 0 glueballs may decay preferentially to K mesons, because in perturbation theory

$$M(gg \rightarrow \bar{q}q)_{J=0} \propto m_8 \qquad (4.24)$$

so that $s\bar{s}$ dominates. Furthermore, the rate for $\psi \rightarrow \gamma\pi\pi$ in this mass region is extremely small. The upper limit[26] $B(\psi \rightarrow \gamma S^* \rightarrow \gamma\pi\pi) < 0.7 \cdot 10^{-4}$ also applies to S_1 and is two orders of magnitude less than $B(\psi \rightarrow \gamma\iota)$. Nor does S_1 have a larger double pomeron coupling than the other two states, though the conjecture that double pomeron exchange might favor glueball production is one of the motivations of such a search. I wonder whether the three states could instead be explained in terms of "cryptoexotic" $\overline{qq}qq$ states expected[30] in this region and p–wave $\bar{q}q$ states?

According to the 0IZ rule a resonance in $\pi\pi \rightarrow g_T \rightarrow \phi\phi$ could not be a $\bar{q}q$ meson but might be a glueball or an $\bar{s}sg$ meikton. I am again not competent to judge the sophisticated partial wave analysis needed to see the three g_T states[69]. Both Mark III and DM2 data imply stringent upper limits[4] on $\psi \rightarrow \gamma g_T$. For the entire 2.1–2.4 GeV region, the rate according to the Mark III is

$$B(\psi \rightarrow \gamma\phi\phi) \simeq 3 \cdot 10^{-4}. \qquad (4.24)$$

There is a peak at 2200 but with $J^P = 0^-$. Furthermore, there is no $\psi \rightarrow \gamma\phi\phi$ signal at the mass corresponding to $g_T(2010)$, the most prominent state in the $\pi\pi \rightarrow \phi\phi$ analysis.

5 E (1420) AND ALL THAT.....AGAIN: NEW PHYSICS IN THE $J^C = 1^+$ CHANNEL

In the study of $\iota(1460)$ the J = 1 $\overline{K}K\pi$ channel in the E/ι mass region was regarded as background to the new physics emerging in the J = 0 channel. Now that we are close to understanding the new physics of the J = 0 channel (see section 4), we begin to find signs of new physics in the J = 1 channel. It seems that the E/ι region does not want to go away!

I will begin with a capsule summary of the developments from 1966 to the present. In 1966 a $J^{PC} = 0^{-+}$ $\overline{K}K\pi$ resonance was observed[71] in $\bar{p}p$ annihilation at rest with the CERN hydrogen bubble chamber. Named E for Europe, the parameters were M = 1425 $\pm$7 and Γ = 80 $\pm$10. The statistical level was high for the time, 800 total events in the signal over virtually no background, and the 0^{-+} determination was made by two methods. In the years until 1980 the spin–parity was not confirmed, though there were also no experiments that matched the original one in statistical power.

In 1980 at the CERN PS the channel $\pi^- p \rightarrow \overline{K}K\pi n$ was studies[72] with a 4 GeV Π beam. With 100 events over comparable background, the analysis indicated that the E is a $J^{PC} = 1^{++}$ state, confirming a previous experiment[72]; the Particle Data Group[73] then incorporated the E into the meson table as an established 1^{++} resonance. Together with other developments in the 1^{++} nonet, the E was plausibly the $\bar{s}s$ member of the nonet.

Just as order seemed to emerge from chaos, chaos struck again with the discovery of a large "E" signal by the Mark II[74] in $\psi \rightarrow \gamma\overline{K}K\pi$. For several reasons – including the Landau–Yang theorem[75] which suggests that a 1^{++} state should not be copiously produced in the two gluon $\psi \rightarrow \gamma X$ channel – Ishikawa and I suggested[44] that the object seen by the Mark II was a pseudoscalar and probably a glueball. The pseudoscalar hypothesis was confirmed by the Crystal Ball[76] in 1982, and it seemed that both 1^{++} and 0^{-+} $\overline{K}K\pi$ resonances had been established.

As discussed at the end of section 2, the E(1420) seemed a good candidate for the $\bar{s}s$ state to complete the 1^{++} nonet, assuming the LASS–ACCMOR analysis for the "unmixing" of Q_1/K_1 (1280) and

Q_2/K_1 (1400) which implies Q_a and Q_b masses at 1340 MeV. The 1^{++} nonet would then consist of $A_1(1270)$, D(1285), $Q_A(1340)$ and E(1420). The dominance of E ⟶ K^{1*}K and the small branching ratio B(D ⟶ $\overline{K}K\pi$) $\tilde{=}$ 0.1 are both as expected for ideal mixing and the OIZ rule. The mass relations expected for ideal mixing also work very well,

$$1270 \tilde{=} 1285 \quad (5.1)$$

$$2 \cdot 1340 \tilde{=} 1285 + 1420, \quad (5.2)$$

good to 1%!

However there are also serious problems. E has never been observed in Kp scattering, as would be expected for an $\overline{s}s$ state. The A_1 mass is controversial, with data from $\tau \longrightarrow \pi\pi\pi\nu$ favoring lower values as discussed in section 2. Also discussed in section 2, the determination of the Q_A mass is indirect and requires theoretical assumptions. The existence and/or interpretation of E(1420) was also challenged in 1982 by the observation[77] of D$'$(1530) as a 1^{++} resonance in Kp ⟶ $K^*K\Lambda$, as expected for the $\overline{s}s$ member of the nonet.

The existence of the 1^{++} E(1420) was called into question again in 1985 by the high statistics πp ⟶ $\overline{K}K\pi$n BNL experiment that observed the η(1400) (see section 4). This experiment reported[78] $\eta(1420) \rightarrow K^*K$ in the $J^{PC} = 0^{-+}$ partial wave but no 1^{++} resonance at the E mass. However another high statistics experiment, WA76 at CERN[79], studying central production of $\overline{K}K$ at high energy with π and p beams, <u>did</u> report a 1^{++} E in the K^*K channel (1000 events)! They measured M = 1425 $\pm$2 MeV and = 62 $\pm$5 MeV. At this point we could only cry out for help!

In 1986 help arrived (or so it seemed at first) from the TPC collaboration. In untagged $\gamma\gamma \longrightarrow \overline{K}K\pi$ scattering they obtained the upper bound on $\iota \longrightarrow \gamma\gamma$ discussed in section 4, but in tagged events $\gamma\gamma^* \longrightarrow \overline{K}K\pi$ they discovered an unmistakeable "E" signal[24], confirmed by the Mark II[25]. The Landau–Yang theorem[75] now enters the story for the second time (now read in the other direction): since a J = 1 particle cannot couple to two massless gauge bosons, the data requires – as shown quantitatively by fits to the q^2 dependence – that the $\gamma\gamma^*$ signal be J = 1. The signal was observed by the TPC at $M = 1425^{+10}_{-18}$ and by the Mark II at 1423 $\pm$4 MeV, in good agreement with the 1^{++} K^*K signal of WA76 at 1425 $\pm$2 MeV. This seemed to be very strong confirmation of a 1^{++} E. Well, maybe.... The data suggested two or three problems. First, the initial TPC signal[24] was too much of a good thing. In the Renard[81] convention (a factor 2 larger than that used by the TPC), which corresponds to the physical partial width for real decays such as 1^{++} the TPC result was

$$\frac{m_E^2}{Q^2}\Gamma(E \rightarrow \gamma\gamma^*)\cdot B(E \rightarrow \overline{K}K\pi) = 12 \pm 4 \pm 4\text{keV} \qquad (5.3)$$

Correcting an error in the prescient paper of Renard, we can use the nonrelativistic quark model for a crude estimate of the D and E partial widths in terms of the known $\gamma\gamma$ widths of other p–wave mesons. Assuming D and E to be the ideally mixed 1^{++} states, we find comparing to $\Gamma(f \rightarrow \gamma\gamma)$ that

$$\frac{m_E^2}{Q^2}\Gamma(E \rightarrow \gamma\gamma^*) \sim 0.4\text{ keV} \qquad (5.4)$$

$$\frac{m_D^2}{Q^2}\Gamma(D \rightarrow \gamma\gamma^*) \sim 4\tfrac{1}{2}\text{ keV} \qquad (5.5)$$

Equations (5.3) and (5.4) differ by at least an order of magnitude. It seemed very difficult to understand the data in terms of an $\bar{s}s$ state.

A second problem, with a similar message, emerges from the Mark III study of hadronic J/ψ decays[4 26]. An "E" signal is seen in $\psi \rightarrow \omega$"E" but not in $\psi \rightarrow \phi$"E":

$$B(\phi \rightarrow \omega\text{"E"})B(\text{"E"} \rightarrow \overline{K}K\pi) = (6.8 \pm 2.4)\cdot 10^{-4} \qquad (5.6)$$

$$B(\psi \rightarrow \phi\text{"E"})B(\text{"E"} \rightarrow \overline{K}K\pi) < 1.1\cdot 10^{-4}(90\%\text{ CL}) \qquad (5.7)$$

Again we have data that cannot be understood in terms of an $\bar{s}s$ state.

The J^P of the "E" in (5.6) is not known.

Motivated by these two problems I proposed[27] a radical solution: that the state seen in $\gamma\gamma^* \bar{K}K\pi$ and $\psi \rightarrow \omega\bar{K}K\pi$ is an isoscalar exotic with $J^{PC} = 1^{-+}$ that could be interpreted as a $(\bar{u}u + \bar{d}d)g_{TE}$ ground state meikton. The most attractive feature of the exotic hypothesis is that if it is confirmed it points unmistakeably to new physics. We would not need to repeat the kind of analysis needed in the 0^{-+} channel to decide whether iota represents new physics. But we would still need to decide whether a 1^{-+}X(1420) is a meikton or something else. According to the meikton hypothesis, X(1420) would be the state predicted in table 3.1 at ~ 1500 MeV.

Most recently, new results were presented at the HADRON '87 meeting by the TPC, Mark II, and LASS. The TPC quoted a somewhat smaller $\gamma\gamma^*$ width than eq. (5.3),

$$\frac{m_E^2}{Q^2}\Gamma(E \rightarrow \gamma\gamma^*)B(E \rightarrow K^*K) = (7.0 \pm 2 \pm 1.4)\text{keV} \qquad (5.8)$$

Equation (5.8) assumes a ρ form faster fit to the q_2 dependence of the γ^*. Smaller values are reported by the Mark II[25],

$$\frac{m_E^2}{Q^2}\Gamma(E \rightarrow \gamma\gamma^*)B(E \rightarrow \bar{K}K\pi) = \left[\frac{2.7 \pm 1.2 \pm 1.5 \text{ keV } \rho}{1.7 \pm 0.8 \pm 0.3 \text{ keV } \phi}\right] \qquad (5.9)$$

where the smaller value (with a ϕ form factor fit) would be appropriate for an $\bar{s}s$ state. The Mark II also reports a measurement of $\gamma\gamma^* \rightarrow D(1280) \rightarrow \eta\pi\pi$. Using $B(D \rightarrow \eta\pi\pi)$ from the PDG[8] it implies

$$\frac{m_D^2}{Q^2}\Gamma(D \rightarrow \gamma\gamma^*) = 8.2 \pm 2.2 \pm 1.5 \text{ keV} \qquad (5.10)$$

If we assume that E(1420) and D(1285) are the I = 0 members of the A_1 nonet, then (5.10) and (5.8) require mixing far from ideal whereas (5.10) and (5.9) can be accommodated by a small, negative mixing angle not far from ideal.

As discussed in section 2, LASS presented results[32] at HADRON '87 confirming the existence of the 1^{++} $D'(1530)$, observed in $K\bar{p} \to D'\Lambda \to K^*K\Lambda$. In additon the BNL experiment which previously reported[78] the absence of a 1^{++} E signal presented new data[41] which seems to have an enhancement in the 1^{++} K^*K intensity at the E mass in the small t region, $|t| < 0.2$ GeV^2. (Phase motion was not displayed for this data.)

All together this makes a very confusing picture. However one point is clear: <u>there are too many $J^C = 1^+$ states for the $\bar{q}q$ model</u>. Both D(1285) and $D'(1530)$ are now established $J^{PC} = 1^{++}$ states, and the $\gamma\gamma^*$ data clearly establishes a $J^C = 1^+$ state at 1420 MeV. Whether we accept the evidence from hadron scattering for a 1^{++} E or not, there is convincing evidence for <u>three</u> $J^C = 1^+$ states where the $\bar{q}q$ model can only accommodate the two isoscalars of the 1^{++} nonet.

Since it is possible that a 1^{++} E(1420) exists but is not the $J^C = 1^+$ state seen in $\gamma\gamma^* \to \bar{K}K\pi$, I will refer to the $\gamma\gamma^* \to \bar{K}K\pi$ state as X(1420). There are then two critical questions:

1) Does a 1^{++} E(1420) exist?
2) Are X(1420) and E(1420) identical, or could X(1420) be a P = – state?

The critical tests are to determine the parity of the $\gamma\gamma^* \to \bar{K}K\pi$ resonance and to measure the spin and parity of the $\bar{K}K\pi$ signal in $\psi \to \omega\bar{K}K\pi$ at the E/X mass.

At HADRON '87 the TPC announced[82] a parity measurement for X favoring P = + but stated that P = – could not be excluded. The Mark II presented data[25] verifying P = + for the D signal in $\gamma\gamma^* \to \eta\pi\pi$ but stated that no conclusion is possible for the parity of $\gamma\gamma^* \to X \to \bar{K}K\pi$. The TPC parity determination is not definitive because of very low statistics and because it assumes that only one of the possible amplitudes dominates. This assumption would be justified[82] for $q^2 << M_X$ but the data sample used is dominated by larger q^2 values. Much higher statistics (one or more orders of magnitude) would be needed to have a large enough sample of event with

$q_2 << M_X^2$.

The E/X signal in $\psi \to \omega\overline{K}K\pi$, eq. (5.6), is at $M = 1444 \pm 5^{+10}_{-20}$ MeV with $\Gamma = 40^{+17}_{-13} \pm 10$ MeV. An E/X signal is also seen[26] in $\eta\pi\pi$,

$$B(\psi \to \omega"E")B("E" \to \eta\pi\pi) = (9.2 \pm 2.4 \pm 2.8)\cdot 10^{-4} \quad (5.11)$$

with $M = 1421 \pm 8 \pm 10$ MeV and $\Gamma = 45^{+32}_{-23} \pm 15$. The J^P of both signals, (5.6) and (5.11), is not known. However, the Mark II reports an upper limit[25] from $\gamma\gamma^* \to X \to \eta\pi\pi$,

$$\frac{B(X \to \eta\pi\pi)}{B(X \to \overline{K}K\pi} < 0.56 \begin{smallmatrix}+0.44\\-0.18\end{smallmatrix} \; (90\% \text{ CL}) \qquad (5.12)$$

which is inconsistent with the nominal ratio of eq. (5.11) to eq. (5.6). If $\psi \to \omega"E" \to \overline{K}K\pi$ is due to X, then $\psi \to \omega"E" \to \eta\pi\pi$ is probably not X but perhaps $\eta(1400)$. To understand these signals we require J^P analysis of both the two $\psi \to \omega"E"$ signals ($\eta\pi\pi$ and $\overline{K}K\pi$) and of the corresponding signals in the D(1285)/η(1275) region.

If $X = (\bar{u}u + \bar{d}d)g$, then why is $B(X \to \overline{K}K) > B(X \to \eta\pi\pi)$ as in (5.12)? Naively we would expect the opposite inequality. The answer[27] may be found in the quantum numbers of X in the meikton hypothesis: I = 0 and $J^{PC} = 1^{-+}$. They imply that if the dipion is in a J eigenstate, then the lowest partial wave is

$$X \to [(\pi\pi)_{L=2} + \eta] \qquad (5.13)$$

which is strongly suppressed since the d–wave dipion is negligible below 1 GeV. On the other hand,[83] if $\eta\pi$ is in a J eigenstate the decay proceeds by

$$X \to [(\eta\pi)_{L=1} + \pi]_{L=0} \qquad (5.14)$$

But $(\eta\pi)_{L=1}$ corresponds precisely to the I=1 $J^{PC} = 1^{-+}$ exotic expected in table 3.1 at ~1400 MeV and possibly seen[28] in that mass region. This decay is also suppressed if the splitting between X and its I=1 partner is less than m_π.

According to the $\bar{q}qg_{TE}$ hypothesis the dominant decays of X are expected to be

$$\begin{array}{llll} X & K^*K & L=1 & Q=3D\ \text{MeV} \\ X & A_1\pi & L=0 & Q=15\ \text{MeV} \end{array}$$

The preliminary indication is that the $\gamma\gamma^* \to X \to \bar{K}K\pi$ signal[25] is dominated by K^*K. The $\rho\pi\pi$ mode with $\pi\pi$ in a J eigenstate is also suppressed, since the lowest such partial wave is

$$[\rho + (\pi\pi)_{L=1}]_{L=1} \tag{5.15}$$

The L = 1 dipion is usually dominated by the ρ which is however forbidden in (5.15) by Bose statistics. Similarly
$X \to \rho\rho$ is forbidden by Bose statistics in the narrow width approximation for the ρ and is therefore suppressed in general.
$X \to \delta\pi \to \bar{K}K\pi$ or $\eta\pi\pi$ is strictly forbidden by conservation of J and P.

If we normalize the masses of the 1^{-+} nonet in table 3.1 downward by 90 MeV to fit X(1420) to the "ω" = $(\bar{u}u + \bar{d}d)g$, then we expect the other members of the nonet to be $X\rho$ (1320), $X_\phi(1700)$, and $X_K(1500)$. Some possible decay modes are

$$\begin{array}{l} X_\rho \to (\pi\eta, \pi\eta', \pi\rho, K^*K)_{L=1} (\pi b, \pi D)_{L=0} \\ X_\phi \to (\eta\eta', K^*K)_{L=1} \\ X_K \to (\pi K, \eta K, \phi K)_{L=1}, (\pi Q_1)_{L=0} \end{array}$$

Of these $(\pi\eta)_{L=1}$, $(\pi\eta')_{L=1}$, and $(\eta\eta')_{L=1}$ are especially interesting since they uniquely imply $J^{PC} = 1^{-+}$. As discussed in section 3, hadronic J/ψ decay is a very good place to look for these states, either in

$$\psi \to (\omega X, \phi X_\phi, \rho X_\rho, K^* X_K)_{L=1}$$

or

$$\psi \to (hX, h'X_{\phi}, b_1X_{\rho}, Q_1X_K)_{L=0}$$

The preliminary report[28] of a p–wave $\eta\pi$ resonance in the 1300–1400 MeV region is encouraging. The natural channel to seek this object is

$$\psi \to \rho\, X_{\rho} \to \rho\pi\eta$$

To summarize, if ,meiktons exist at m $\geq$ 1200 MeV, as in table 3.1, then the spectrum is even more complicated than previously imagined. In this case there could be as many as five I=0 $J^{PC} = 0^{-+}$ states between 1200 and 1500 MeV: two $\bar{q}q$ mesons, two $\bar{q}qg$ meiktons and one glueball. They could overlap with one another and mix. In the face of such complexity, the best chance to establish the existence of meiktons is to find the $J^{PC} = 1^{-+}$ nonet which cannot be confused with ordinary $\bar{q}q$ physics.

SUMMARY AND CONCLUSIONS

The long distance properties of QCD are not understood quantitatively or qualitatively. One of the most fundamental qualitative questions is whether the hadronic spectrum contains states with gluon constituents, glueballs and $\bar{q}qg$ mixed states or meiktons. If so, are they characterized by valence gluon modes as the bag model naively suggests? Are there also excitations of the $\bar{q}q$ spectrum involving collective modes of the soft confining quanta which are different from gluonic states made of valence gluons? There are opinions about these questions but no real answers.

It is interesting to think hard about how well QCD is really established. The evidence is convincing but there are no precise tests, as there have been for QED, which challenge the theory at a deep quantitative level. Therefore despite the "conventional wisdom" in favor of QCD, we should be open to the possiblility of fundamental surprises. Such surprises could emerge from careful study of the hadron spectrum, since the spectrum is the central consequence of any theory.

I have stressed that only crude theoretical models are available today. As was discussed extensively at the workshop which preceded us in Beijing in May, in the 1990's we can look forward to serious lattice calculations of the spectrum from future generations of computers – mainframe and special purpose. This would be well synchronized with high statistics data from BEPC. Theory will be in a better position to guide experiment, to interpret the data and to suggest important questions. Precise, high statistics experiments will be needed to guide the theory, to test the improved calculations and to learn which approximations are adequate. This could be the best kind of physics, which does not happen often in a physics lifetime: strong theory and strong experiment working effectively together.

While theory today cannot provide reliable quantitative predictions, we are still not operating completely in the dark. We have a very successful theory of the hadrons in QCD, and we do understand its short distance properties and the qualitative features of the spectrum of "ordinary" mesons and baryons. We therefore have good chances to find and recognize examples of gluonic states. Each case must be analyzed individually, but some possible guidelines include the following:

1) Glueballs and meiktons are not $\bar{q}q$ mesons, that is, they are <u>extra</u> states from the viewpoint of the nonrelativistic $\bar{q}q$ spectrum. This is the most reliable property (it could be deduced by the philosophy department without building an accelerator) but it is also the most difficult to put into practice, since it requires a detailed understanding of the "ordinary" $\bar{q}q$ spectrum. Physicists interested in gluonic states must understand what is already known about the $\bar{q}q$ spectrum and what important questions still need answers.

2) Sticky dynamics can help identify glueballs: small $\gamma\gamma$ widths and large gg couplings suggested by large $\Gamma(\psi \rightarrow \gamma G)$ partial widths. Stickiness, (see section 1) is the ratio of the square of these couplings with phase space factors removed.

3) Meiktons might give rise to apparent 0IZ rule violations of a kind that are strongly suppressed for $\bar{q}q$ mesons, as discussed in section 3. The reliability of this suggestion needs to be tested by experience. Some examples from recent data were discussed in section 4.

4) Exotic quantum numbers are as definite a sign of new physics as the existence of extra states with nonexotic quantum numbers, and it might be an easier sign to put in to insist on right to American idiom or guaranteed by US constitution! (Freedom of Speech). In section 5 I discussed clues that there may be $J^{PC} = 1^{-+}$ states in the mass region predicted by the bag model for meiktons.

All these remarks were illustrated in the discussion of sections 4 and 5. Our knowledge of QCD at short distances suggests that $J^{PC} = 0^{++}, 0^{-+}$, and 2^{++} glueballs are produced prominently in radiative J/ψ decay. With improved upper limits on their $\gamma\gamma$ decay widths, both the 0^{-+} $\iota(1460)$ and the 2^{++} $\theta(1730)$ are quite sticky,

$$\iota : \eta' : \eta \sim (>65) : 4 : 1$$

and

$$\theta : f' : f \sim (>20) : 3 : 1$$

Since $\eta(1275)$ is established, the question of whether iota is "extra" depends on whether $\eta(1400)$ exists: it is seen by two experiments but with discrepancies that could leave a skeptic unconvinced. It might be verified by establishing $J^P = 0^-$ for the clear enhancement seen in $\psi \to \gamma\eta\pi\pi$ with the right mass and width. According to my Monte Carlo, verification of $\eta(1400)$ would establish $\iota(1460)$ as a glueball at the 95% confidence level. As discussed in section 4 and also by Toki[4], the structure and mixing of iota will require careful study even if the glueball interpretation is confirmed.

In addition to its stickiness, the failure to observe in the high statistics Kp scattering experiment of LASS implies (see section 4) $B(\theta \to \overline{K}K) << 1$ which in turn implies $B(\psi \to \gamma\theta) >> 1 \cdot 10^{-3}$. Since it has long been clear that θ is "extra", the large rate for $\psi \to \gamma\theta$ would confirm the glueball interpretation. The challenge is to find the missing decay modes of $\theta(\gtrsim 70\%)$ in radiative ψ decay and therefore confirm the large value of $\Gamma(\psi \to \gamma\theta)$ directly. High statistics partial wave analysis in quasi two body decay channels is required.

It is probable from the absence of signals in $\psi \to \gamma(\pi\pi, \overline{K}K, \eta\eta,)$ that theoretical estimates of a light scalar glueball mass, < 1300 MeV, cannot be correct. If the 0^{++} glueball is heavier, it is not surprising that it has not been discovered since low partial waves with flat angular distributions are the most difficult to observe over the typically flat background. Precise, high statistics partial wave analysis is again essential.

The experimental search for meiktons is less advanced but several clues exist that correspond roughly to the mass regions of tables 3.1 and 3.2. If $\eta(1390)$ seen at KEK and $\eta(1420)$ at BNL are both real and different from each other (and if the 0^{-+} $\rho\rho$ enhancement seen by the Mark III at 1530 is not due to iota), then there are too many states for even a glueball plus a $\overline{q}q$ nonet. We might then need a 0^{-+} $\overline{q}qg$ nonet in addition. The $J^{PC} = 1^{--}$ $C(1480) \to \phi\pi$ decay might be an example of an apparently OIZ violating meikton decay. The same explanation could apply to the possible $\xi(2230) \to \phi\omega$ decay; to confirm it and measure the spin of ξ will require at least 25,000,000 J/ψ decays. Finally these are hints of I = 0 and I = 1 $J^{PC} = 1^{-+}$ exotic meiktons in the mass region expected in table 3.1: the $\gamma\gamma^* \to \overline{K}K\pi$ and $\psi \to \gamma\overline{K}K\pi$ enhancements at 1420 MeV and the possible p–wave $\pi\eta$ resonance at ~1400 MeV which would necessarily have $J^{PC} = 1^{-+}$ if confirmed. If these are the 1^{-+} exotic meiktons they should also be produced in hadronic J/ψ decay.

Because the search for gluonic states is so difficult, the prospects are excellent that BEPC will be able to make important contributions in the early 1990's. The goal is to improve on present data samples by an order of magnitude, meaning 50,000,000 to 100,000,000 J/ψ decays. This is certainly an ambitious goal. To accomplish it, several major challenges must be met. BEPC must run at near the design parameters in luminosity and beam spread. BES must obtain high quality data and be able to handle instantaneous data acquisition rates much greater than 10 Herz. And the offline analysis may require a parallel processor farm, as discussed in Witherell's lectures.

While data from many different experiments are necessary, J/ψ decay is expecially important for the discovery and analysis of gluonic states. BEPC will therefore be one of the most important world facilities for this physics, arguably <u>the</u> most important.

Acknowledgements

I wish to thank Professor Ye Ming Han and the members of the Institute of High Energy Physics for their magnificent hospitality.

REFERENCES

1. M. Gell Mann, *Phys. Lett*, **8**, 214, 1964; G. Zweig, CERN–8419/TH 412, 1964 (unpublished).

2. M. Chanowitz and S. Sharpe, *Nuc. Phys.* **B222**, 211, 1983; T. Barnes, F. Close, and F. de Viron, *Nuc. Phys.* **B224,** 241, 1983; M. Flensburg, C. Peterson, and L. Skold, *Z. Phys.* **C22**, 293, 1984.

3. J. D. Bjorken, Proc. SLAC Summer Inst. (ed. A. Mosher), 1979.

4. W. Toki, these proceedings.

5. M. Chanowitz, Proc. of the Second Intl. Conf. on Hadron Spectroscopy, April 16–18, 1987, KEK (and LBL – 23437)

6. M. Chanowitz, *Phys. Rev* **D12**, 918, 1975; L. Okun and M. Voloshin, ITEP–95, 1976 (unpublished). Also, with a factor 3 error: T. Appelquist *et al.*, *Phys. Rev. Lett.* **34**, 365,1975.

7. A. Billoire *et al*; *Phys. Lett.* **80B**, 381, 1979.

8. Particle Data Group. *Phys. Lett.* **170B**, 1, 1986.

9. D. Scharre *et al*; *Phys. Rev.* **D23**: 43 1981.

10. E. Bloom and F. Gilman, *Phys. Rev.* Lett. **25**: 1140 (1970).

11. R. Jaffe and K. Johnson, *Phys. Lett.* **60B**, 201, 1976; J. Donoghue, K. Johnson and B. A. Li, *Phys. Lett.***99B**, 416, (1981).

12. L. T. Hansson, K. Johnson, and C. Peterson, *Phys. Rev.* **D26**: 2069 (1982).

13. M. Chanowitz and S. Sharpe, *Phys. Lett.* **132B**, 413, 1983.

14. N. Isgur and J. Paton, *Phys. Lett.* **124B**, 247 1983.

15. I. Balitski, D. Dyakanov, and A. Yung, *Phys. Lett.* **B112**: 71 (1982); J. Govaerts *et. al.*, *Nucl. Phys.* **B248**: 1 (1984); J. Lattorre, P. Pascual, and S. Narison, *Zeit. fur Physik* **C34**: 347 1987.

16. C. Dominguez and N. Paver, *Zeit. fur Physik* **C31**: 591 (1982).

17. S. Sharpe, R. Jaffe, and M. Pennington, *Phys. Rev.* **D30**: 1013 (1984).

18. S. Narison, Proc. XXIII Intl. Conf. on High Energy Physics, ed. S. Loken, Berkeley, July, 1986 (World Scientific, Singapore, 1986).

19. K. Iwasaki, Proc. II Intl. Conf. on Hadron Spectroscopy, *op cit.*

20. Y. Prokoshkin, *ibid.*

21. T. Suzuki, *ibid.*

22. S. Godfrey, R. Kokoski, and J. Paton, *Phys. Lett.* **54B**, 869, 1985.

23. C. S. Gao *et al.*, Proc. of the Workshop on Colliding Beam Physics, eds. W. Chen *et al*, Beijing, June 12–21, 1984.

24. H. Aihara *et al*; *Phys. Rev. Lett.* **57**, 2500, 1986.

25. G. Gidal *et al*; LBL – 22690, 1987 and Proc XXIII Intl. Conf. on H.E.P., *op cit*

26. L. Kopke *et al*; Proc. XXIII Intl. Conf. on H.E.P; *op cit.*

27. M. Chanowitz, *Phys. Lett.* **187B**: 409 (1987).

28. J. Stroot, Proc. II Intl. Conf. on Hadron Spectroscopy, *op cit.*

29. R. Longacre, *ibid.*

30. For a recent summary, see V. Obraztsov, Proc. XXIII Intl. Conf. on H.E.P., *op cit.*

31. K. Au, D. Morgan, and M. Pennington, *Phys. Rev.* **D35**, 1633, 1987.

32. T. Suzuki, Proc. II Intl. Conf. on Hadron Spectroscopy, *op cit*

33. D. Aston, Proc. XXIII Intl. Conf. on H.E.P., *op cit.*

34. M. Chanowitz, Proc. SLAC Summer Inst., ed. A. Mosher, 1981.

35. F. Gilman and R. Kaufmann, SLAC–PUB, 1987.

36. R. Cahn and M. Chanowitz, *Phys. Lett.* **59B**, 277, 1975.

37. H. Haber and J. Perrier, *Phys. Rev.* **D32**, 2961, 1985.

38. A. Seiden, H. Sadrozinski, and H. Harber, UCSC preprint, 1987.

39. R. Jaffe, *Phys. Rev.* **D15**, 267, 1977.

40. S. Protopopescu, Proc. II Intl. Conf. on Hadron Spectroscopy, *op cit.*

41. F. Gilman, these proceedings.

42. R. Carnegie *et al*; *Phys. Lett.* **68B**, 287, 1977.

43. M. Mazzucato, *Nuc. Phys.* **B156**, 532, 1979.

44. K. Ishikawa, *Phys. Rev. Lett.* **46**, 978, 1981; M. Chanowitz, *Phys. Rev. Lett.* **46**, 981, 1981.

45. T. D. Lee, *Phys. Rev.* **D19**, 1802, 1979.

46. M. Chanowitz and S. Sharpe, ref. 2.

47. T. Barnes, F. Close, and S. Monaghan, *Nuc. Phys.* **B198**, 380, 1982.

48. C. Carlson, T. Hansson, and C. Peterson, *Nuc. Phys.* **B198**, 380, 1982.

49. M. Chanowitz, Proc. XIV Intl. Symposium on Multiparticle Dynamics, Tahoe, (1983), ed. P. Yagen and J. Gunion (World Scientific, Singapore, 1983).

50. K. Lin, Submitted to II Intl. Conf. on Hadron Spectroscopy.

51. V. Novikov *et al.*, *Nucl. Phys.* **B191**: 301 (1981).

52. S. Sharpe, Proc. Nashville Conf. 1984:1 (QCD161, I41:1984).

53. S. Sharpe, *ibid*

54. M. Chanowitz, Proc. VI Intl. Workshop on Photon–Photon Collisions, ed. R. Launder, Tahoe, 1984 (World Scientific, Singapore, 1984).

55. V. Novikov *et al.*, *Nuc. Phys.* **B165**, 55, 1980.

56. H. Aihara *et al.*, *Phys. Rev. Lett.* **57**, 51, 1986.

57. K. Tsuru, Proc. II Intl. Conf. on Hadron Spectroscopy, *op cit.*

58. S. Cooper, *ibid.*

59. N. Wermes, Physics in Collison, 1985, Proc. 6'th Intl. Conf. Autun, eds. B. Aurbert and L. Montanet, p. 221 and SLAC–PUB–3730, 1985.

60. R. Longacre, Proc. XXIII Intl. Conf. on H.E.P., *op cit.*

61. H. Aihara *et al.*, *Phys. Rev. Lett.* **57**, 404, 1986.

62. T. Armstrong *et al.*, *Phys. Lett.* **167B**, 133, 1986.

63. R. Longacre, private communication.

64. Y. Prokoshkin, Proc II Intl. Conf. on Hadron Spectroscopy, *op cit.*

65. B. J. Marie, Proc. XXIII Intl. Conf. on H.E.P., *op cit.*

66. V. Obraztsov, *ibid.*

67. C. Edwards *et al.*, *Phys. Rev. Lett.* **48**, 458,1982.

68. S. Gerstein, submitted to XXIII Intl. Conf. on H.E.P., Berkeley, 1986.

69. R. Longacre, Proc. II Intl. Conf. on Hadron Spectroscopy, *op cit.*

70. P. Baillon *et al.*, *Nuovo Cim.* **A50**, 393, 1967.

71. C. Dionisi, *et al.*, *Nuc. Phys.* **B169**, 1,1980.

72. O. Dahl *et al.*, *Phys. Rev.* **163**, 1377, 1967.

73. Particle Data Group, *Rev. Mod. Phys.* **53**, No. 2, part 1, S1, 1980.

74. D. Scharre, G. Trilling *et al.*, *Phys. Lett.* **97B**, 329,1980.

75. C. N. Yang, *Phys. Rev.* **77**, 243, 1950.

76. C. Edwards, *Phys. Rev Lett.* **49**, 259, 1982.

77. P. Gavillet *et al.*, *Z. Phys.* **C16**, 119, 1982.

78. S. Chung *et al.*, *Phys. Rev. Lett.* **55**, 779, 1985.

79. T. Armstrong *et al.*, *Phys. Lett.* **146B**, 273, 1984.

80. F. Renard, *Nuovo Cim.* **80A**, 1, 1984.

81. D. Bauer, Proc. II Intl. Conf. on Hadron Spectroscopy, *op cit.*

82. R. Cahn, LBL–22555, 1986 (to be published in *Phys. Rev.* D).

83. J. Rosner, private communication.

WEAK DECAYS OF CHARMED QUARKS AN EXPERIMENTAL REVIEW

DAVID G. HITLIN
California Institute of Technology, Pasadena, CA 91125

Abstract The experimental status of the study of weak decays of charmed mesons and baryons is reviewed.

INTRODUCTION

We now have a great deal of experimental information on the weak decays of charmed quarks, and a considerable body of calculations which serve to organize and explain the data. These lectures will treat the subject of weak decays of charmed quarks in some detail, summarizing the current experimental situation and briefly discussing theoretical ideas. As the literature in this area has now become quite vast, no attempt will be made to treat the subject historically.

STATES CONTAINING HEAVY QUARKS

We will be concerned with the weak decays of hadrons containing charmed quarks. There are three pseudoscalar charmed mesons, the $(D^0(c\bar{u}), D^+(c\bar{d}))$ isodoublet and the $D_s^+(c\bar{s})$ isosinglet. All are $^1S_0$ combinations of light and charmed quarks; all have weak decays. Figure 1 shows the lowest-lying $J^P = 0^-$ charmed mesons. Their vector D^* partners ($^3S_1$) decay through electromagnetic transitions or single pion emission. The low Q value of the $D^* \to D\pi$ transition has proven particularly valuable in isolating weak hadronic D decays.

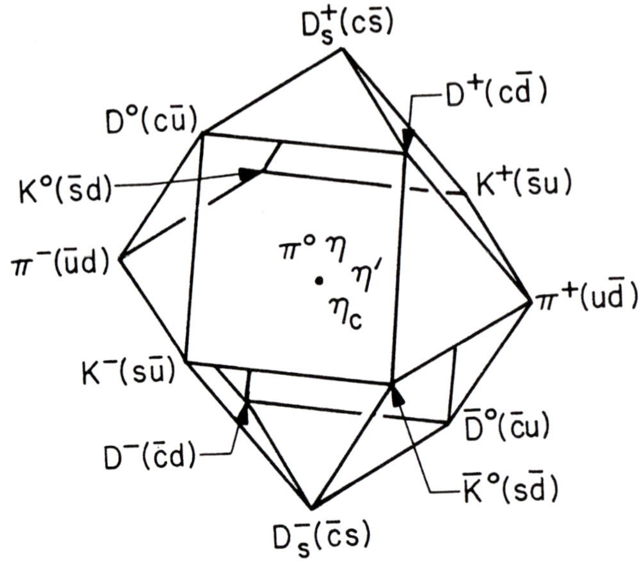

Fig. 1 The $J^P = 0^-$ mesons in a four quark world.

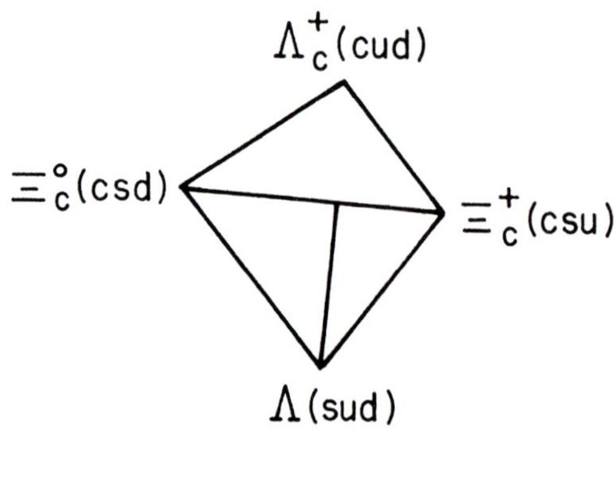

Fig. 2 The $J^P = \frac{1}{2}^-$ baryons.

TABLE I

Measurements of the D_s^+ Mass

Experiment	Mode	Mass (MeV)
CLEO	$\phi\pi^+$	$1970 \pm 5 \pm 5$
TASSO	$\phi\pi^+$	$1975 \pm 9 \pm 10$
ARGUS	$\phi\pi^+$	$1970.4 \pm 2.3 \pm 3.0$
ARGUS	$\phi 3\pi$	$1975.7 \pm 4.7 \pm 3.0$
NA-11	$KK\pi$	1972.8 ± 2.1
NA-32	$KK\pi$	1972.6 ± 2.1
NA-32 ccd	$KK\pi$	1969.1 ± 1.2
TPC	$KK\pi$	1948 ± 30
HRS	$\phi\pi^+$	$1963 \pm 3 \pm 3$
Mark III	$\phi\pi^+$	$1973 \pm 4 \pm 4$
E691	$\phi\pi^+$	$1968.1 \pm 0.8 \pm 0.6$
Average		1969.5 ± 0.7

In systems with one heavy and one light quark, the ${}^3S_1 - {}^1S_0$ mass splitting is governed by the long distance behavior of the potential. For a linear potential, the splitting should be essentially independent of the meson mass.[1] This is indeed the case, as the mass-squared differences of the $\rho - \pi$, $K^* - K$, $D^* - D$, $D_s^+ - D_s$ and $B^* - B$ systems all fall in the range[2,3,4] 0.55-0.57 GeV^2.

Figures 2, 3 and 4 show the lowest-lying $J^P = \frac{1}{2}^-, \frac{3}{2}^+$ and $\frac{1}{2}^+$ baryons in a four quark scheme. Heavy baryon mass splittings have been calculated in detail in bag and potential models.[5] Four charmed baryons (Λ_c^+, Ξ_c^+, Ξ_c^0 and Ω_c^0) decay weakly.

The D^0 and D^+ masses[2] are well established:

$$M_{D^0} = 1864.6 \pm 0.6 \text{ MeV}$$

$$M_{D^+} = 1869.3 \pm 0.6 \text{ MeV} \quad .$$

Measurements of the D_s^+ mass are summarized in Table I[6] and Figure 2.

Measurements of the Λ_c^+ mass are summarized in Table II[7] and Figure 6.

The Hadronic Weak Current

Within the context of the Standard Model, an anomaly-free, renormalizable theory is ensured by placing the charged leptons in doublets with their associated neutrinos and the quarks in left-handed weak isospin doublets and right-handed singlets. The hadronic current is not then described by these left-handed doublets, but by a generalization of the Cabibbo prescription due to Kobayashi and Maskawa.[8] For N quark generations, mixing is described by an $N \times N$ unitary matrix. In the six quark case, three angles and one phase suffice for a full description of this matrix. There are several definitions of these parameters in general use.[9] We will employ the original parametrization. There have also been extensive discussions of the generalization of the K-M matrix to four or more quark generations.[10] The parametrization we will employ

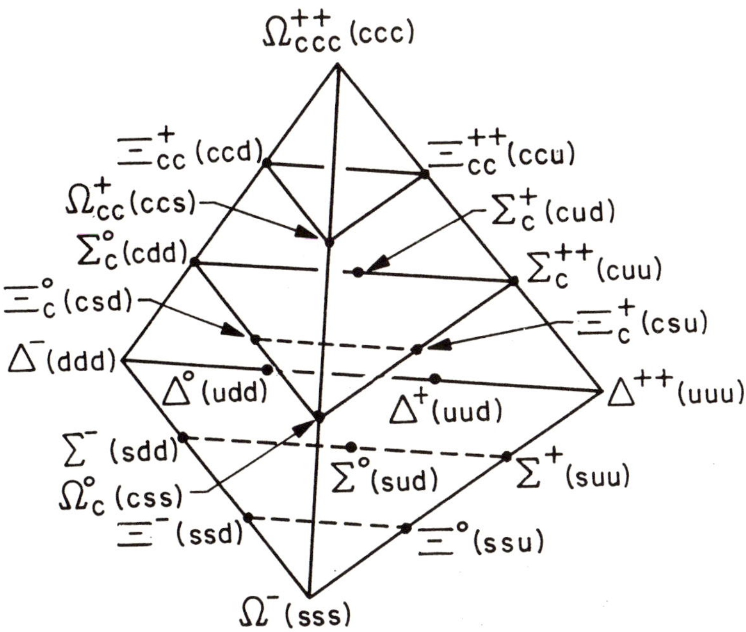

Fig. 3 The $J^P = \frac{3}{2}^+$ baryons.

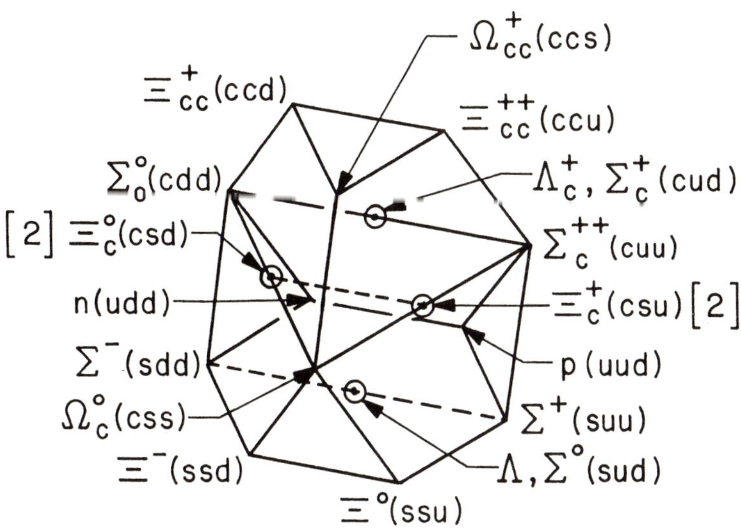

Fig. 4 The $J^P = \frac{1}{2}^+$ baryons.

TABLE II

Measurements of the Λ_c^+ Mass

Experiment	Mass (MeV)
Cazzoli	2260 ± 20
Knapp	2260 ± 20
Baltay	2257 ± 10
Cnops	2254 ± 12
Giboni	2262 ± 10
Mark II	2285 ± 6
Allasia	2250 ± 20
Calicchi	2290 ± 3
Kitagaki I	2275 ± 10
Russell	2284 ± 10
Bosetti	2270 ± 15
Kitagaki II	2270 ± 15
WA-58	2301 ± 17
Aleev	2268 ± 6
CLEO	$2287 \pm 11 \pm 5$
ACCMOR	2287 ± 0.6
LEBC-EHS	$2284 \pm 2.3 \pm 0.5$
ARGUS	$2283.1 \pm 1.7 \pm 2.0$
E691	$2284 \pm 2 \pm 1$
R-608	$2305 \pm 3 \pm 6$
Average	2286.3 ± 0.5

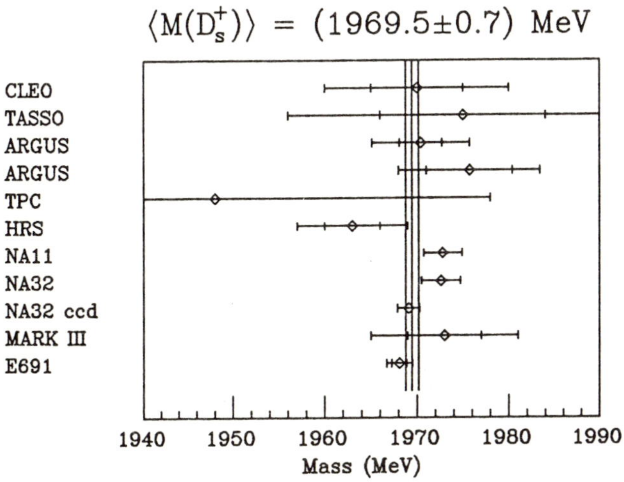

Fig. 5 Comparison of measurements of the D_s^+ mass.

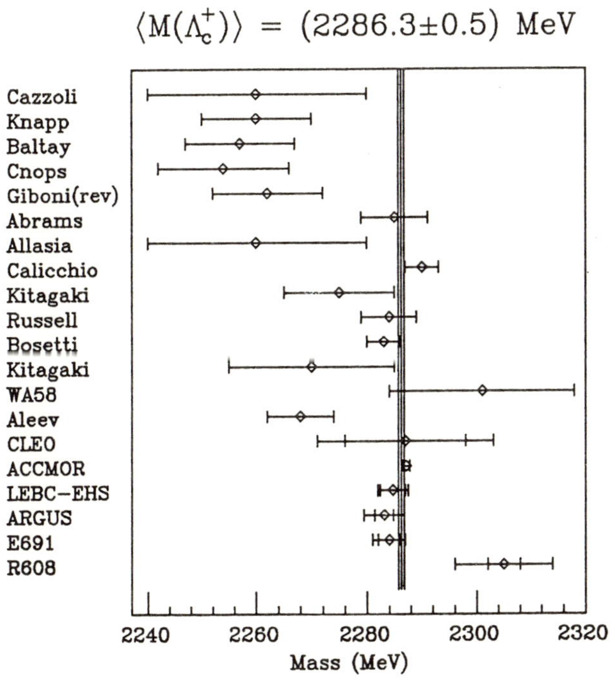

Fig. 6 Comparison of measurements of the Λ_c^+ mass.

contains 3 angles: θ_1, θ_2, θ_3, and one phase: δ, which, in a natural way, admits the possibility of CP violation.

The hadronic charged weak current thus has the form

$$J_u^T = \left(\bar{u}, \bar{c}, \bar{t}\right)\gamma_\mu\left(1-\gamma_5\right)\begin{pmatrix} V_{ud} & V_{us} & V_{ub} \\ V_{cd} & V_{cs} & V_{cb} \\ V_{td} & V_{ts} & V_{tb} \end{pmatrix}\begin{pmatrix} d \\ s \\ b \end{pmatrix},$$

with the K-M matrix completely described as:

$$\begin{pmatrix} V_{ud} & V_{us} & V_{ub} \\ V_{cd} & V_{cs} & V_{cb} \\ V_{td} & V_{ts} & V_{tb} \end{pmatrix} = \begin{pmatrix} c_1 & s_1c_3 & s_1s_3 \\ -s_1c_2 & c_1c_2c_3 - s_2s_3e^{i\delta} & c_1c_2s_3 + s_2c_3e^{i\delta} \\ -s_1s_2 & c_1s_2c_3 + c_2s_3e^{i\delta} & c_1s_2s_3 - c_2c_3e^{i\delta} \end{pmatrix}.$$

Here, $c_i \equiv \cos\theta_i$, $s_i \equiv \sin\theta_i$.

The values of the K-M angles and phase are determined by a fitting to a number of experimental inputs.

We will use the recent evaluation of K-M matrix elements by Kleinknecht and Renk.[11] The couplings are obtained from the following types of measurements:

V_{ud}: comparison of muon decay with nuclear beta decay, incorporating radiative corrections.

V_{us}: rates of K_{e3} decay or semileptonic hyperon decay. The V_{us} obtained by these two methods differs by more than the experimental error, perhaps due to SU(3) breaking effects, which are thought to be larger in hyperon decays. The K_{e3} data are therefore used.

V_{cd}: comparison of ratios of single and di-muon production in ν and $\bar{\nu}$ reactions, which can be analyzed to yield the rate of single charm quark production by the charged weak current. Extraction of V_{cd} then requires a knowledge of $B(c \to \mu X)$, which is obtained from the measurements of $B(D \to eX)$ on the $\psi(3770)$ resonance in e^+e^- annihilation. Note that the recent Mark III measurement[12] of $B(D^0 \to \pi^- e^+ \nu_e)$ provides, in principle, a determination of $|V_{cd}/V_{cs}|$ with many fewer assumptions than the technique presently employed, but that the statistics are, at present, quite limited.

V_{cs}: rate for the exclusive decays $D \to Ke\nu$, with the assumption that $f_+(0)$, which differs from 1 due to SU(4) breaking, can be calculated by sum rules, and neglecting the q^2 dependence of f_+.

V_{ub}: study of the end point spectrum for electrons produced in semileptonic B meson decay has been the classical method. The analysis of the endpoint spectrum has been controversial; the limit set is model dependent. Kleinknecht uses $|V_{ub}|/|V_{cb}| < 0.19$ at 90% confidence level. The recent ARGUS observation of charmless hadronic B meson decays allows[13] extraction of a (model-dependent) value of $|V_{bu}/V_{bc}|$ compatible with this bound.

V_{cb}: uses the measured b quark lifetime (a differing mixture of B containing hadrons in different experiments) together with the average B meson semileptonic branching ratio.

With these experimental inputs, the K-M matrix elements can be found by a least-squares adjustment. On the assumption of the existence of only three quark generations, unitarity allows a reasonably precise extraction of the matrix elements, despite our paucity of knowledge on V_{td}, V_{ts} or V_{tb}. Kleinknecht obtains:

$$\begin{array}{c} \\ u \\ c \\ t \end{array} \begin{array}{c} \begin{array}{ccc} \quad d \quad & \quad s \quad & \quad b \quad \end{array} \\ \begin{pmatrix} 0.9743-0.9757 & 0.219-0.225 & 0.000-0.008 \\ 0.219-0.225 & 0.9733-0.9748 & 0.039-0.050 \\ 0.002-0.017 & 0.037-0.048 & 0.9987-0.9993 \end{pmatrix} \end{array},$$

where the numbers quoted are the 90% confidence level allowed range.

A similar analysis, allowing the possibility of four quark generations, provides a possible window for new physics. Using the Bose-Paschos[7] parametrization, Kleinknecht finds:

$$\begin{array}{c} \\ u \\ c \\ t \\ t' \end{array}\begin{array}{c} \begin{array}{cccc} d & s & b & b' \end{array} \\ \begin{pmatrix} 0.9708-0.9750 & 0.218-0.224 & 0.000-0.008 & 0.028-0.092 \\ 0.169-0.237 & 0.66-0.98 & 0.039-0.050 & 0.00-0.78 \\ 0.00-0.17 & 0.00-0.77 & 0.030-0.999 & 0.000-0.997 \\ 0.00-0.15 & 0.00-0.35 & 0.000-0.999 & 0.05-0.9993 \end{pmatrix} \end{array},$$

using the constraint $f_+(0) < 1$ for D semileptonic decay. Clearly, we have little evidence for or against a fourth generation at the current level of precision. Observation of a t quark and measurement of its couplings will provide the next major step in this area of analysis.

We will be concerned henceforth with experimental results involving only the four quark sector of the $K-M$ matrix. We will refer to $c \to s$ transitions as "Cabibbo-allowed", for obvious historical reasons, and to $c \to d$ transitions as "Cabibbo-suppressed".

CHARMED QUARK DECAY MODES

In the current-current picture, we can isolate three distinct classes of heavy quark decay:

Purely Leptonic Decay

For pseudoscalar mesons, purely leptonic decays $P \to l\bar{\nu}_l$ (Figure 7) provide a clean measure of the axial-vector coupling of the hadronic current to vacuum, i.e., of the matrix element:

$$\langle 0|J_\mu|P\rangle = -if_P q_{p_\mu} .$$

The decay rate is

$$\Gamma(P(Q\bar{q}) \to l\bar{\nu}_l) = \frac{G_F^2}{8\pi}|V_{Qq}|^2 f_P^2 m_P m_l^2 \left[1 - \frac{m_l^2}{m_P^2}\right]^2 .$$

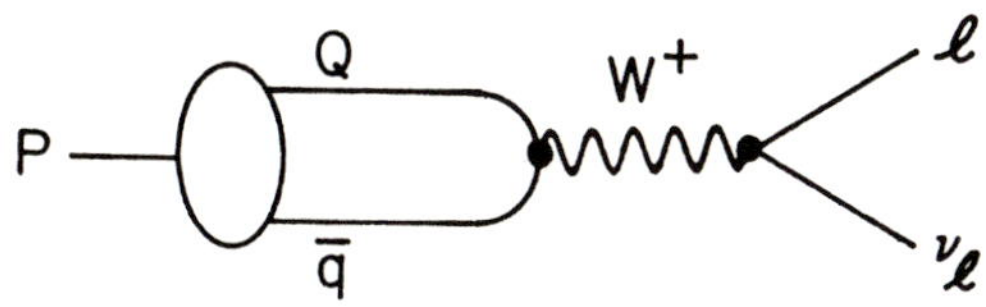

Fig. 7 Purely leptonic decay of a heavy $Q\bar{q}$ meson.

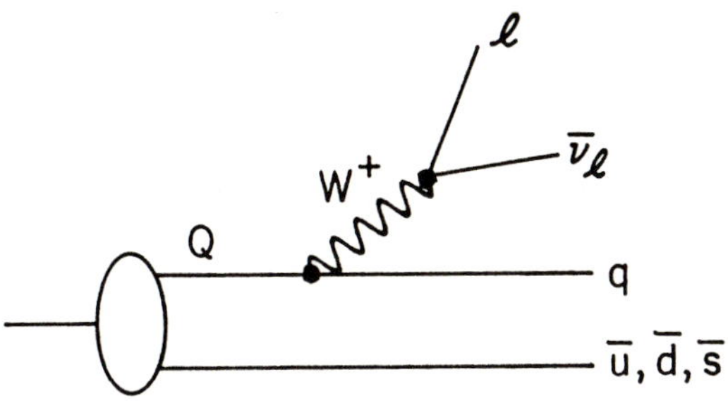

Fig. 8 Semileptonic decay of a heavy $Q\bar{q}$ meson.

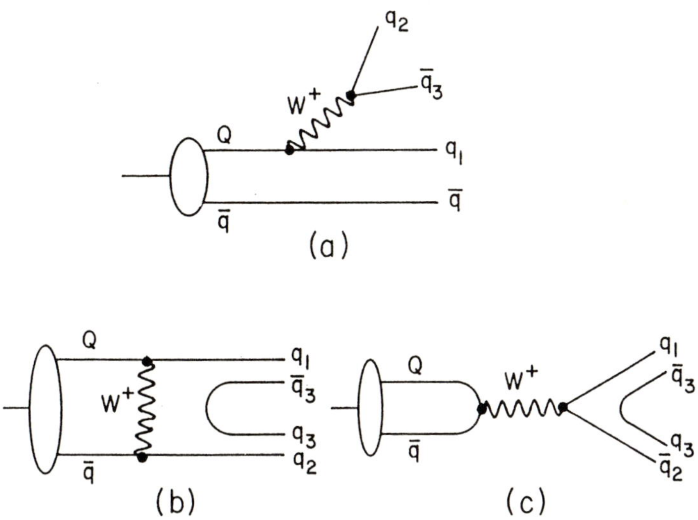

Fig. 9 Hadronic decays of a heavy $Q\bar{q}$ meson in the valence quark approximation. a) light quark spectator decay; b) W exchange; c) annihilation. The exchange and annihilation amplitudes (generically "non-spectator amplitudes") are helicity-suppressed at the light quark vertex.

TABLE III Estimates of Purely Leptonic Branching Fractions		
l \ P $B(P \to l\nu_l)$	D^+	D_s^+
$e^+\nu_e$	3.7×10^{-9}	3.1×10^{-8}
$\mu^+\nu_\mu$	1.6×10^{-4}	1.3×10^{-3}
$\tau^+\nu_\tau$	3.6×10^{-4}	1.2×10^{-2}

While f_π and f_K are, of course, well measured, no such measurements exist for heavy-quark-containing mesons. The reason is that the helicity suppression is so great that purely leptonic branching ratios are quite small. Thus, leptonic decays contribute in only a negligible way to heavy meson lifetimes. Table III shows estimates of D^+, D_s^+ meson purely leptonic branching ratios, using current lifetimes (see below), for $f_P \equiv f_K = .133$ GeV. Direct observation of purely leptonic decay is clearly difficult. Theoretical predictions for the size of f_D, f_{D_s} and f_B have varied over the years, but most estimates have now converged to the region of .10-.25 GeV.[14] The only available experimental information comes from Mark III,[15] which has analyzed its large sample of $\psi(3770)$ to set a limit on $B(D^+ \to \mu^+\nu_\mu)$. This is done by searching a sample of events in which D^+ hadronic decays have been reconstructed for a muon in the recoil spectrum and zero missing mass. No such events have been found, leading to a limit of $B(D^+ \to \mu^+\nu_\mu) < 7.2 \times 10^{-4}$ at 90% confidence level. When theoretical uncertainties in the D^+ lifetime and V_{cd} are included, this leads to a limit of $f_D < 290$ MeV at 90% Confidence Level.

Semileptonic Decays

Heavy meson semileptonic decays occur primarily through beta decay of the heavy quark (Figure 8). In this so-called spectator approximation, the rate for semileptonic decay can be scaled from muon decay:

$$\Gamma(P \to Xl\bar{\nu}_l) = \Gamma(Q \to ql\bar{\nu}_l) = \frac{G_F^2}{192\pi^3}|V_{Qq}|^2 m_Q^5 .$$

The decay matrix element is a product of terms involving the weak hadronic current and the leptonic current. A more detailed treatment of semileptonic decays, taking into account final state hadrons and form factors, will be given below.

Hadronic Decays

In the valence quark approximation, hadronic decays of heavy mesons can occur through three types of processes, shown in Figure 9 for the

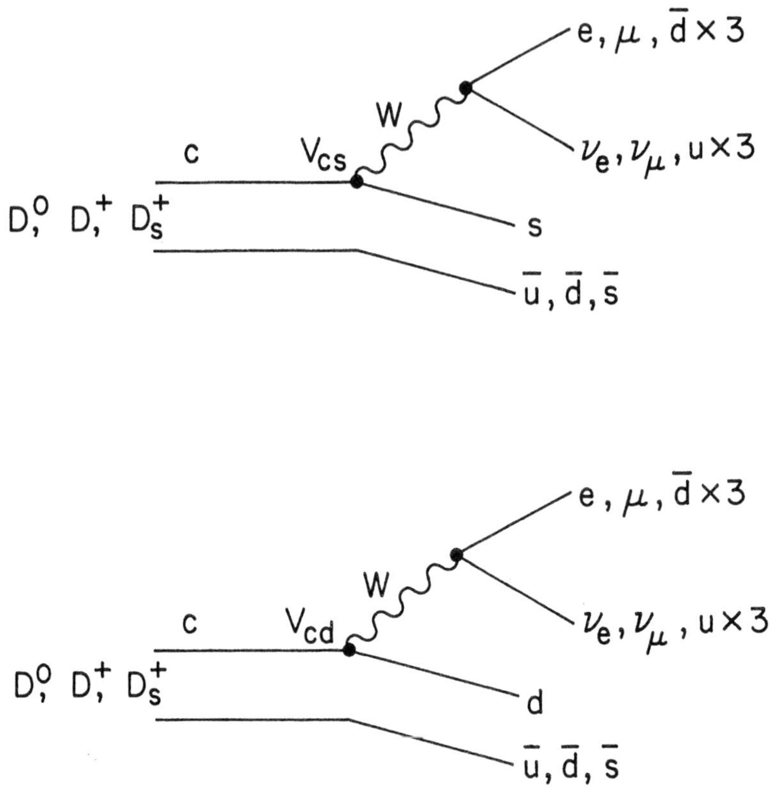

Fig. 10 Spectator amplitudes for charmed meson decay: a) Cabibbo-allowed, and b) Cabibbo-suppressed. Note that in this picture the lifetime of all charmed particles is predicted to be identical, as are all semileptonic branching fractions, which should be $\sim 20\%$.

case of charm decay. The dominant diagram is flavor decay, the W exchange and annihilation diagrams being suppressed by helicity conservation at the light quark vertex. As the matrix element here is a product of terms each involving the weak hadronic current, calculations are non-trivially model dependent. Within the valence quark picture, it is consistent to retain only the flavor decay contribution, leading to the spectator model. This naïve picture, as we shall see, must be considerably refined, but we shall employ it for the time being, in order to estimate the lifetimes of heavy quark states.

Figure 10 illustrates the spectator diagrams for charmed meson decay. Clearly, the expected lifetime for charmed mesons can be scaled from muon decay:

$$\tau_o(D^0, D^+, D_s^+) = \frac{1}{\Gamma_o} \simeq \frac{1}{5}\left(\frac{m_\mu}{m_c}\right)^5 [\Im|V_{cs}|^2 + |V_{cd}|^2]$$
$$\simeq 3 \times 10^{-13}\text{sec for } m_c = 1.5 \text{ GeV},$$

with a phase space factor $\Im \simeq 0.5$.

Two immediate conclusions may be drawn from this simple picture. First, while the correct order of magnitude for charmed particle lifetimes is obtained, the prediction that all charmed particle lifetimes are **identical** is not experimentally verified. Second, inspection of Figure 8 tells us that the semileptonic branching ratio for all charmed particles should be $\sim 20\%$. This is also not the case.

LIFETIME MEASUREMENTS

A great deal of experimental effort has gone into the measurement of charmed particle lifetimes, in both fixed target and e^+e^- annihilation experiments. The experimental situation is summarized in Tables IV[16] and V[17] and Figures 11-14. A clear pattern of charmed particle lifetimes has now emerged:

$$\tau(D^+)/\tau(D^0) = 2.45 \pm 0.09 \quad ,$$

$$\tau(D_s^+)/\tau(D^0) = 1.01 \pm 0.08 \quad .$$

While naïve extrapolation from muon decay gives a good estimate of the average charmed particle lifetime, an understanding of the detailed ordering of lifetimes:

$$\tau(D^+) > \tau(D^0) \simeq \tau(D_s^+) > \tau(\Lambda_c^+) \quad ,$$

perforce must take us beyond the simple spectator model. We must therefore examine in detail our overly-simplified initial premises.

The Heavy Quark Weak Decay Hamiltonian

Summing hadronic and semileptonic decay processes, the bare spectator quark-parton Hamiltonian for weak decays of charmed mesons has the following form:

$$\begin{aligned} H_W^{(0)}(\Delta c = -1) = \frac{G_F}{\sqrt{2}} & [V_{cs}(\bar{s}c)_L + V_{cd}(\bar{d}c)_L] \\ & \times [V_{ud}^*(\bar{d}u)_L + V_{us}^*(\bar{s}u)_L \\ & + \bar{e}\nu_e + \bar{\mu}\nu_\mu + \bar{\tau}\nu_\tau] \, . \end{aligned}$$

where, e.g., $(\bar{s}c)_L = \sum_{i=1}^{3} \bar{s}_i \gamma^\mu (1-\gamma_5) c_i$. As $q^2 \ll M_W^2$, this pointlike interaction approximation is clearly valid.

These bare Hamiltonia generate a hierarchy of decays characterized by the size of the relevant K-M elements. For example, in hadronic charmed particle decays, Cabibbo-allowed terms involving $V_{cs}V_{ud}^*$ are dominant. Strong interaction effects modify this simple picture in several significant ways:

- weak couplings are renormalized
- the color structure of the interaction may be distorted by octet currents
- new chiral structures, e.g., $V + A$ currents may appear

TABLE IV

Measurements of Lifetimes of Charmed Mesons

Experiment	D^+ $\tau(10^{-13}\text{s})$	D^0 $\tau(10^{-13}\text{s})$	D_s^+ $\tau(10^{-13}\text{s})$
E-531	$11.1\pm^{4.4}_{2.9}$	$4.3 \pm^{0.7}_{0.5} \pm^{0.1}_{0.2}$	$2.6\pm^{1.6}_{1.1}$
WA-58	$5.0 \pm^{1.5}_{1.0} \pm 1.9$	$3.6 \pm^{1.2}_{0.8} \pm 0.7$	
SHF	$8.6 \pm 1.3\pm^{0.7}_{0.3}$	$6.1 \pm 0.9 \pm 0.3$	
NA-16	$8.4\pm^{3.5}_{2.2}$	$4.1\pm^{1.3}_{1.0}$	
NA-18	$6.3 \pm^{4.9}_{2.3} \pm 1.5$	$4.1 \pm^{2.6}_{1.3} \pm 0.5$	
NA-27	$11.2\pm^{1.4}_{1.1}$	$4.6\pm^{0.6}_{0.5}$	
NA-1	$9.5\pm^{3.1}_{1.9}$	$3.4\pm^{0.6}_{0.5}$	
NA-11	$10.6\pm^{3.6}_{2.4}$	$3.7\pm^{1.0}_{0.7}$	$3.2\pm^{1.5}_{1.0}$
NA-11	$10.9 \pm 1.4 \pm 0.8$		
NA-32	$10.9\pm^{1.9}_{1.5}$	4.2 ± 0.5	$3.4\pm^{1.3}_{0.9}$
NA-32			$4.5\pm^{1.9}_{1.1}$
NA-14		$3.5\pm^{1.3}_{0.8}$	
NA-14′		4.43 ± 0.37	
E-691	$10.9 \pm 0.3 \pm 0.25$	$4.22 \pm 0.08 \pm 0.1$	$4.7 \pm^{0}.4 \pm 0.2$
DELCO		$4.6 \pm 1.5\pm^{0.7}_{0.6}$	
MARK II	$8.9 \pm^{3.8}_{2.7} \pm 1.3$	$4.7 \pm^{0.9}_{0.8} \pm 0.5$	
MARK II upgrade		$4.4 \pm^{1.2}_{1.1} \pm 0.6$	
HRS	$8.1 \pm 1.2 \pm 1.6$	$4.2 \pm 0.9 \pm 0.6$	$3.5 \pm^{2.4}_{1.8} \pm 0.9$
TASSO		$4.8 \pm^{1.0}_{0.9} \pm^{0.5}_{0.7}$	$5.7 \pm^{3.6}_{2.6} \pm 0.9$
CLEO	$11.4 \pm 1.6 \pm 0.7$	$5.0 \pm 0.7 \pm 0.4$	$4.7 \pm 2.2 \pm 0.5$
ARGUS	$10.5 \pm 0.8 \pm 0.7$	$4.8 \pm 0.4 \pm 0.3$	$5.6 \pm^{1.3}_{1.2} \pm 0.8$
Average	$10.45\pm^{0.31}_{0.29}$	4.27 ± 0.10	$4.31\pm^{0.36}_{0.32}$

TABLE V

Measurements of Lifetimes of Charmed Baryons

Experiment	Λ_c^+ $\tau(10^{-13}\text{s})$	Ξ_c^+ $\tau(10^{-13}\text{s})$	Ω_c^0 $\tau(10^{-13}\text{s})$
E-531	$2.0\pm^{0.7}_{0.6}$		
WA-58	$2.3\pm^{0.9}_{0.6}\pm0.4$		
NA-27	$1.2\pm^{0.5}_{0.3}$		
NA-1	$1.1\pm^{0.8}_{0.4}$		
NA-32	$1.6\pm^{0.4}_{0.3}\pm0.3$		
E-691	$2.2\pm0.3\pm0.2$		
E-400		$6.2\pm^{1.8}_{1.6}$	
WA-62		$4.8\pm^{2.1}_{1.0}\pm^{2.0}_{1.0}$	$7.9\pm2.8\pm2.0$
Average	$1.60\pm^{0.24}_{0.18}$	$5.7\pm^{1.6}_{1.1}$	7.9 ± 3.4

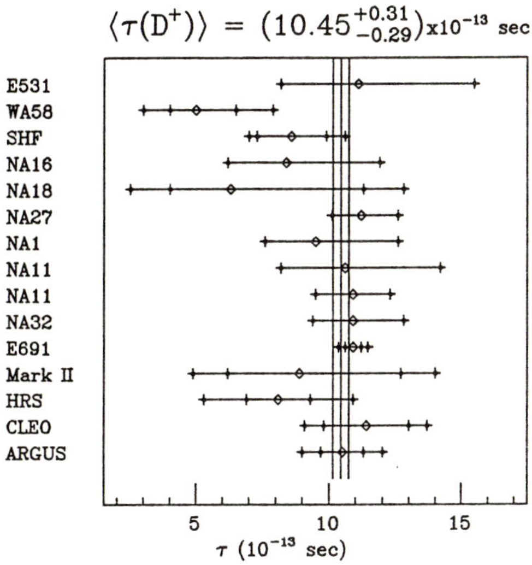

Fig. 11 Comparison of measurements of the D^+ lifetime. Where given, systematic errors are also shown. For purposes of calculating a weighted average, statistical and systematic uncertainties are combined in quadrature, although this procedure has no compelling justification in statistical theory.

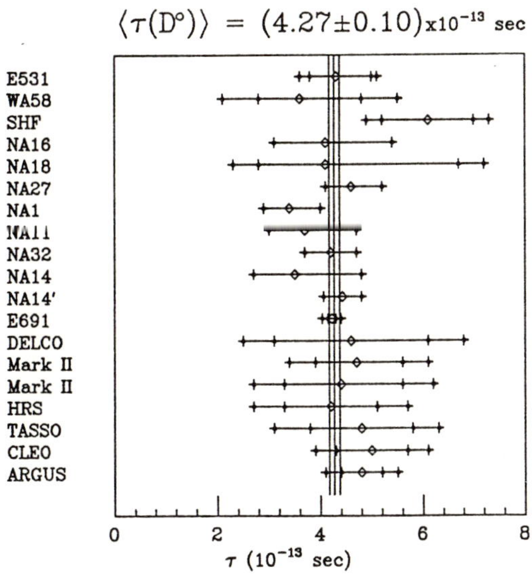

Fig. 12 Comparison of measurements of the D^0 lifetime.

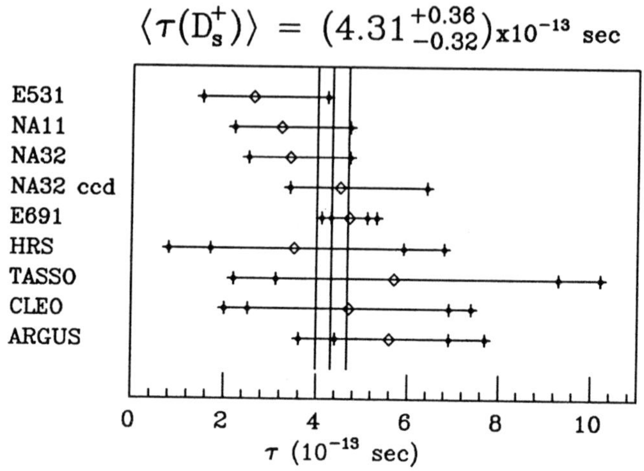

Fig. 13 Comparison of measurements of the D_s^+ lifetime.

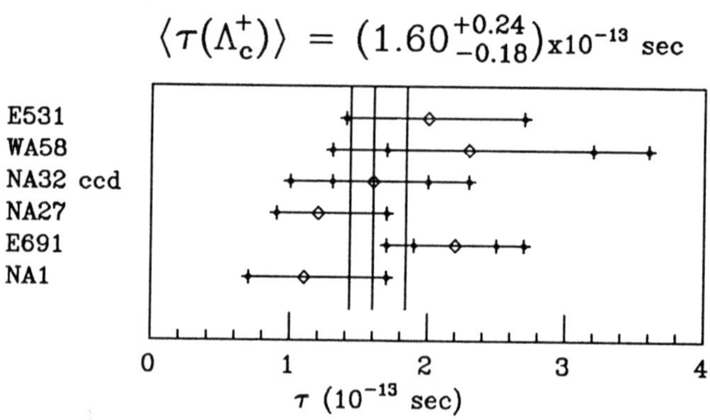

Fig. 14 Comparison of measurements of the Λ_c^+ lifetime.

- there may be induced S and T interactions

In practice, the last two effects are small, but the first two are quite significant. QCD provides a framework in which these strong interaction effects can be estimated.[18] The lowest order corrections, i.e., $O(\alpha_s)$ hard gluon contributions, fall into two classes. Those which involve self energy or vertex corrections to the four quark operators are absorbed in the renormalization of G_F, while those involving gluon exchange along with the $W^{\pm}$ are logarithmically divergent. Cutting the latter off at M_W yields a first order correction to $H_W^{(0)}$:

$$H_W^{(1)} = -\frac{G_F V_{cs} V_{ud}^*}{\sqrt{2}} \frac{3\alpha_s}{8\pi} \log\left(\frac{M_W^2}{\mu^2}\right) (\bar{s}\lambda_a c)_L (\bar{d}\lambda^a u)_L$$

for Cabibbo-allowed hadronic charm decays, with analogous corrections for the other terms. Thus the hard gluon corrections induce a new operator, which has the same chiral and flavor structure, but involves color octet currents, λ_a. Semileptonic decays clearly receive no such corrections.

The physical effects of hard gluon exchanges are most easily seen if we make a Fierz transformation:

$$(\bar{s}\lambda_a c)_L (\bar{d}\lambda_a u)_L = -\frac{2}{3}(\bar{s}c)_L(\bar{d}u)_L + 2(\bar{s}d)_L(\bar{u}c)_L\,.$$

If we then sum the contributions to the Hamiltonian:

$$\begin{aligned} H_W &= H_W^{(0)} + H_W^{(1)} + \ldots \\ &\simeq \frac{G_F}{\sqrt{2}} V_{cs} V_{ud}^* \left[\left(1 + \frac{\alpha_s}{4\pi} \ln \frac{M_W^2}{\mu^2}\right) (\bar{s}c)_L (\bar{d}u)_L \right. \\ &\qquad \left. - \frac{3\alpha_s}{4\pi} \ln \frac{M_W^2}{\mu^2} (\bar{s}d)_L (\bar{u}c)_L \right], \end{aligned}$$

we see that the hard gluon exchange has two effects:

- the strength of the charged current interaction is renormalized
- there is an induced effective neutral current interaction.

Similar effects are expected, of course, for the Cabibbo-suppressed terms.

It is useful to define the symmetric and antisymmetric linear combinations

$$O_{\pm}(\Delta c = -1) = \frac{1}{2}[(\bar{s}c)_L(\bar{u}d)_L \pm (\bar{s}d)_L(\bar{u}c)_L] .$$

The bare Hamiltonian for Cabibbo-allowed hadronic charm decays then has the form:

$$H_W^{(0)} = \frac{G_F}{\sqrt{2}} V_{cs} V_{ud}^* [O_+ + O_-] ,$$

while the first order corrected version is:

$$H_W = \frac{G_F}{\sqrt{2}} V_{cs} V_{ud}^* [c_+^{(1)} O_+ + c_-^{(1)} O_-] ,$$

where

$$c_+^{(1)} = 1 - \frac{\alpha_s}{2\pi} \ln \frac{M_W^2}{\mu^2}$$
$$c_-^{(1)} = 1 + \frac{\alpha_s}{2\pi} \ln \frac{M_W^2}{\mu^2} .$$

Clearly, the bare light quark spectator model is recovered in the limit $c_+^{(1)} = c_-^{(1)} = 1$.

The effective four quark operators $O_{\pm}$ may usefully be classified by their color SU(3) transformation properties. The current-current structure is $\underline{3} \times \underline{3} = \underline{3}^* + \underline{6}$. The $\underline{3}^*$ representation of SU(3) is antisymmetric, while the $\underline{6}$ is symmetric. Hard gluon corrections thus enhance the $\underline{3}^*$ and suppress the $\underline{6}$. In the case of the $\Delta s = -1$ Hamiltonian which governs K decay, a similar decomposition obtains. Since O_- transitions are $\Delta I = \frac{1}{2}$, while O_+ transitions can be either $\Delta I = \frac{1}{2}$ or $\frac{3}{2}$, the origin of the $\Delta I = \frac{1}{2}$ rule is likely to be found in QCD corrections. Thus, the $\Delta I = \frac{1}{2}$ rule is often referred to as "$\underline{6}$ dominance". These QCD modifications to the structure of the current-current Hamiltonian governing inclusive weak hadronic decays can be calculated to all orders in α_s using renormalization group methods.[19] In terms of the anomalous dimensions d_+ and d_-:[20]

$$c_\pm = \left[\frac{\alpha_s(\mu^2)}{\alpha_s(M_W^2)}\right]^{\frac{d_\pm}{2b}} = \left[1 + \alpha_s(\mu^2)\frac{b}{4\pi}\ln\left(\frac{M_W^2}{\mu^2}\right)\right]^{\frac{d_\pm}{2b}} .$$

Here, $b = 11 - \frac{2N_f}{3}$, where N_f is the number of fermionic degrees of freedom.

In the leading log (LL) approximation, the value of $c_\pm$ depends on the effective coupling constant g_s, defined by:

$$c_\pm\left(g_s, \frac{M_W}{\mu}\right) = c_\pm\left(0, \frac{M_W}{\mu}\right)\left[\frac{\ln(M_W^2/\Lambda_{QCD}^2)}{\ln(\mu^2/\Lambda_{QCD}^2)}\right]^{\frac{d^\pm}{2b}} .$$

Note that since $d_- = -2d_+$, we have the relation $c_- = 1/c_+^2$.

The calculation has, in fact, been carried out[21] including two loop (next to leading log) corrections to the β function. These "NLL" corrections are small, lending confidence to the prescription:

$$c_\pm\left(g_s, \frac{M_W}{\mu}\right) = \left[\frac{\alpha_s(\mu^2)}{\alpha_s(M_W^2)}\right]^{\frac{d_\pm}{b}}\left[1 + \frac{\alpha_s(\mu^2) - \alpha_s(M_W^2)}{\pi}\rho_\pm\right],$$

where

$$\rho_+ = \frac{1}{b}\left[\frac{-221}{24} + \frac{5}{9}N_f\right] + \frac{1}{b^2}\left[51 - \frac{19}{3}N_f\right]$$

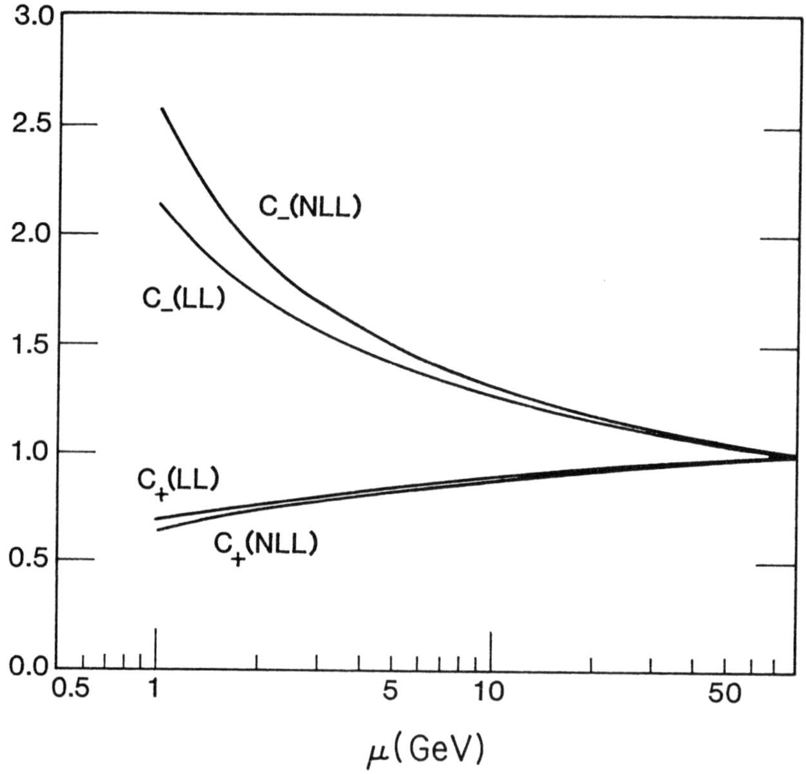

Fig. 15 Scale dependence of the operator coefficients c_- and c_+ as calculated in the leading log (LL) approximation, and to the next-to-leading-log (NLL) order.

$$\rho_- = \frac{1}{b}\left[\frac{263}{12} - \frac{10}{9}N_f\right] + \frac{1}{b^2}\left[-102 + \frac{38}{3}N_f\right].$$

The c_+ and c_- coefficients for $N_f = 6$ are shown in Fig. 15 as a function of the scale μ. Clearly, hard gluon exchanges modify the operator coefficients of the charm Hamiltonian to a greater degree than they do those for the bottom and top sectors.

We can most clearly visualize the affect of QCD corrections on heavy quark weak processes by rewriting the Hamiltonian once more. Defining

$$c_1 = \frac{c_+ + c_-}{2} \quad , \quad c_2 = \frac{c_+ - c_-}{2} ,$$

we have, for the $\Delta c = -1$ Hamiltonian:

$$H_W(\Delta c = -1) = \frac{G_F}{\sqrt{2}} V_{cs} V_{ud}^* [c_1(\bar{s}c)_L(\bar{u}d)_L + c_2(\bar{s}d)_L(\bar{c}u)_L] ,$$

where c_1, c_2 are evaluated at the appropriate scale μ. In the absence of QCD corrections, $c_1 = 1$, $c_2 = 0$. Thus, hard gluon exchange is responsible for:

- Shortening the charmed quark lifetime, as the non-leptonic decay width is enhanced:

$$\Gamma_{NL} = (2c_+^2 + c_-^2)\Gamma_0 .$$

- Inducing an "effective neutral current" interaction, proportional to c_2^2. This term also often referred to as a "color-suppressed" interaction.

- Reducing semileptonic branching ratios, the so-called "non-leptonic enhancement". Since

$$\Gamma(c \to su\bar{d}) = (2c_+^2 + c_-^2)\Gamma(c \to sl\nu_l) ,$$

the semileptonic branching fraction becomes

$$B(c \to eX) = \frac{1}{2 + 2c_+^2 + c_-^2} \simeq 15\% \,,$$

rather than the naïve value of 20%. The reduction in phase space due to non-zero quark masses, also, of course, slightly reduces the naïve estimate.

- The $\Delta I = \frac{1}{2}$ rule in non-leptonic K decay and its generalizations to the heavy quark sector.

These modifications are all born out by experiment, providing striking support for Quantum Chromodynamics. In one important area, however, we must go farther. The framework we have developed thus far ensures, up to small phase space corrections, that the lifetimes and semileptonic branching ratios of all charmed mesons are identical. As D^0, D^+ and D_s^+ lifetimes are manifestly not identical, we must extend our investigation to a more detailed study of non-leptonic decays. As experimental data improve, measurements of the separate lifetimes of the individual B mesons will undoubtedly be made. With appropriate scaling, the results discussed here apply equally to the b quark sector.

Comparison of the lifetimes and semileptonic branching ratios of the D^0 and D^+ mesons provides the first as to the origin of the differing lifetimes. Neglecting small Cabibbo-suppressed semileptonic decays of the D^+, isospin invariance tells us that $\Gamma(D^0 \to e^+X) = \Gamma(D^+ \to e^+X)$, as both are purely spectator decays: $c \to se\nu_e$. We can then relate the ratios of lifetimes and semileptonic branching fractions:

$$\begin{aligned}\frac{B(D^+ \to e^+X)}{B(D^0 \to e^+X)} &= \frac{\Gamma(D^+ \to e^+X)}{\Gamma(D^+ \to \text{all})} \cdot \frac{\Gamma(D^0 \to \text{all})}{\Gamma(D^0 \to e^+X)} \\ &\simeq \frac{\Gamma(D^0 \to \text{all})}{\Gamma(D^+ \to \text{all})} = \frac{\tau(D^+)}{\tau(D^0)} \,.\end{aligned}$$

Mark III has measured[22] the separate D^0 and D^+ inclusive semileptonic branching fractions:

$$B(D^+ \to e^+X) = (17.0 \pm 1.9 \pm 0.7)\%\,,$$

$$B(D^0 \to e^+X) = (7.5 \pm 1.1 \pm 0.4)\%\,.$$

The ratio is

$$\frac{B(D^+ \to e^+X)}{B(D^0 \to e^+X)} = 2.3^{+0.5+0.1}_{-0.4-0.1}\,,$$

quite close to the world average directly measured lifetime ratio:

$$\frac{\tau(D^+)}{\tau(D^0)} = 2.45 \pm 0.09\,.$$

The lifetime and semileptonic branching ratios can be combined to extract the partial widths for semileptonic decay:

$$\Gamma_{SL}(D^0) = 1.75 \pm 0.28 \times 10^{11}\ \mathrm{sec}^{-1}$$

$$\Gamma_{SL}(D^+) = 1.65 \pm 0.20 \times 10^{11}\ \mathrm{sec}^{-1}\quad ,$$

which are, in fact, identical. Thus, the source of the differing charmed meson lifetimes must be sought in the hadronic decay sector.

As $B(D^+ \to e^+X)$ is slightly larger than the QCD corrected estimate, we are still, however, faced with a puzzle. Are the hadronic decays of the D^0 enhanced by a factor of two, or are the D^+ decays inhibited? Scaling from muon decay is not helpful due to the fifth power dependence of the lifetime on the (unknown) charmed quark mass. There is a direct experimental approach, however. It so happens, perhaps *a fortiori*, that the hadronic decays of the D meson are dominated by two body and quasi-two body decay modes. It has been possible to reconstruct a large number, in fact the great majority of such exclusive decays, thus allowing us to approach the matter of the charmed meson lifetime, a manifestly *inclusive* matter, by summing over a large number of measured *exclusive* channels.

J

CHARMED QUARK DECAYS – EXPERIMENT

D Hadronic Decays

A full understanding of the mechanism of weak decays of charmed mesons requires a detailed study of exclusive decay modes. We now have a great deal of information on the exclusive decays of the D^0 and D^+, and are beginning to learn some important facts about D_s^+ decays. For clarity, we will discuss Cabibbo-allowed and -suppressed decays separately. The classic technique for reconstruction of exclusive states is of course, the invariant mass technique.

Charmed particles are produced in hadronic reactions, photoproduction, neutrino reactions and high energy e^+e^- annihilation. Charm production cross sections range from $\sim 10^{-3}$ of the total cross section in hadronic reactions, to $\sim 10^{-2}$ in photoproduction, to $\sim 10^{-1}$ in neutrino reactions to 30% or more in e^+e^- annihilation. Much progress has been made in recent years in sophisticated triggering schemes to enrich event samples, and recently, in the use of high precision vertex detectors, which greatly enhance signal-to-background ratios by allowing reconstruction of detached secondary vertices.

In order to reduce combinatoric background, several methods are employed. The GIM mechanism ensures that Cabibbo-allowed charm hadronic decays produce either K_s^0 or charged kaons of a specific sign (i.e., $D^0 \to K^-$, etc.). Semileptonic decays similarly produce leptons of a specific sign ($D^0 \to e^+$). Such correlations are important both in isolating a signal and demonstrating its validity. Given sufficient spatial resolution, the relatively long D lifetime allows isolation of a separated vertex, substantially enhancing signal-to-noise ratios. Since at high energy, $D^*(1^-)$ production is three times larger than $D(0^-)$ production, a requirement that the mass of a reconstructed D and a slow pion form a D^* mass can be a powerful method of suppressing background (see Figure 16).

Most information on D^0 and D^+ exclusive decay modes, however, comes from low energy e^+e^- annihilation at SPEAR. Due to the existence of the $\psi(3770)$ resonance, which decays predominantly to $D\bar{D}$

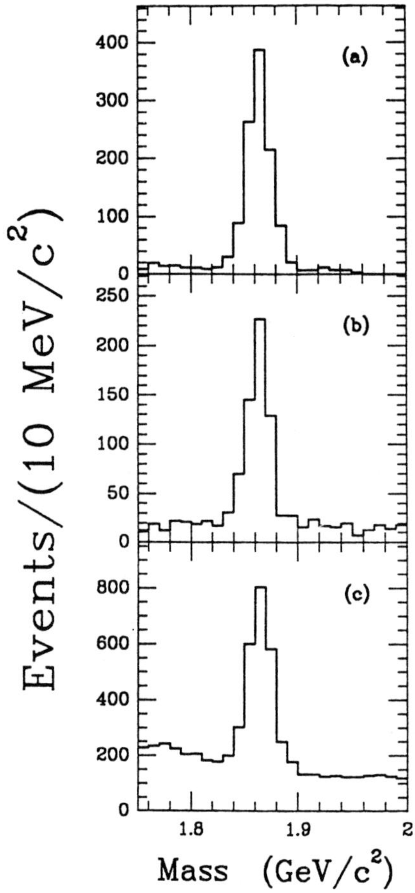

Fig. 16 Three D^0 mass spectra from experiment E691.[23] a) $D^0 \to K^-\pi^+$, with the requirement that there be an additional π^+ in the event which fits the reaction $D^{*+} \to \pi^+ D^0$; b) $D^0 \to K^-\pi^+\pi^+\pi^-$, also with the D^* requirement, and c) $D^0 \to K^-\pi^+$ without the D^* requirement. The minimum decay length is 5 standard deviations.

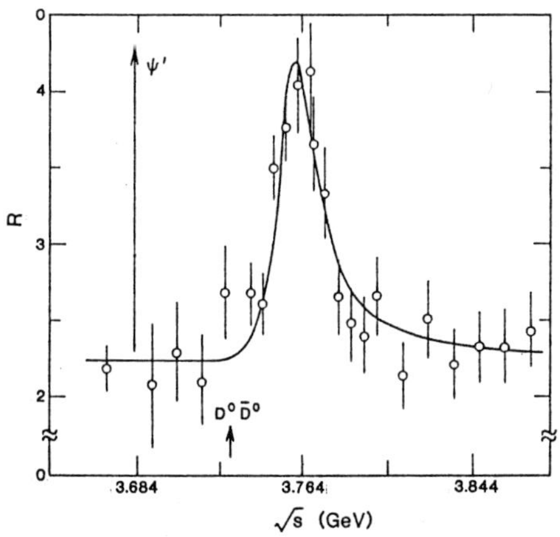

Fig. 17 The Mark II measurement[24] of the resonance parameters of the $\psi(3770)$ resonance in e^+e^- annihilation (radiatively corrected).

pairs, the experimental situation is greatly simplified, making it possible to study exclusive decays in great detail. The $\psi(3770)$ resonance (Fig. 17) is a very prominent feature in the e^+e^- total cross section. The $\psi(3770)$ decays almost entirely to $D\bar{D}$, making it possible to measure absolute branching ratios for exclusive D decays by simultaneously measuring the $\psi(3770)$ cross section. The only direct measurement which bears on this point[25] shows that $B(\psi(3770) \to D\bar{D}) = 1.09 \pm 0.23$. As will be discussed further below, it is in fact possible to measure absolute branching ratios at the $\psi(3770)$ without knowing the cross section. Other $\psi(3770)$ decay modes are expected from conventional mechanisms: a) the OZI violating decay $\psi(3770) \to J/\psi\pi\pi$, b) decays to the χ states, the $^3P_{0,1,2}$ charmonium levels, and c) direct hadronic decays such as $\psi(3770) \to \pi^+\pi^-\pi^0$. Preliminary evidence for such decays, at a small level, has been found.[26] Lipkin has advanced the idea that nearby resonances can enhance OZI violation in $\psi(3770)$ decay,[27] although no direct evidence for this exists.

The final states produced in $\psi(3770) \to D\bar{D}$ decay are particularly simple. Decays to other charmed states or decays involving even so much as an additional slow pion are kinematically forbidden. Further, the D's have a unique momentum: $P_{D^0} = 274$ MeV/c and $P_{D^+} =$ 237 MeV/c. This will prove particularly useful in the reconstruction of Cabibbo-suppressed hadronic decays and of semileptonic decays.

Another unique feature of D production at the $\psi(3770)$ allows us to improve the resolution in the reconstruction of masses for hadronic decay modes. Rather than plotting the conventional invariant mass of particle combinations, we define the "beam-constrained mass":

$$M_{BC} = \sqrt{E^2_{\text{beam}} - \sum_i P_i^2}\,,$$

exploiting the fact that the total energy of all decay products must sum to the energy of each beam. As the uncertainty in the beam energy is far smaller than the uncertainty in the total reconstructed energy of the decay tracks, this approach yields much improved mass

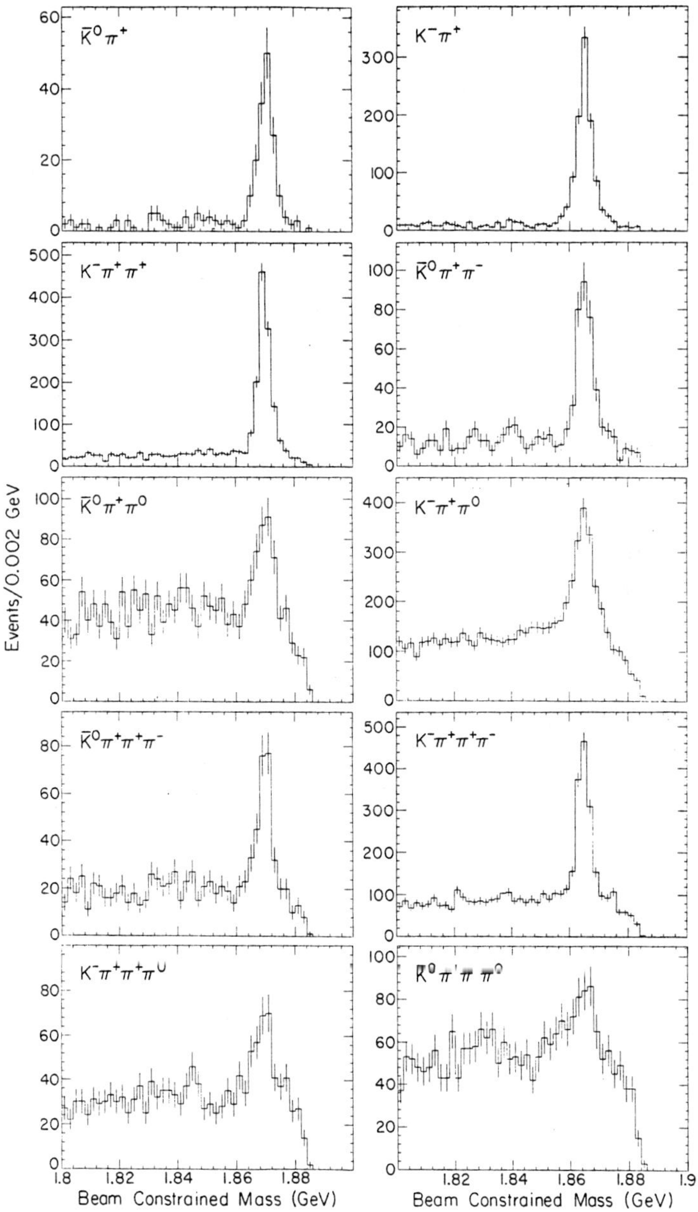

Fig. 18 Beam constrained mass plots from Mark III for five D^+ and five D^0 Cabibbo-allowed decays.

TABLE VI

Cabibbo-Allowed D^0 Hadronic Branching Ratio Measurements

Decay Mode	Experiment	Measurement	Branching Ratio (%)	Comments
$D^0 \to K^-\pi^+$	LGW	$\sigma \cdot B = (0.25 \pm 0.05)$ nb	4.3 ± 1.0	a,j
	Mark II	$\sigma \cdot B = (0.24 \pm 0.02)$ nb	4.1 ± 0.6	b,j
	Mark III	$\sigma \cdot B = (0.248 \pm 0.009 \pm 0.014)$ nb	$4.2 \pm 0.4 \pm 0.4$	c,h
	LEBC-EHS	topological	$4.0 \pm^{2.1}_{1.0} \pm 0.2$	
$D^0 \to \bar{K}^0\pi^0$	Mark II	$\sigma \cdot B = (0.18 \pm 0.08)$ nb	3.1 ± 1.4	b,j
	Mark III	$\sigma \cdot B = (0.108 \pm 0.020 \pm 0.010)$ nb	$1.9 \pm 0.4 \pm 0.2$	c
$D^0 \to \bar{K}^0\phi$	ARGUS	$\frac{B(D^0 \to \bar{K}^0\phi)}{B(D^0 \to \bar{K}^0\pi^+\pi^-)} = 0.155 \pm 0.033$	$0.99 \pm 0.24 \pm 0.14$	d
	CLEO	$\frac{B(D^0 \to \bar{K}^0\phi)}{B(D^0 \to \bar{K}^0\pi^+\pi^-)} = 0.142 \pm 0.049$	$0.91 \pm 0.33 \pm 0.13$	d
	Mark III	$\sigma \cdot B = (0.05 \pm^{0.03}_{0.02} \pm^{0.02}_{0.01})$ nb	$0.86 \pm^{0.50}_{0.41} \pm^{0.31}_{0.17}$	c,j
$D^0 \to \bar{K}^0\omega$	Mark III	$\sigma \cdot B = (0.187 \pm 0.073 \pm 0.047)$ nb	$3.2 \pm 1.3 \pm 0.8$	c,j
$D^0 \to \bar{K}^0\eta$	Mark III	$\sigma \cdot B = (0.088 \pm 0.039 \pm 0.012)$ nb	$1.5 \pm 0.7 \pm 0.2$	c,j
$D^0 \to K^-\rho^+$	Mark III		$10.8 \pm 0.4 \pm 1.7$	e,j
$D^0 \to \bar{K}^0\rho^0$	Mark III		$0.75 \pm 0.09 \pm 0.47$	e,j
$D^0 \to K^{*-}\pi^+$	Mark III		$5.2 \pm 0.3 \pm 1.5$	e,f,j
$D^0 \to \bar{K}^{*0}\pi^0$	Mark III		$2.6 \pm 0.3 \pm 0.7$	e

TABLE VI Continued

Decay Mode	Experiment	Measurement	Branching Ratio (%)	Comments
$D^0 \to (K^-\pi^+\pi^0)_{\text{total}}$	LGW	$\sigma \cdot B = (1.4 \pm 0.6)$ nb	24.1 ± 10.8	a,j
	Mark II	$\sigma \cdot B = (0.68 \pm 0.23)$ nb	11.7 ± 4.2	b,j
	Mark III	$\sigma \cdot B = (0.759 \pm 0.044 \pm 0.083)$ nb	$13.3 \pm 1.2 \pm 1.3$	c,h
	LEBC-EHS	topological	$10.6 \pm^{6.1}_{2.8} \pm 0.6$	
$(K^-\pi^+\pi^0)_{\text{nonres}}$	Mark III		$1.2 \pm 0.2 \pm 0.6$	c,e,h
$D^0 \to (\bar{K}^0\pi^+\pi^-)_{\text{total}}$	LGW	$\sigma \cdot B = (0.46 \pm 0.12)$ nb	7.9 ± 2.3	a,j
	Mark II	$\sigma \cdot B = (0.30 \pm 0.08)$ nb	5.2 ± 1.5	b,j
	Mark III	$\sigma \cdot B = (0.372 \pm 0.030 \pm 0.031)$ nb	$6.4 \pm 0.8 \pm 0.8$	c,j
	Avery *et al.*	$\dfrac{B(D^0 \to \bar{K}^0\pi^+\pi^-)}{B(D^0 \to K^-\pi^+)} = 1.7 \pm 0.8$	7.1 ± 3.5	g
	LEBC-EHS	topological	$4.5 \pm^{5.9}_{1.4} \pm 0.3$	
$(\bar{K}^0\pi^+\pi^-)_{\text{nonres}}$	Mark III		$2.1 \pm 0.3 \pm 0.7$	c,e,j
$D^0 \to (\bar{K}^0K^-K^+)_{\text{nonres}}$	Mark III	$\sigma \cdot B = (0.05 \pm^{0.02}_{0.01} \pm 0.01)$ nb	$0.85 \pm^{0.27}_{0.24} \pm^{0.19}_{0.18}$	c,d
$D^0 \to K^-\pi^+\pi^+\pi^-$	LGW	$\sigma \cdot B = (0.36 \pm 0.10)$ nb	6.2 ± 1.9	a,j
	Mark II	$\sigma \cdot B = (0.68 \pm 0.11)$ nb	11.7 ± 2.4	b,j
	Mark III	$\sigma \cdot B = (0.525 \pm 0.026 \pm 0.054)$ nb	$9.1 \pm 0.8 \pm 0.8$	c,h
	LEBC-EHS	topological	$6.5 \pm^{1.7}_{1.1} \pm 1.9$	
	Bailey *et al.*	$\dfrac{B(D^0 \to K^-\pi^+\pi^+\pi^-)}{B(D^0 \to K^-\pi^+)} = 2.0 \pm 1.0$	8.4 ± 4.3	g
	ARGUS	$\dfrac{B(D^0 \to K^-\pi^+\pi^+\pi^-)}{B(D^0 \to K^-\pi^+)} = 2.17 \pm 0.36$	9.1 ± 1.7	g
$D^0 \to \bar{K}^0\pi^+\pi^-\pi^0$	Mark III	$\sigma \cdot B = (0.666 \pm 0.113 \pm 0.153)$ nb	$11.5 \pm 2.2 \pm 2.8$	c,j
$D^0 \to \bar{K}^0\pi^+\pi^0\pi^0$	Mark III		$14.9 \pm 3.7 \pm 3.0$	c,h

resolution ($\sigma \sim 3$ MeV) as compared to the invariant mass technique ($\sigma \sim 12$ MeV).

Charged particles, identified as K or π by time-of-flight or dE/dx measurements, are required to originate at the primary vertex, as the spatial precision of the Mark III is insufficient to allow direct sensitivity to D lifetimes. Decays containing K^0's are reconstructed through the $\pi^+\pi^-$ channel. For modes involving a π^0 or η, a $2C$ fit is performed: the total energy of the charged particles and photons is constrained to E_{beam}, while the photons are constrained to the π^0 or η mass. A large number of D^0 and D^+ Cabibbo-allowed decay modes have been reconstructed in this way. Some of these[28] are shown in Fig. 18. The various measurements of Cabibbo-allowed D^0 and D^+ branching ratios are summarized in Tables VI and VII.[29] Where possible, the primary measurement ($\sigma \cdot B$ or ratio of branching fractions) is given. The normalization of these branching ratios will be discussed below.

Several hadronic D meson decay modes have also been observed in higher energy experiments. Except in isolated cases, the measurements of $\sigma \cdot B$, are primarily useful in studying charm production cross sections, using the value of B for a particular decay mode, as obtained from $\psi(3770)$ results, to extract the production cross section in various processes. The recent advances in statistics in photoproduction now also allow meaningful measurements of the ratio of decay branching fractions into interesting modes. These are also shown in Tables VI and VII.

A number of Cabibbo-suppressed hadronic decay modes have also been reconstructed. These are expected at a rate of $\sim tan^2\theta_c$ or 5% of comparable allowed modes. Since the signals are smaller, the problem of $K - \pi$ misidentification assumes greater significance. The unique momentum of D's at the $\psi(3770)$ can be exploited to display explicitly background due to incorrect particle assignment. For most Cabibbo-suppressed modes, this involves plotting the invariant mass for events in which the reconstructed D momentum is within 50 MeV of the expected value. Background is measured using sidebands of reconstructed "D" momentum. Figures 19 and 20 show an example of this

TABLE VII

Cabibbo-Allowed D^+ Hadronic Branching Ratio Measurements

Decay Mode	Experiment	Measurement	Branching Ratio (%)	Comments
$D^+ \to \bar{K}^0\pi^+$	LGW	$\sigma \cdot B = (0.14 \pm 0.05)$ nb	3.3 ± 1.3	a,k
	Mark II	$\sigma \cdot B = (0.14 \pm 0.03)$ nb	3.3 ± 0.9	b,k
	Mark III	$\sigma \cdot B = (0.135 \pm 0.012 \pm 0.010)$ nb	$3.2 \pm 0.5 \pm 0.2$	c,h
$D^+ \to \bar{K}^0\rho^+$	Mark III		$6.9 \pm 0.8 \pm 2.3$	e,k
$D^+ \to \bar{K}^{*0}\pi^+$	Mark III		$5.9 \pm 1.9 \pm 2.5$	e,k
$D^+ \to (K^-\pi^+\pi^+)_{\text{total}}$	LGW	$\sigma \cdot B = (0.36 \pm 0.06)$ nb	8.6 ± 2.0	a,k
	Mark II	$\sigma \cdot B = (0.38 \pm 0.05)$ nb	$9.1 \pm 1.3 \pm 0.4$	b,k
	Mark III	$\sigma \cdot B = (0.388 \pm 0.013 \pm 0.029)$ nb	$9.1 \pm 1.3 \pm 0.4$	c,h
	LEBC-EHS	topological	$6.3 \pm^{2.8}_{1.4} \pm 1.1$	
$(K^-\pi^+\pi^+)_{\text{nonres}}$	Mark III		$7.2 \pm 0.6 \pm 1.8$	e,k
$D^+ \to (\bar{K}^0\pi^+\pi^0)_{\text{total}}$	Mark II	$\sigma \cdot B = (0.78 \pm 0.48)$ nb	18.6 ± 11.8	a,k
	Mark III	$\sigma \cdot B = (0.417 \pm 0.081 \pm 0.075)$ nb	$10.2 \pm 2.5 \pm 1.6$	c,k
$(\bar{K}^0\pi^+\pi^0)_{\text{nonres}}$	Mark III		$1.3 \pm 0.7 \pm 0.9$	e,k
$D^+ \to K^-\pi^+\pi^+\pi^0$	Mark III	$\sigma \cdot B = (0.177 \pm 0.042 \pm 0.042)$ nb	$4.2 \pm 1.2 \pm 1.2$	c,k
	LEBC-EHS	topological	$2.2 \pm^{4.7}_{0.6} \pm 0.4$	
$D^+ \to \bar{K}^0\pi^+\pi^+\pi^-$	Mark II	$\sigma \cdot B = (0.51 \pm 0.18)$ nb	12.1 ± 4.7	a,k
	Mark III	$\sigma \cdot B = (0.305 \pm 0.031 \pm 0.030)$ nb	$6.6 \pm 1.5 \pm 0.5$	c,h
	LEBC-EHS	topological	$24.3 \pm^{6.4}_{4.1} \pm 4.1$	
	CLEO	$\dfrac{B(D^+ \to \bar{K}^0\pi^+\pi^+\pi^-)}{B(D^+ \to \bar{K}^0\pi^+)} = 2.3 \pm 0.5$	7.4 ± 2.0	i
$D^+ \to K^-\pi^+\pi^+\pi^+\pi^-$	ACCMOR		SEEN	
$D^+ \to K^-\pi^+\pi^+\pi^0\pi^0$	LEBC-EHS	topological	$2.2 \pm^{4.7}_{0.6} \pm 0.4$	
$D^+ \to \bar{K}^0\pi^+\pi^+\pi^-\pi^0$	LEBC-EHS	topological	$4.4 \pm^{5.2}_{0.6} \pm 0.4$	

Comments for Cabibbo-Allowed D^0 and D^+ Hadronic Decay Measurement Tables

a Measurement at $\sqrt{s} = 3.774$ GeV

b Measurement at $\sqrt{s} = 3.771$ GeV

c Measurement at $\sqrt{s} = 3.768$ GeV

d Derived using Mark III measurements of $\sigma \cdot B(D^0 \to \bar{K}^0\pi^+\pi^-)$ and σ_{D^0} at the $\psi(3770)$

e Refer to Table XIV for further details

f Average of determinations from $D^0 \to K^-\pi^+\pi^0$ and $D^0 \to \bar{K}^0\pi^+\pi^-$ Dalitz plot analyses

g Derived using Mark III measurement of $B(D^0 \to K^-\pi^+)$

h Branching ratio derived from global fit double tab measurement

i Derived using Mark III measurement of $B(D^+ \to \bar{K}^0\pi^+)$

j Branching ratio derived using Mark III value $\sigma_{D^0} = (5.8 \pm 0.5 \pm 0.5)$ nb

k Branching ratio derived using Mark III value $\sigma_{D^+} = (4.2 \pm 0.6 \pm 0.3)$ nb

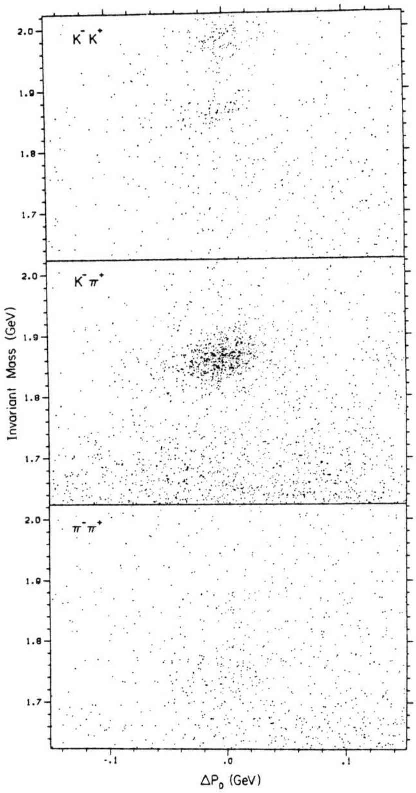

Fig. 19 Scatter plots of invariant mass *vs* difference from expected D^0 momentum for the Cabibbo-allowed decay $D^0 \to K^-\pi^+$ and two Cabibbo-suppressed channels, $D^0 \to K^-K^+$ and $\pi^-\pi^+$ (Mark III).

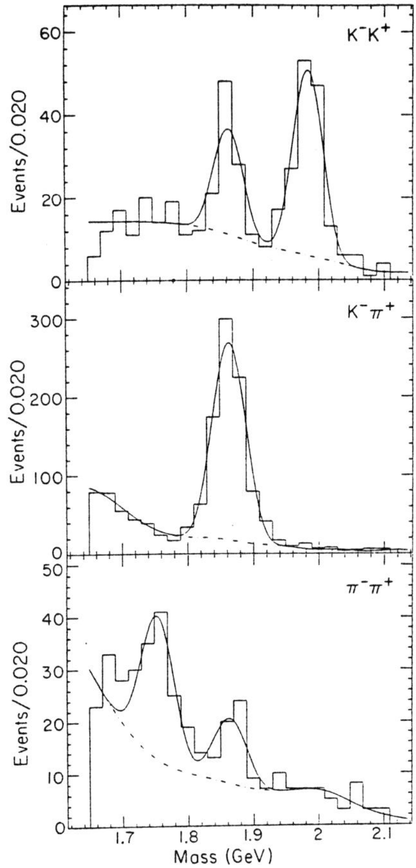

Fig. 20 Invariant mass spectra for the Cabibbo-allowed decay $D^0 \to K^-\pi^+$ and the two Cabibbo-suppressed decays $D^0 \to K^-K^+$ and $D^0 \to \pi^-\pi^+$, for events within ± 50 MeV of the expected D^0 momentum (Mark III). Note that peaks due to $\pi - K$ misidentification are clearly separated from the actual signal.

technique. The invariant mass projections shown in Fig. 20 for events within 50 MeV of the correct D momentum, display clear D signals, as well as cleanly separated misidentification peaks. Branching ratios are expressed as ratios to closely related Cabibbo-allowed modes, to minimize systematic errors and facilitate comparison with models. These are summarized in Tables VIII and IX.[30] The decays $D^+ \to \pi^+\pi^0$, $D^0 \to \pi^+\pi^-\pi^0$ and $D^0 \to \pi^+\pi^-\pi^+\pi^-$ are subject to large combinatorial backgrounds when analyzed by this technique. These modes, also shown in Tables VIII and IX, are, therefore, reconstructed by the beam-constrained method in the recoil spectrum against a sample of reconstructed Cabibbo-allowed decays: branching ratios so obtained are thus absolutely normalized. E691 and LEBC-EHS have now also reconstructed a number of Cabibbo-suppressed D hadronic decays using the invariant mass technique. The ability of the experiments to isolate cleanly the decay vertex results in small backgrounds even with branching ratios well below the 1% level. These are also summarized in Tables VIII and IX. The E691 $K^+K^-\pi^+$ mass distributions are shown in Figure 21. Note the clear Cabibbo-suppressed D^+ signal together with the allowed D_s^+ signal.

Normalization of D Hadronic Branching Ratios

Two methods are available to extract branching fractions at the $\psi(3770)$ from the primary $\sigma \cdot B$ measurements. The most common method has been to measure the cross section at the peak of the $\psi(3770)$ and normalize the results on the assumption that the $\psi(3770)$ decays solely to $D\bar{D}$ pairs. The various measurements of the $\psi(3770)$ resonance are shown in Fig. 22 and Table X. Unfortunately, these do not agree very well, although there is a tendency for later measurements to yield smaller values of the resonance height.

This has led the Mark III group to develop a new method which provides an absolute normalization of D meson branching ratios, independent of the production cross section. The technique consists of comparing the number of events in which a single D hadronic decay is reconstructed ("single tags") with the number in which a pair of D's is reconstructed ("double tags"). If the appropriate reconstruc-

TABLE VIII

Cabibbo-Suppressed D^0 Hadronic Branching Ratio Measurements

Decay Mode	Experiment	Measurement	Branching Ratio (%)	Comments
$D^0 \to \pi^-\pi^+$	Mark II	$\frac{B(D^0 \to \pi^-\pi^+)}{B(D^0 \to K^-\pi^+)} = 0.033 \pm 0.015$	0.14 ± 0.07	a
	Mark III	$\frac{B(D^0 \to \pi^-\pi^+)}{B(D^0 \to K^-\pi^+)} = 0.033 \pm 0.010 \pm 0.006$	$0.14 \pm 0.04 \pm 0.03$	a
	LEBC-EHS	topological	$0.5 \pm^{1.2}_{0.2} \pm 0.4$	
$D^0 \to K^-K^+$	Mark II	$\frac{B(D^0 \to K^-K^+)}{B(D^0 \to K^-\pi^+)} = 0.133 \pm 0.030$	0.56 ± 0.15	a
	Mark III	$\frac{B(D^0 \to K^-K^+)}{B(D^0 \to K^-\pi^+)} = 0.122 \pm 0.018 \pm 0.012$	$0.51 \pm 0.09 \pm 0.06$	a
$D^0 \to \bar{K}^0K^0$	Mark III	$\frac{B(D^0 \to \bar{K}^0K^0)}{B(D^0 \to K^-\pi^+)} < 0.11$ @ 90% CL	< 0.46 @ 90% CL	a
$D^0 \to \bar{K}^{*0}K^0 + \text{cc}$	Mark III	$\frac{B(D^0 \to \bar{K}^{*0}K^0 + \text{cc})}{B(D^0 \to K^{*-}\pi^+ + K^-\rho^+)} < 0.034$ @ 90% CL	< 0.55 @ 90% CL	b
$D^0 \to K^{*-}K^+ + \text{cc}$	Mark III	$\frac{B(D^0 \to K^{*-}K^+ + \text{cc})}{B(D^0 \to K^{*-}\pi^+ + K^-\rho^+)} = 0.05 \pm 0.03$	0.8 ± 0.5	b
$D^0 \to \pi^-\pi^+\pi^0$	Mark III		$1.1 \pm 0.4 \pm 0.2$	c
$D^0 \to \pi^-\pi^+\pi^+\pi^-$	Mark III		$1.5 \pm 0.6 \pm 0.2$	c
	LEBC-EHS	topological	$0.5 \pm^{1.1}_{0.1} \pm 0.1$	
$D^0 \to \pi^0\pi^0$	Crystal Ball	$\sigma \cdot B < 0.019$ nb @ 90% CL	< 0.3 @ 90% CL	d
$D^0 \to \eta^0\pi^0$	Crystal Ball	$\sigma \cdot B < 0.5$ nb @ 90% CL	< 0.9 @ 90% CL	d
$D^0 \to \eta^0\eta^0$	Crystal Ball	$\sigma \cdot B < 0.07$ nb @ 90% CL	< 1.2 @ 90% CL	d
$D^0 \to \pi^0\pi^0\pi^0$	Crystal Ball	$\sigma \cdot B < 0.11$ nb @ 90% CL	< 1.9 @ 90% CL	d
$D^0 \to \eta^0\pi^0\pi^0$	Crystal Ball	$\sigma \cdot B < 0.45$ nb @ 90% CL	< 7.8 @ 90% CL	d

Comments

a Branching ratio derived using Mark III value for $B(D^0 \to K^-\pi^+)$

b Branching ratio derived using Mark III value for $B(D^0 \to K^{*-}\pi^+ + K^-\rho^+)$

c Absolute measurement

d Branching ratio limit using Mark III value $\sigma_{D^0} = (5.8 \pm 0.5 \pm 0.5)$ nb

TABLE IX

Cabibbo-Suppressed D^+ Hadronic Branching Ratio Measurements

Decay Mode	Experiment	Measurement	Branching Ratio (%)	Comments
$D^+ \to \bar{K}^0 K^+$	Mark II	$\frac{B(D^+ \to \bar{K}^0 K^+)}{B(D^+ \to \bar{K}^0 \pi^+)} = 0.25 \pm 0.15$	0.80 ± 0.48	a
	Mark III	$\frac{B(D^+ \to \bar{K}^0 K^+)}{B(D^+ \to \bar{K}^0 \pi^+)} = 0.317 \pm 0.086 \pm 0.048$	$1.01 \pm 0.32 \pm 0.18$	a
$D^+ \to \pi^+ \pi^0$	Mark III	$\frac{B(D^+ \to \pi^+ \pi^0)}{B(D^+ \to \bar{K}^0 \pi^+)} < 0.15$ @ 90% CL	< 0.48 @ 90% CL	a
$D^+ \to \phi \pi^+$	Mark III	$\frac{B(D^+ \to \phi \pi^+)}{B(D^+ \to K^- \pi^+ \pi^+)} = 0.084 \pm 0.021 \pm 0.011$	$0.77 \pm 0.22 \pm 0.11$	b
	E691	$\frac{B(D^+ \to \phi \pi^+)}{B(D^+ \to K^- \pi^+ \pi^+)} = 0.103 \pm 0.014 \pm 0.009$	$0.94 \pm 0.18 \pm 0.09$	b
$D^+ \to \bar{K}^{*0} K^+$	Mark III	$\frac{B(D^+ \to \bar{K}^{*0} K^+)}{B(D^+ \to K^- \pi^+ \pi^+)} = 0.048 \pm 0.021 \pm 0.011$	$0.44 \pm 0.20 \pm 0.11$	b
	E691	$\frac{B(D^+ \to \bar{K}^{*0} K^+)}{B(D^+ \to K^- \pi^+ \pi^+)} = 0.074 \pm 0.012 \pm 0.007$	$0.67 \pm 0.15 \pm 0.07$	b
$D^+ \to \rho^0 \pi^+$	E691	$\frac{B(D^+ \to \rho^0 \pi^+)}{B(D^+ \to K^- \pi^+ \pi^+)} = 0.022 \pm 0.008 \pm 0.003$	$0.20 \pm 0.08 \pm 0.03$	b
$D^+ \to \pi^- \pi^+ \pi^+$	Mark III	$\frac{B(D^+ \to \pi^- \pi^+ \pi^+)}{B(D^+ \to K^- \pi^+ \pi^+)} = 0.042 \pm 0.016 \pm 0.010$	$0.38 \pm 0.15 \pm 0.09$	b
	E691	$\frac{B(D^+ \to \pi^- \pi^+ \pi^+)}{B(D^+ \to K^- \pi^+ \pi^+)} = 0.041 \pm 0.008 \pm 0.004$	$0.37 \pm 0.09 \pm 0.04$	b
$D^+ \to (K^- K^+ \pi^+)_{total}$	LEBC-EHS	topological	$0.8 \pm^{1.7}_{0.2} \pm 0.1$	
$D^+ \to (K^- K^+ \pi^+)_{nonres}$	Mark III	$\frac{B(D^+ \to K^- \pi^+ \pi^+)}{B(D^+ \to K^- K^+ \pi^+)} = 0.059 \pm 0.026 \pm 0.009$	$0.54 \pm 0.25 \pm 0.09$	b
	E691	$\frac{B(D^+ \to K^- K^+ \pi^+)}{B(D^+ \to K^- \pi^+ \pi^+)} = 0.050 \pm 0.009 \pm 0.006$	$0.46 \pm 0.10 \pm 0.06$	b
$D^+ \to \phi \pi^+ \pi^- \pi^+$	E691	$\frac{B(D^+ \to \phi \pi^+ \pi^- \pi^+)}{B(D^+ \to K^- \pi^+ \pi^+)} = 0.01$ @ 90% CL	< 0.09 @ 90% CL	b

Comments

a Branching ratio derived using Mark III value for $B(D^+ \to \bar{K}^0 \pi^+)$

b Branching ratio derived using Mark III value for $B(D^+ \to K^- \pi^+ \pi^+)$

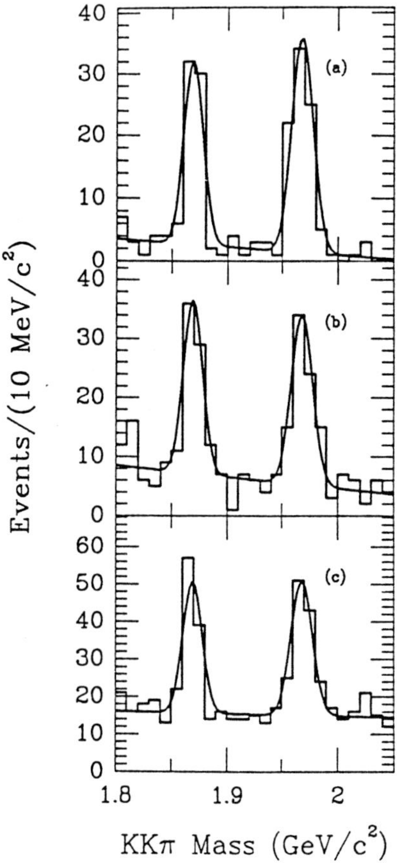

Fig. 21 $K^+K^-\pi^+$ mass distributions from E691:
a) $\phi\pi^+$, b) $\bar{K}^{*0}K^+$ and c) non-resonant $K^+K^-\pi^+$. Clear signals are seen for Cabibbo-suppressed D^+ decay, as well as for allowed D_s^+ decay.

Fig. 22 Comparison of three measurements of the $\psi(3770)$ resonance of SPEAR. Note that the Mark II and Crystal Ball (CB) results are τ-subtracted, while the earlier Lead Glass Wall (LGW) measurement is not.

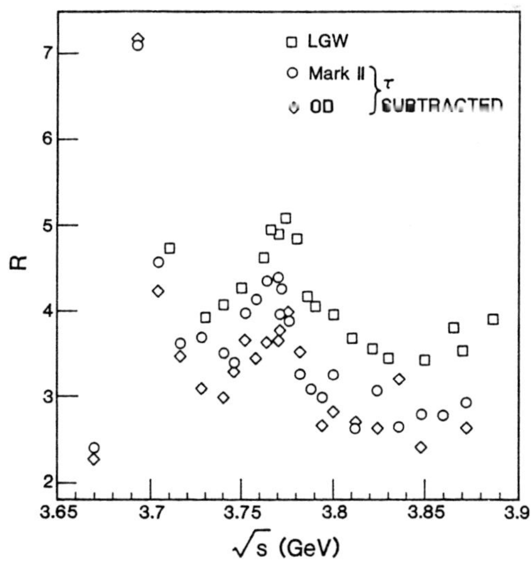

tion efficiencies for single and double tag combinations are known, the individual branching ratios can be extracted without reference to the production cross section.[34–36]

In the conventional, indirect method of extracting branching ratios, the number of reconstructed D mesons into a particular mode, i, is given by:

$$S_i = 2N_{D\bar{D}}B_i\epsilon_i \quad ,$$

where S_i represents the number of "single tags"; B_i, the branching ratio; ϵ_i, the reconstruction efficiency; and $N_{D\bar{D}}$, the number of produced $D\bar{D}$ pairs. $N_{D\bar{D}}$ is found by a measurement of the $\psi(3770)$ cross section, on the assumption that the $\psi(3770)$ decays only to $D\bar{D}$.

In the direct or "double tag" method, comparison of the number of single tags with the number of observed double tags, *i.e.*, the number of events in which a $D\bar{D}$ pair is reconstructed:

$$D_{ij} = N_{D\bar{D}}B_iB_j\epsilon_{ij} \quad ,$$

yields the branching ratio into the mode j without reference to the production cross section. In practice, since the number of single tags is much larger than the number of double tags, the S_i serve to determine relative branching ratios, while the D_{ij} set the absolute scale of the branching ratios. The double tag sample is determined by complete kinematic fitting of events, requiring that the mass of the two reconstructed D mesons (M_x) can be identical. In order to preserve the overall efficiency, only a loose χ^2 cut is placed on the quality of the fit. Background is subtracted by study of the sideband below $M_x = M_D$.

The original publication[35] of absolutely normalized branching ratios using this procedure suffered from a contamination due to two backgrounds, which have now been removed.[36] When the original version of this analysis led to surprisingly high values for the D branching ratios, further study was undertaken, which pointed to two sources of background in the double tag sample which were not removed by

TABLE X

Direct Measurements of $\psi(3770)$ Cross Section

	Lead Glass Wall[31]	Mark II[32]	Crystal Ball[33]
$\sqrt{s}$ (GeV)	3.774	3.771	3.771
σ (nb)	10.3 ± 1.6	$7.0 \pm 0.8 \pm 0.7$	6.4 ± 0.9

TABLE XI

Mark III D^0 and D^+ Branching Fractions Derived from Global Fit

Decay Mode	Branching Fraction (%)
(a) Results of Global Fits	
$D^0 \to K^-\pi^+$	$4.2 \pm 0.4 \pm 0.4$
$D^0 \to K^-\pi^+\pi^-\pi^+$	$9.1 \pm 0.8 \pm 0.8$
$D^0 \to K^-\pi^+\pi^0$	$13.3 \pm 1.2 \pm 1.3$
$D^+ \to K^-\pi^+\pi^+$	$9.1 \pm 1.3 \pm 0.4$
$D^+ \to \bar{K}^0\pi^+$	$3.2 \pm 0.5 \pm 0.2$
$D^+ \to \bar{K}^0\pi^+\pi^0$	$10.2 \pm 2.5 \pm 1.6$
$D^+ \to \bar{K}^0\pi^+\pi^-\pi^+$	$6.6 \pm 1.5 \pm 0.5$
(b) New Double Tag Measurement	
$D^0 \to K^-\pi^+\pi^0\pi^0$	$14.9 \pm 3.7 \pm 3.0$

the sideband subtraction. These may easily to understood by referring to Fig. 23, which shows Monte Carlo distributions for the fitted mass, M_x, for a) $M_{\bar{K}^0\pi^+}$ in the double tag $K^-\pi^+$ *vs* $K^+\pi^-$ sample and b) $M_{\bar{K}^0\pi^+}$ in the sample $K^+\pi^-$ *vs* K^-K^+, $\pi^-\pi^+$ and $K^-\pi^+\pi^0$. All three of the latter show peaks at $M_x \simeq M_D$. In the case of K^-K^+ and $\pi^-\pi^+$, the source is particle misidentification, which allows the Cabibbo-suppressed channels to act as a background for a Cabibbo-allowed decay. In the case of $K^-\pi^+\pi^0$, certain kinematic configurations in which the π^0 is not reconstructed also produce a $K\pi$ mass close to M_D. Such backgrounds could be removed by a tighter χ^2 cut, with substantial loss of efficiency. It has been possible instead, however, to remove the background without substantial loss of efficiency by use of an additional cut. Figure 24 shows the distribution of $\Delta M \equiv M_{bc} - M_{inv}$, the difference of the beam constrained and invariant masses for all $K\pi$ double tags. Figure 24a shows the data, while Fig. 24b shows Monte Carlo distributions for $K\pi$ *vs* $K\pi$ and the three potential background modes. Clearly, a cut of $|\Delta M| < 60$ MeV for all charged modes ($-120 < \Delta M < 100$ MeV for modes with neutrals) removes most background with little loss of signal. Such a cut has been used to derive the double tag sample shown in Figure 25. Table XI shows the D^0 and D^+ branching ratios derived from this analysis. The fit χ^2 is 3.5 for 5 degrees of freedom for D^0 modes and 1.8 for 3 degrees of freedom for D^+ modes. Note that this revised analysis has also uncovered a new decay mode, $D^0 \to K^-\pi^+\pi^0\pi^0$, which is shown in Figure 24. The branching ratios obtained from the global fit are 19 to 25% smaller than those derived from the original analysis. The global fit also determines the D^0 and D^+ production cross sections at the $\psi(3770)$:

$$\sigma_{D^0} = (5.8 \pm 0.5 \pm 0.6)\ \text{nb}$$

$$\sigma_{D^+} = (4.2 \pm 0.6 \pm 0.3)\ \text{nb}$$

$$\sigma_{D^0}/\sigma_{D^+} = 1.36 \pm 0.23 \pm 0.14 \quad .$$

These D meson production cross sections may be compared with the direct measurements of the $\psi(3770)$ cross section in Table X. Values

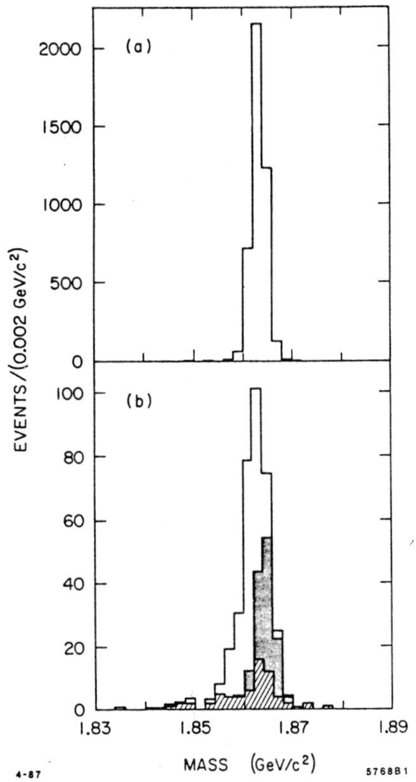

Fig. 23 Monte Carlo distributions for M_x from fits to the hypothesis $K^-\pi^+$ vs $K^+\pi^-$ for the Mark III double tag analysis.

a) $K^-\pi^+$ *vs* $K^+\pi^-$

b) $K^-\pi^+$ *vs* $\pi^+\pi^-$ (shaded), K^+K^- (solid) and $K^+\pi^-\pi^0$ (cross-hatched), arising from particle misidentification or loss of the π^0.

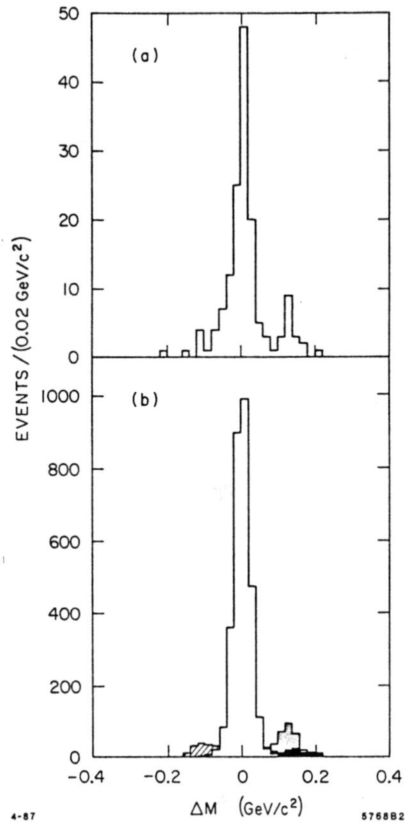

Fig. 24 Distribution of $\Delta M \equiv M_{bc} - M_{inv}$ for a) $K^-\pi^+$ *vs* $K^+\pi^-$ signal, and b) Monte Carlo simulations of the signal and backgrounds due to $K^-\pi^+$ *vs* $\pi^-\pi^+$ (cross-hatched), K^-K^+ (shaded) and $K^+\pi^-\pi^0$ (solid) modes.

of σ_D obtained in the Mark III double-tag analysis are clearly lower than values of σ_D obtained from the $\psi(3770)$ on the assumption that the $\psi(3770)$ decays only to $D\bar{D}$ pairs. This difference is likely to reflect experimental differences, rather than to support the conclusion that the $\psi(3770)$ has large decay modes to other than $D\bar{D}$ final states.[37]

Tables VI through IX show the values of D meson branching ratios obtained using the normalization provided by the Mark III global double-tag analysis. Note that the LEBC-EHS D branching ratios, independently normalized by a topological technique, are in good agreement with the other measurements. As these branching ratios are used to extract B inclusive and exclusive branching ratios, as well as charm production cross sections, care should be exercised to ensure that the corrected normalization is henceforth employed. The previous normalization has been corrected downward by slightly more than twenty percent, now yielding absolute branching ratios which are consistent with branching ratios extracted using the conventional technique, when the most recent $\psi(3770)$ cross section measurements are employed. Note also that branching ratios quoted for Mark II and Lead Glass Wall measurements are derived using the new Mark III normalization, not the $\psi(3770)$ cross sections determined by these experiments.

Do we have other evidence to support the current normalization? There are several areas we can investigate. Figure 27 shows the charged particle multiplicity measured in D^0 and D^+ decays. The Lead Glass Wall,[38] Mark II[39] and LEBC-EHS[40] (topological) measurements are inclusive. The Mark III measurements are exclusive, obtained from the $(90 \pm 5 \pm 9)\%$ of D^0 decays and $(80 \pm 4 \pm 8)\%$ of D^+ decays (hadronic and semileptonic) which have been reconstructed. [Note that with the LEBC-EHS observation of $D^+ \to K^-\pi^+\pi^+\pi^+\pi^-$ and $D^+ \to K^-\pi^+\pi^+\pi^0\pi^0$, 87% of D^+ decays have now been accounted for.] The excellent agreement between the multiplicities obtained from the inclusive and exclusive approaches indicates that the large majority of D decay modes have indeed been found.

A second check is the fraction of charm produced in high energy e^+e^- annihilation which appears as D mesons, which is expected to be

TABLE XII

***D* Meson Fraction of Continuum Charm Production**

Experiment	D Fraction (%)
ARGUS[41]	$69 \pm 6 \pm 18$
CLEO[42]	$60 \pm 6 \pm 14$
HRS[43]	$60 \pm 6 \pm 5$

TABLE XIII

Inclusive Charm Production in *B* Meson Decay

Charmed Particle per B Decay	ARGUS[44]	CLEO[45]
D^0	$0.63 \pm 0.08 \pm 0.08$	$0.50 \pm 0.06 \pm 0.06$
D^+	$0.25 \pm 0.05 \pm 0.04$	$0.21 \pm 0.05 \pm 0.03$
D_s^+	$0.12 \pm 0.03 \pm 0.04$	$0.11 \pm 0.03 \pm 0.03$
Λ_c^+	0.074 ± 0.03	0.074 ± 0.029
J/ψ	$2 \times (0.0107 \pm 0.0016 \pm 0.0022)$	$2 \times (0.0117 \pm 0.0017 \pm 0.0021)$
ψ'	$2 \times (0.0046 \pm 0.0017 \pm 0.0011)$	$2 \times (0.0046 \pm 0.0017 \pm 0.0011)^*$
Total	1.10 ± 0.14	0.93 ± 0.11

* ARGUS value

about 2/3. Clearly, the D cross sections depend on the normalization of D branching ratios, especially $D^0 \to K^-\pi^+$ and $D^+ \to K^-\pi^+\pi^+$. With the current normalization, the fraction of continuum charm production appearing as D mesons from three event experiments is shown in Table XII. The agreement is satisfactory.

A final check comes from measurements of inclusive charm production in B meson decay. Since charmed quarks can be produced either directly from the $b \to c$ transition or via W decay, we expect 1.15 charmed quarks per B meson decay. Table XIII summarizes the measurements of ARGUS and CLEO. The ARGUS measurements are in excellent agreement with expectations, while the CLEO results are something more than 2 standard deviations low. The discrepancy is entirely in the inclusive D^0 and D^+ fractions. Note that care must be taken in averaging the results of the two groups, which are correlated through use of common charmed particle branching ratios.

Resonant Substructure in D Hadronic Decay Modes

The four $D \to K\pi\pi$ decay channels $D^0 \to K^-\pi^+\pi^0$, $D^0 \to \bar{K}^0\pi^+\pi^-$, $D^+ \to \bar{K}^0\pi^+\pi^0$ and $D^+ \to K^-\pi^+\pi^+$ measured by Mark III provide an interesting opportunity to study D meson couplings to pseudoscalar-vector (PV) channels.[46] The fractions of PV and non-resonant three-body contributions to each mode are determined by a maximum-likelihood fit of the Dalitz plot distributions to a coherent sum of amplitudes. The K^* and ρ amplitudes are parametrized by a relativistic Breit-Wigner function (BW), which includes the vector decay matrix element and a mass-dependent width. The three-body amplitude (A_{3b}) is assumed constant over the Dalitz plot. For example, the likelihood function for the $D^0 \to K^-\pi^+\pi^0$ mode has the form:

$$L^{-+0} = |f_1 e^{i\phi_1} A_{3b} + f_2 e^{i\phi_2} BW_{\rho^+} + f_3 e^{i\phi_3} BW_{K^{*-}} + f_4 e^{i\phi_4} BW_{\bar{K}^{*0}}|^2 \quad .$$

A background term, found by analyzing a sample of events with beam-constrained mass below the D mass, is also included.

The results of the fits are shown in Table XIV, and Figures 28-31. The two D^0 modes and $D^+ \to \bar{K}^0\pi^+\pi^0$ are seen to be dominated by

Table XIV

Mark III $D \rightarrow K\pi\pi$ Pseudoscalar-Vector Branching Fractions

Decay Mode	Fraction (%)	Phase (degrees)	BR (%)
$D^0 \rightarrow K^-\pi^+\pi^0$			$13.3 \pm 1.2 \pm 1.3$
$K^-\rho^+$	$81 \pm 3 \pm 6$	0.0	$10.8 \pm 0.4 \pm 1.7$
$K^{*-}\pi^+$	$12 \pm 2 \pm 3$	154 ± 11	$4.9 \pm 0.7 \pm 1.5$
$\bar{K}^{*0}\pi^0$	$13 \pm 2 \pm 3$	7 ± 7	$2.6 \pm 0.3 \pm 0.7$
non-resonant	$9 \pm 2 \pm 4$	52 ± 9	$1.2 \pm 0.2 \pm 0.6$
$D^0 \rightarrow \bar{K}^0\pi^+\pi^-$			$6.4 \pm 0.5 \pm 1.0$
$\bar{K}^0\rho^0$	$12 \pm 1 \pm 7$	93 ± 30	$0.8 \pm 0.1 \pm 0.5$
$K^{*-}\pi^+$	$56 \pm 4 \pm 5$	0.0	$5.3 \pm 0.4 \pm 1.0$
non-resonant	$33 \pm 5 \pm 10$	—	$2.1 \pm 0.3 \pm 0.7$
$D^+ \rightarrow \bar{K}^0\pi^+\pi^0$			$10.2 \pm 2.5 \pm 1.6$
$\bar{K}^0\rho^+$	$68 \pm 8 \pm 12$	0.0	$6.9 \pm 0.8 \pm 2.3$
$\bar{K}^{*0}\pi^+$	$19 \pm 6 \pm 6$	43 ± 23	$5.9 \pm 1.9 \pm 2.5$
non-resonant	$13 \pm 7 \pm 8$	250 ± 19	$1.3 \pm 0.7 \pm 0.9$
$D^+ \rightarrow K^-\pi^+\pi^+$			$9.1 \pm 1.3 \pm 0.4$
$\bar{K}^{*0}\pi^+$	$13 \pm 1 \pm 7$	105 ± 8	$1.8 \pm 0.2 \pm 1.0$
non-resonant	$79 \pm 7 \pm 15$	0.0	$7.2 \pm 0.6 \pm 1.8$

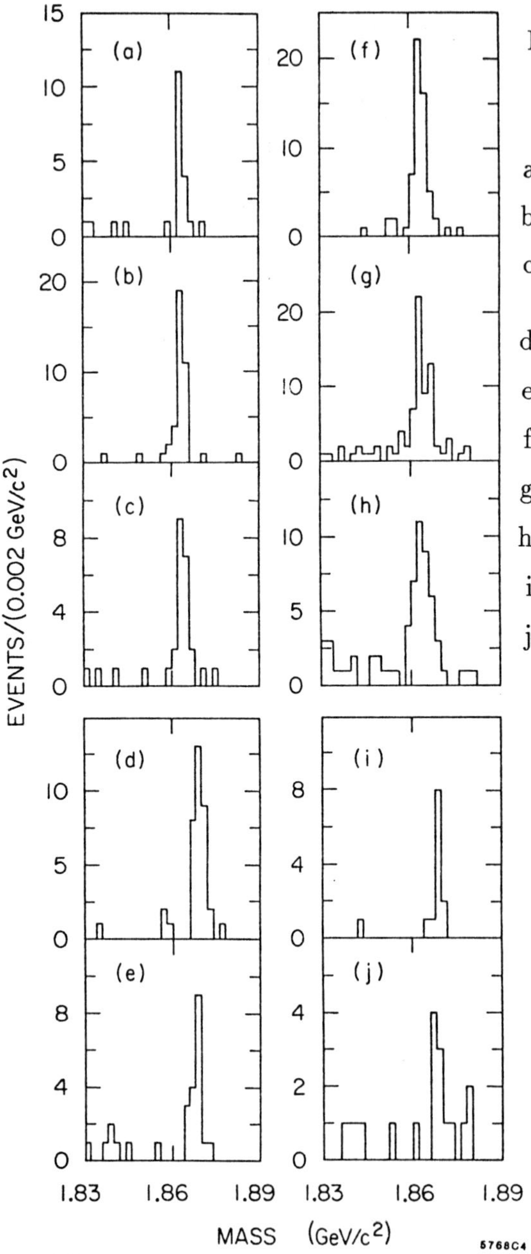

Fig. 25

M_x distribution for double tags

a) $K^-\pi^+$ *vs* $K^+\pi^-$

b) $K^-\pi^+$ *vs* $K^+\pi^-\pi^-\pi^+$

c) $K^-\pi^+\pi^+\pi^-$ *vs* $K^+\pi^-\pi^-\pi^+$

d) $K^-\pi^+\pi^+$ *vs* $K^+\pi^-\pi^-$

e) $K^-\pi^+\pi^+$ *vs* $K^0\pi^-\pi^0$

f) $K^-\pi^+$ *vs* $K^+\pi^-\pi^0$

g) $K^-\pi^+\pi^+\pi^-$ *vs* $K^+\pi^-\pi^0$

h) $K^-\pi^+\pi^0$ *vs* $K^+\pi^-\pi^0$

i) $K^-\pi^+\pi^+$ *vs* $K^0\pi^-$

j) $K^-\pi^+\pi^+$ *vs* $K^0\pi^-\pi^-\pi^+$.

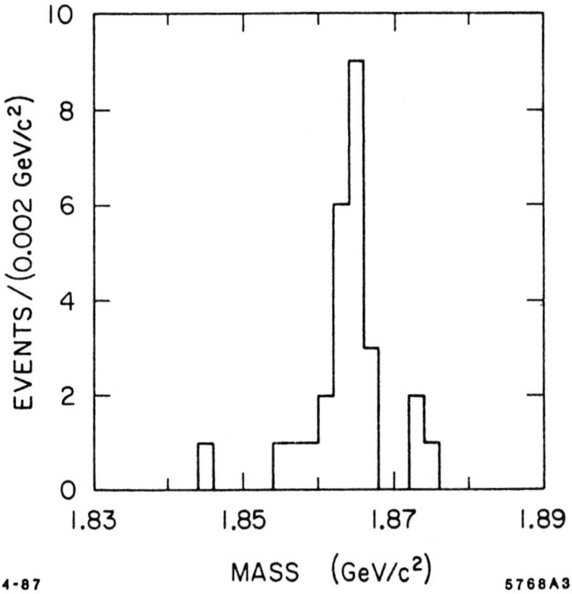

Fig. 26 Fitted mass M_x for $K^-\pi^+\pi^0\pi^0$ from the double tag sample $K^+\pi^-$ *vs* $K^-\pi^+\pi^0\pi^0$.

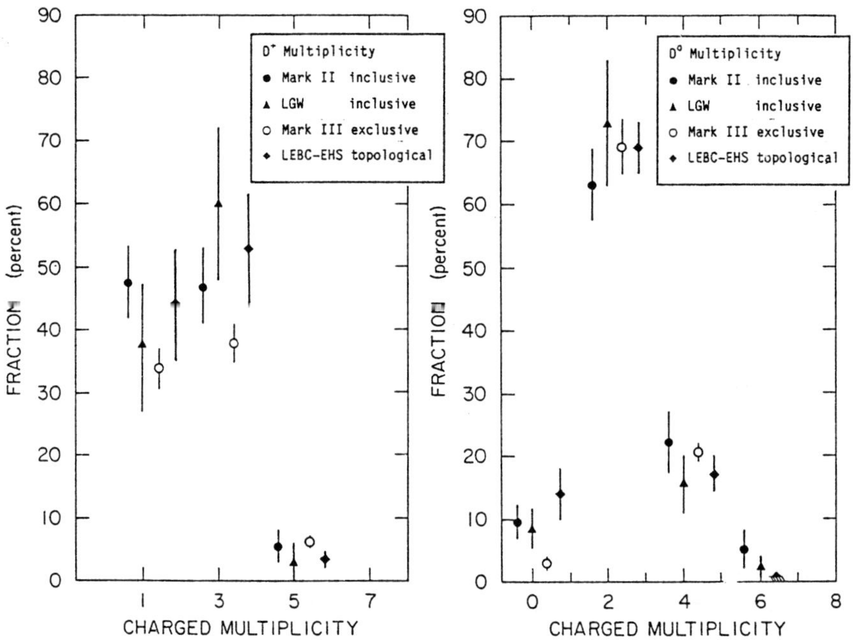

Fig. 27 Charged particle multiplicities for a) D^+ decay and b) D^0 decay derived from inclusive, topological and exclusive experiment.

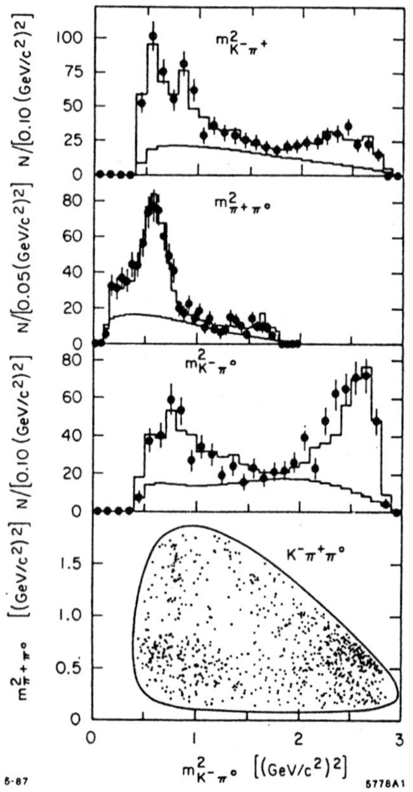

Fig. 28 Dalitz plot analysis by Mark III of the decay channel $D^0 \to K^-\pi^+\pi^0$. The Dalitz plot shows the $L = 1$ structure characteristic of a decay of the type $P \to PV$. The data is also shown in the three projections with the results of the fit described in the text superimposed as histograms. The background determined from lower mass sidebands is the lower histogram in each projection. The $K^-\pi^+$, $K^-\pi^0$ and $\pi^+\pi^0$ projections show prominent K^* and ρ signals.

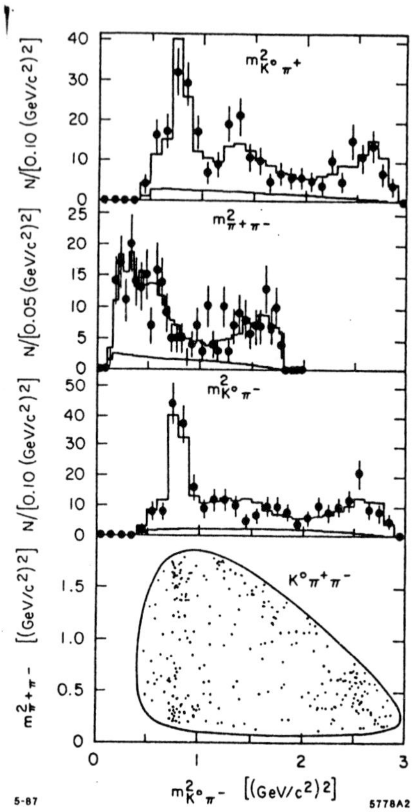

Fig. 29 The Dalitz plot for $D^0 \to \bar{K}^0\pi^+\pi^-$ and the three projections.

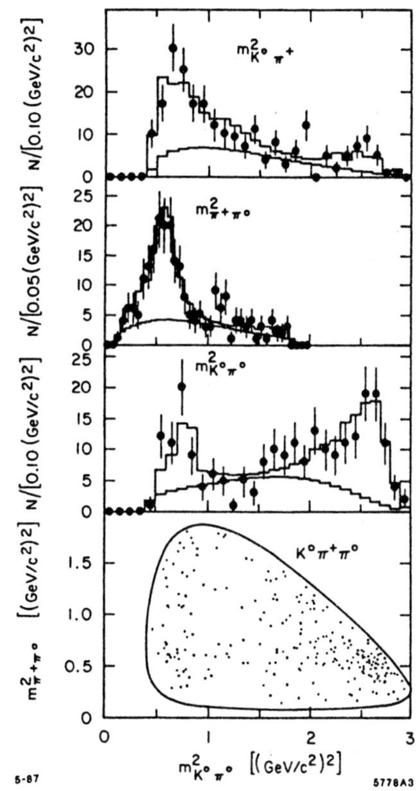

Fig. 30 The Dalitz plot for $D^+ \to \bar{K}^0\pi^+\pi^0$ and the three projections.

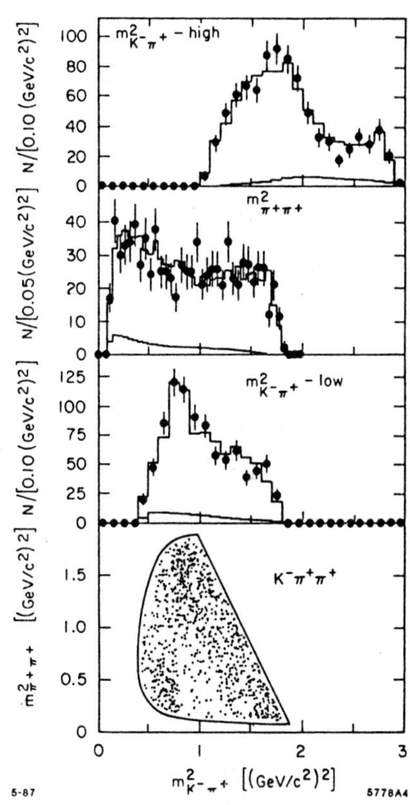

Fig. 31 The Dalitz plot for $D^+ \to K^-\pi^+\pi^+$ and the three projections.

pseudoscalar-vector channels. As the $D^+ \to K^-\pi^+\pi^+$ mode (Fig. 31) contains identical pions, an additional Bose-symmetric amplitude of the form $B_{3b} = |m^2_{K\pi_1} - m^2_{K\pi_2}|$ has been included in the fit. The fit attributes much to the apparent K^{*0} enhancement seen in the $K\pi$ projection to interference with this term, so that only a small resonant term results. Both $K^{*-}\pi^+$ and $\bar{K}^{*0}\pi^+$ appear in two channels. For the former, seen in $K^-\pi^+\pi^0$ and $\bar{K}^0\pi^+\pi^-$, the agreement is excellent. As one of the $\bar{K}^{*0}\pi^+$ channels is $K^-\pi^+\pi^+$, the agreement here is not satisfactory. The determination from the $\bar{K}^0\pi^+\pi^0$ mode is considered more reliable.

The decay $D^0 \to \bar{K}^0\phi$ has sparked special interest as a possible way of obtaining direct evidence for non-spectator hadronic decays. The ARGUS $\bar{K}^0\phi$ signal is shown in Fig. 32. The three observations of the $\bar{K}^0\phi$ mode are summarized in Table VI. This "smoking gun" mode will be discussed in further detail below, as will related tests in D_s^+ decay.

Both $D^0 \to \bar{K}^0\omega$ and $D^0 \to \bar{K}^0\eta$ modes have also been established by Mark III (see Table VI). Work is now underway to isolate possible vector-vector substructure in $D \to K\pi\pi\pi$ modes.

D Semileptonic Decays

As we have seen, semileptonic processes play an important role in the phenomenology of charmed and bottom meson decay. The ratio of inclusive semileptonic branching ratios of the D^+ and D^0 mesons provided, for a time, the most precise indication of the nonequality of D^+ and D^0 lifetimes. This relationship is based on the fact that these semileptonic decays are spectator processes, as they predominantly are, in both the Cabibbo-allowed and -suppressed sector. There are two semileptonic decay processes (see Fig. 33), Cabibbo-allowed for the D_s^+ and Cabibbo-suppressed for the D^+, which could originate from annihilation diagrams. Both are expected to be small, not only because of wave-function suppression, but because final state hadrons must arise from flavor singlet gluon emission.

Inclusive semileptonic branching ratios, interesting in their own

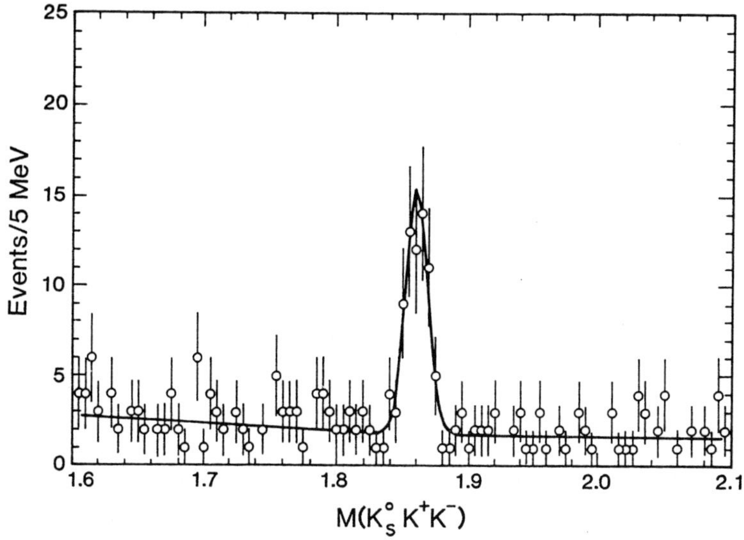

Fig. 32 ARGUS plot of $M(K_s^0 K^+ K^-)$ with the K^+K^- required to be a ϕ, showing evidence for the decay $D^0 \to \bar{K}^0 \phi$.

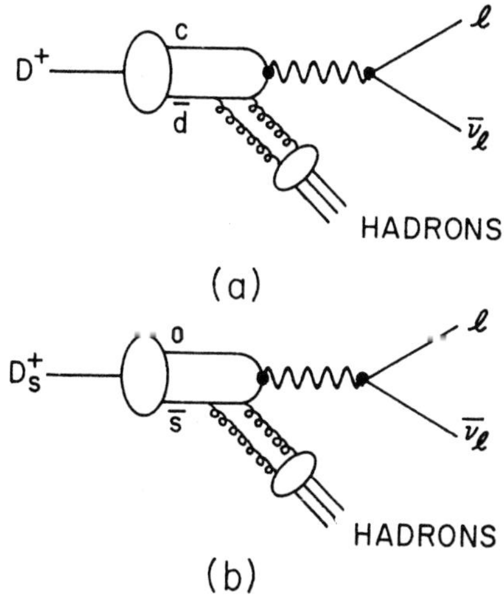

Fig. 33 Semileptonic decays arising from non-spectator amplitudes. a) $D^+ \to l\nu_l$+hadrons, which is Cabibbo-suppressed, and b) $D_s^+ \to l\nu_l$+ hadrons, which is Cabibbo-allowed.

right, are needed to extract information on the charmed sea from dimuon events produced in neutrino reactions, are used to normalize charm production cross sections in hadronic reactions and are central to many limits placed on neutral meson mixing. Measurements in e^+e^- annihilation have heretofore produced values of the average semileptonic branching ratio for c quarks (and b quarks), without distinguishing the individual parent charmed hadron. It has been possible to use the unique characteristics of $D\bar{D}$ production at the $\psi(3770)$, however, to obtain the separate D^0 and D^+ inclusive semileptonic branching ratios, as well as to measure particular exclusive channels and, in the case of $D \to Ke\nu$, to extract the q^2 dependence of the vector form factor.

The key to these studies is the Mark III collection of reconstructed hadronic D decays at the $\psi(3770)$, which provide a large tag sample of 3435 ± 39 $D^0 \to K^-\pi^+$, $K^-\pi^+\pi^0$, $K^-\pi^+\pi^+\pi^-$ and 1729 ± 20 $D^+ \to \bar{K}^0\pi^+$, $K^-\pi^+\pi^+$ events. The rest of the event consists of decay products of the accompanying $\bar{D}^0$ or D^-. (In the above, all references to a particle or decay mode imply the charge conjugate particle or mode as well.) Thus, for example, the fraction of electrons in the recoil spectrum against hadronic tags provides a direct measurement of the semileptonic branching fraction. The astute reader will have noticed, by the way, that the uncertainty in the number of tags quoted above is less than $\sqrt{N_{\rm tags}}$. This is so because each tag provides a single independent experiment in which an electron may be identified. Were there no background to the tags, there would be no uncertainty as to the number of such experiments. Thus, the uncertainty in the number of tag trials is just the uncertainty in the number of background events.

In order to measure the inclusive semileptonic branching ratio, it is necessary to know rather precisely the probability of misidentifying a hadron in the recoil spectrum as an electron. As the GIM current produces, e.g., e^+ from D^0 and D^+, it is possible to directly measure this misidentification probability by counting the number of "wrong sign" electrons. One must also apply small corrections for γ conversions and Dalitz decays, as well as for background under the tag sample. Table XIII summarizes measurements of individual and average D

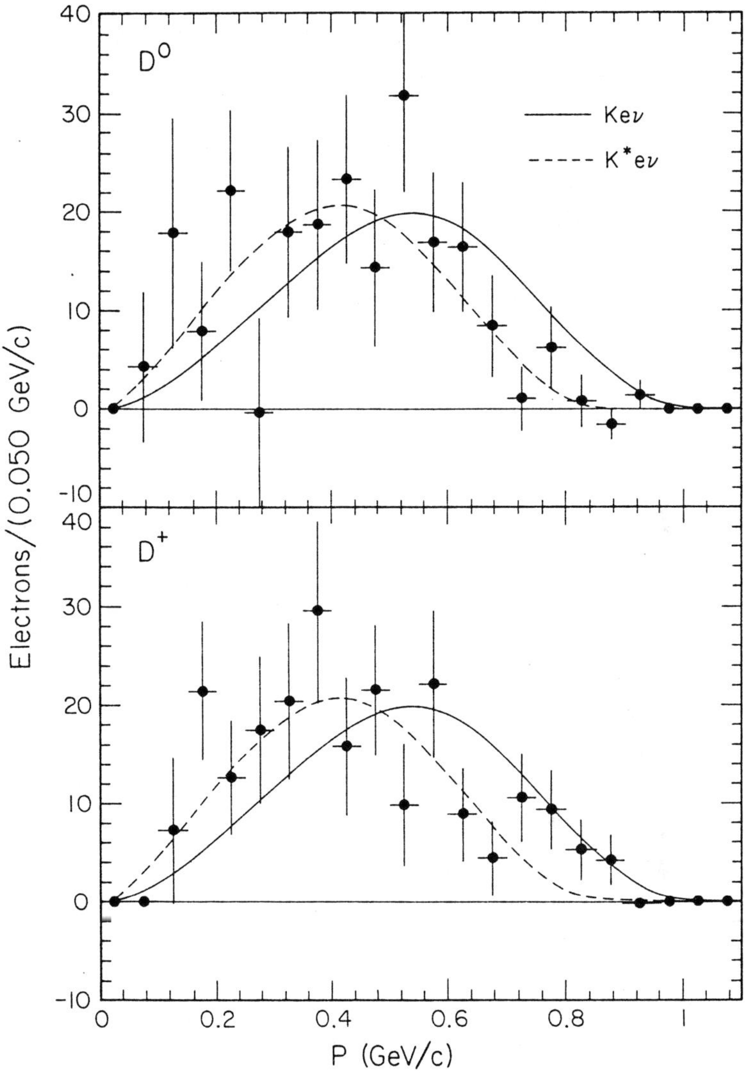

Fig. 34 Inclusive electron spectra for D^0 and D^+ decay at the $\psi(3770)$ from Mark III, with the expected shape for $Ke\nu$ and $K^*e\nu$ superimposed.

inclusive semileptonic branching fractions.

The inclusive lepton spectra obtained separately for D^0 and D^+ by the Mark III[47] are shown in Fig. 34, with expected spectra from $Ke\nu$ and $K^*e\nu$, each separately normalized, superimposed. It is clear that the actual spectra are from not pure $D \rightarrow Ke\nu$ decay, but contain heavier hadron components which soften the electron momentum distribution.

Using the hadronic tag sample, the Mark III has now been able to reconstruct a significant number of exclusive semileptonic decays.[12] This has allowed explicit measurement of the hadron content of the final state. In this case, the tag sample consists of 3130 ± 39 D^0 hadronic decays in the channels $K^-\pi^+$, $K^-\pi^+\pi^+\pi^-$ and $K^-\pi^+\pi^0$, over a background of 1512 ± 39 events, and 1695 ± 28 D^+ decays in the channels $K^-\pi^+\pi^+$, $\bar{K}^0\pi^+$, $\bar{K}^0\pi^+\pi^+\pi^-$ and $\bar{K}^0\pi^+\pi^0$, with a background of 765 ± 28. The $\psi(3770)$ decay to $D\bar{D}$ characteristics are particularly valuable in this case, since both the charm and momentum of the D recoiling against the hadronic tag are known. Thus, both the kaon (in the case of charged K's) and the lepton charge of the recoil can be required to be consistent with the GIM current. Since $\vec{P}_D$ is known, there is no kinematic ambiguity of the type usually encountered in the study of semileptonic K decay; rather than balancing only the transverse momentum of the missing neutrino, we know its total momentum.

The many possible sources of background to semileptonic channels will not be discussed in detail here. The most obvious one will, however, serve to illustrate the method of analysis. $D^0 \rightarrow K^-\pi^+\pi^0$, in which the π^+ is misidentified as an e^+ and both photons from the π^0 are missed appear as candidates for $D^0 \rightarrow K^-e^+\nu_e$. We can define $U \equiv E_{\text{miss}} - |\vec{P}_{\text{miss}}|$, which is 0 for a missing neutrino and positive for a missing π^0. Figure 35 shows the distribution in U *vs* the beam-constrained mass of the tag for D^0 events, showing that it is possible to isolate $K^-e^+\nu_e$ quite clearly. Other backgrounds, such as those from $K^-\pi^+$, $K^-\pi^0e^+\nu_e$ and $K^0_L\pi^-e^+\nu_e$ are removed by cuts on the invariant mass of the Ke system and on the total missing energy. After these cuts and kinematic fitting of the complete event, 47 events are reconstructed in

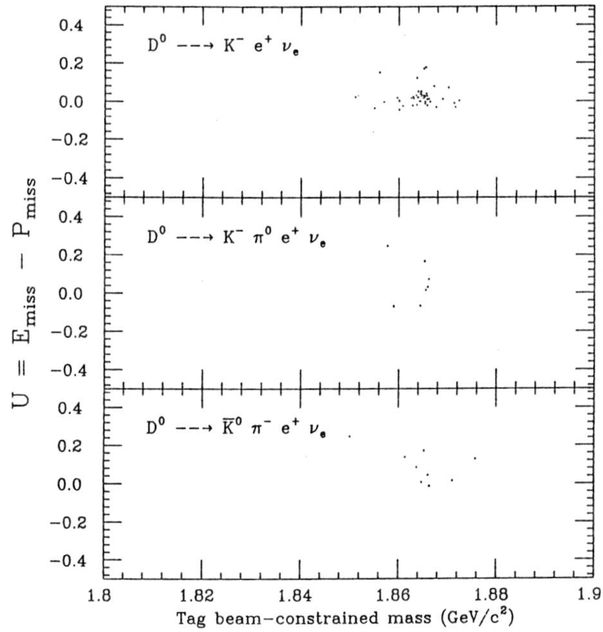

Fig. 35 Scatter plot of $E_{\text{missing}} - |P_{\text{missing}}|$ *vs* beam-constrained mass of the reconstructed hadronic decay (tag) for three D^0 exclusive semileptonic decay modes (Mark III).

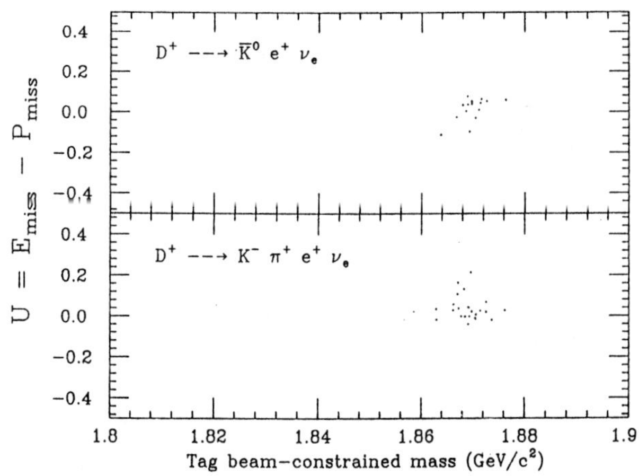

Fig. 36 Scatter plot of $E_{\text{missing}} - |P_{\text{missing}}|$ *vs* beam-constrained mass of the reconstructed hadronic decay (tag) for two D^+ exclusive semileptonic decay modes (Mark III).

the $D^0 \to K^- e^+ \nu_e$ channel, with a background of 2.1 events, yielding a branching ratio $B(D^0 \to K^- e^+ \nu_e) = (3.9 \pm 0.6 \pm 0.6)\%$.

The analysis of channels such as $D^0 \to K^- \pi^0 e^+ \nu_e$, $D^0 \to \bar{K}^0 \pi^- e^+ \nu_e$, $D^+ \to \bar{K}^0 e^+ \nu_e$ and $D^+ \to K^- \pi^+ e^+ \nu_e$ are very similar. Semimuonic channels have also been studied. The problem is again very similar, although here the problem is to identify muons clearly without using the muon detection system, as the muon system lower threshold of 600 MeV/c admits only $\sim 20\%$ of muons from D decay, and the solid angle coverage is only 60% of 4π.

Figures 35 and 36 show the distribution of U versus beam constrained tag mass for all reconstructed semileptonic modes of the D^0 and D^+. Table XVI shows the exclusive branching ratios obtained. The D^0 modes sum to a value close to the inclusive result, while the D^+ exclusive modes are $\sim 2\sigma$ lower than the inclusive result. Table XVII compares the Mark III measurements of inclusive and exclusive semileptonic branching ratios.

E691 has also reported a preliminary result[51] on the branching ratio $B(D^0 \to K^- e^+ \nu_e)$. They isolate the semileptonic decay via the reaction $D_s^+ \to D^0 \pi^+ \to (K^- e^+ \nu_e)\pi^+$, choosing the lower D^* mass combination to resolve the two-fold kinematic ambiguity. The result is:

$$\frac{B(D^0 \to K^- e^+ \nu_e)}{B(D^0 \to K^- \pi^+)} = 0.77 \pm 0.12 \pm 0.13 \quad .$$

When combined with $B(D^0 \to K^- \pi^+) = (4.2 \pm 0.4 \pm 0.3)\%$, this yields

$$B(D^0 \to K^- e^+ \nu_e) = (3.2 \pm 0.6 \pm 0.6)\% \quad ,$$

when statistical and systematic errors are combined in quadrature. The agreement with the Mark III result is thus satisfactory.

It is interesting to plot the $K\pi$ mass in the various $D \to K\pi l \nu_l$ channels. These are shown individually in Figure 37; the sum is plotted in Figure 38, together with a fit to the sum of a K^* Breit-Wigner and

TABLE XV

Inclusive D Semileptonic Branching Ratios

Mode	Experiment	Branching Ratio (%)
$B(D^0 \to e^+X)$	Mark III[47]	$7.5 \pm 1.1 \pm 0.4$
	Mark II[48]	5.5 ± 3.7
	DELCO[49]	< 4.0 @ 95% CL
	LEBC-EHS[40]	15 ± 5
$B(D^+ \to e^+X)$	Mark III	$17.0 \pm 1.9 \pm 0.7$
	Mark II	16.8 ± 6.0
	DELCO	$22.0\pm^{4.4}_{2.2}$
	LEBC-EHS	$20\pm^{9}_{7}$
$B(D \to e^+X)$	Mark III	$11.7 \pm 1.0 \pm 0.5$
(assuming	Mark II	10.0 ± 3.2
$\frac{\sigma(D^0)}{\sigma(D^+)} = 1.27$)	DELCO	8.0 ± 1.5
	LGW[50]	7.6 ± 2.8

Table XVI

Mark III Measurements of Exclusive Semileptonic D Meson Branching Ratios

Decay Mode	Events Signal (Bkgnd)	Branching Ratio (%)
$D^0 \to K^- e^+ \nu_e$	47 (2.1)	$3.9 \pm 0.6 \pm 0.6$
$K^- \pi^0 e^+ \nu_e$	7 (1.1)	$1.7 \pm^{0.9}_{0.7} \pm 0.6$
$\bar{K}^0 \pi^- e^+ \nu_e$	9 (0.7)	$2.2 \pm^{0.9}_{0.7} \pm 0.4$
$\pi^- e^+ \nu_e$	3 (0.9)	$0.4 \pm^{0.4}_{0.3} \pm 0.1$
$D^0 \to K^- \mu^+ \nu_\mu$	56 (9.4)	$4.1 \pm 0.7 \pm 1.2$
$\bar{K}^0 \pi^- \mu^+ \nu_\mu$	20 (8.5)	$2.7 \pm^{1.1}_{1.0} \pm 1.6$
$D^+ \to \bar{K}^0 e^+ \nu_e$	15 (1.1)	$6.3 \pm^{2.0}_{1.6} \pm 1.1$
$K^- \pi^+ e^+ \nu_e$	24 (1.2)	$3.9 \pm^{0.9}_{0.8} \pm 0.7$
$D^+ \to \bar{K}^0 \mu^+ \nu_\mu$	37 (8.9)	$10.2 \pm^{2.2}_{2.1} \pm 3.6$

Table XVII

Comparison of Mark III Inclusive and Exclusive D Meson Semileptonic Branching Ratios

Decay Mode	Branching Ratio (%)
$D^0 \to (K^- + (K\pi)^-) l^+ \nu_l$	7.9 ± 1.5
$e^+ X$	$7.5 \pm 1.1 \pm 0.4$
$D^+ \to (\bar{K}^0 + (K\pi)^0) l^+ \nu_l$	13.3 ± 2.5
$e^+ X$	$17.0 \pm 1.9 \pm 0.7$

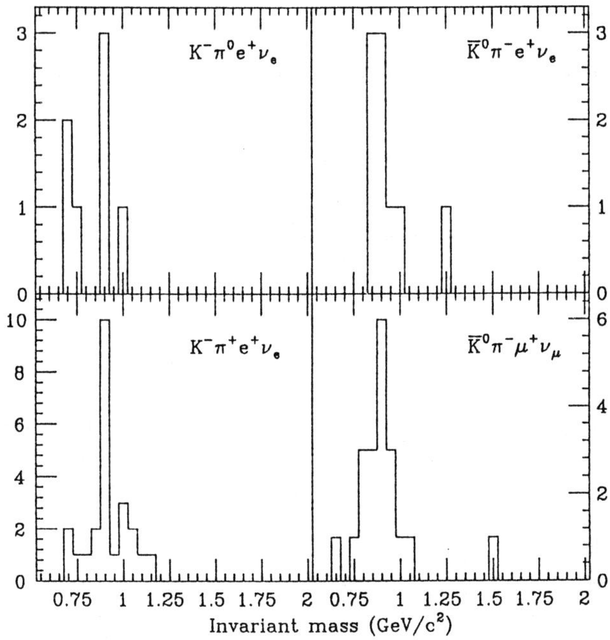

Fig. 37 Distribution of $K\pi$ invariant mass for four D_{l4} exclusive semileptonic decay modes (Mark III).

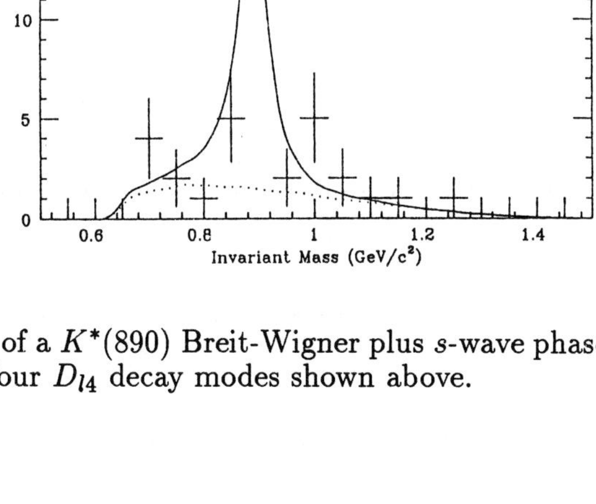

Fig. 38 Fit of a $K^*(890)$ Breit-Wigner plus s-wave phase space for the sum of the four D_{l4} decay modes shown above.

The Dalitz plot distribution for the $D^0 \to K^- e^+ \nu_e$ and $D^0 \to K^- \mu^+ \nu_\mu$ events is shown in Figure 39. The t dependence of f_+ is given by a once-subtracted dispersion relation, dominated by the lowest $1^- c\bar{s}$ resonance, the D_s^{+*}. We thus expect

$$f_+(t) = \frac{f_+(0)}{1 - \dfrac{t}{M^2_{D_s^{+*}}}} .$$

Since t depends only on E_K, a fit to the kaon spectrum suffices to extract $M_{D_s^{+*}}$. Figure 40 shows this spectrum together with a curve for the best fit value of $M_{D_s^{+*}}$:

$$M_{D_s^{+*}} = 2.1^{+1.7}_{-0.4}\ \mathrm{GeV} .$$

Thus, within rather large errors, the t dependence of the f_+ form factor follows the naïve expectation.

In the case of semileptonic D decays to final states containing K^*, $K\pi$, etc., there may be five t dependent form factors. The data is not yet sufficiently precise to allow an attempt to extract them.

Semileptonic decays of the D_s^+ are also of interest, although as yet there is no experimental information in this area. Two points should be noted. First, now that there are reliable lifetime measurements for all three 0^- charmed mesons, it would be very interesting to be able to compare their inclusive semileptonic branching ratios to ascertain whether the relationship seen for D^0 and D^+ decays is maintained. Second, the study of the hadronic component in exclusive decays could be especially rewarding. In the case of the $c\bar{s}$ annihilation amplitude (Fig. 33), the source of hadrons is singlet multiple gluon emission. Thus, along with ϕ and η, which are expected to dominate spectator D_s^+ semileptonic decay, one might expect some amount of glueball production.

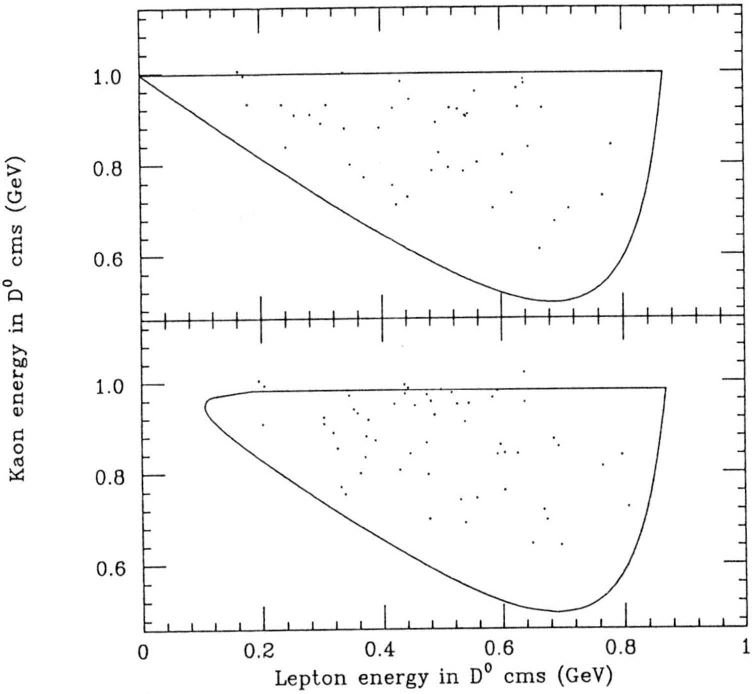

Fig. 39 Dalitz plot for D^0_{e3} and $D^0_{\mu 3}$ exclusive events (Mark III).

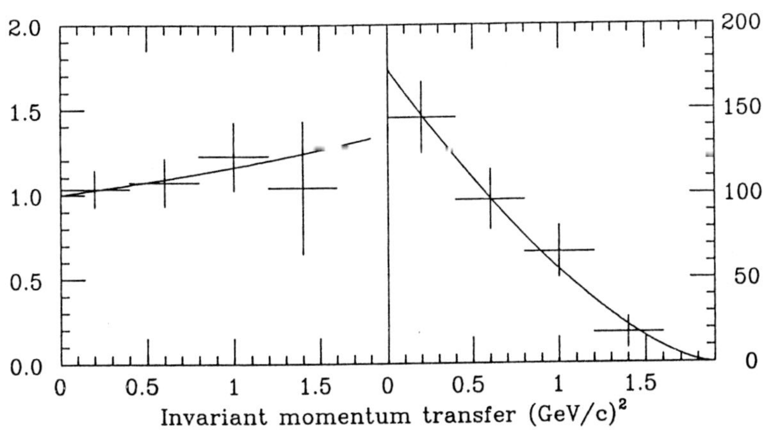

Fig. 40 a) $f_+(t)/f_+(0)$ *vs* t and b) kaon momentum spectrum for the Mark III D_{l3} events. The best fit curves are superimposed.

Rare D Decays

$D^+ \to \mu^+\nu_\mu$

As discussed above, the pseudoscalar coupling constant, f_D, can be measured in the leptonic decay $D^+ \to \mu^+\nu_\mu$. The only extant experimental information on the pseudoscalar coupling constant is a limit on this decay branching ratio, set by Mark III.[15]

The data sample used consists of 9.3 pb^{-1} at the $\psi(3770)$. A sample of single tag events comprising seven D^- hadronic decay modes is first reconstructed. There are $2490 \pm 42 \pm 42$ such tagged events. The recoil spectrum against these tags is then searched for evidence of $D^+ \to \mu^+\nu_\mu$ decay. The signature required is a single positively charged track, identified as a muon either by the muon-detection system for $|\cos\theta| \leq 0.65$, or by an energy deposition in the end-cap shower counters consistent with a non-interacting minimum-ionizing particle. Principal backgrounds come from D^+ decays to a single charged particle plus a π^0 or a $\bar{K}^0$, which produces either a π^0 via K^{*0} decay or a K^0_L. Requiring that there be no neutral particles detected in the recoil spectrum eliminates most such background. A final requirement that the identified muon candidate have momentum $0.775\ \text{GeV} < p_\mu^{lab} < 1.125\ \text{GeV}/c$, and that the missing mass squared in the event be close to zero eliminates all candidates. A maximum likelihood fit to the missing mass squared distribution leads to an upper limit $B(D^+ \to \mu^+\nu_\mu) < 7.2 \times 10^{-4}$ (at 90% confidence levels), including systematic uncertainties. Using the E691 result for the D^+ lifetime and $|V_{cd}|^2 = 0.0493$, the 90% confidence level limit is $f_D < 290$ MeV.

This limit is still above most predictions for f_D, but it is below the value required to enhance non-spectator processes in naïve perturbative calculations[53] to a level which could account for the $D^0 - D^+$ lifetime difference.

$D \to l^+l^-$, l^+l^-X

Limits on charm-changing neutral currents have been obtained in

s-wave $K\pi$ phase space. The $K\pi$ final state is not entirely K^*: there is a significant non-resonant contribution. The fit with s-wave $K\pi$ phase space yields

$$\frac{B(D \to K^* e^+ \nu_e)}{B(D \to K\pi e^+ \nu_e)} = 0.55 \pm 0.13\,.$$

The result using p-wave phase space is 0.57 ± 0.13. The fact that there is a non-resonant component cannot be accommodated in the quark model treatment of semileptonic decay of, for example, Grinstein *et al.*[52] The implications of this problem on using B semileptonic decay spectra to extract the $b \to u/b \to c$ fraction have yet to be addressed.

In the case of the $D^0 \to K^- e^+ \nu_e$ channel, the Dalitz plot distribution has also been analyzed to extract the t dependence of the vector form factor.[12] The matrix element for Cabibbo-allowed semileptonic D decay may be written as

$$M = \frac{G_F}{\sqrt{2}} J_\mu^{\text{lepton}} J^\mu_{\text{hadron}}\,,$$

where

$$J_\mu^{\text{lepton}} = \nu_e \gamma_\mu (1 - \gamma_5) e^+\,,$$

$$\begin{aligned} J^\mu_{\text{hadron}} = \sqrt{2}|V_{cs}|[&f_+(t)(P_D^\mu + P_K^\mu) \\ &+ f_-(t)(P_D^\mu - P_K^\mu)]\,, \end{aligned}$$

and the four momentum transfer $t = (P_D - P_K)^2 - M_D^2 + M_K^2 - 2M_D E_K$.

Since $(P_D - P_K)$ is proportional to M_l, the second term in J^μ_{hadron} may be neglected for any heavy quark semileptonic decay at the present level of precision. Thus, in the D center-of-mass, we can write the Dalitz plot distribution as

$$\frac{d\Gamma}{dx_e dx_k} \propto |f_+|^2 \left(2x_e - x_e x_k + x_e^2 + x_k - \frac{M_K^2}{M_D^2} - 1\right),$$

where $x_e = E_e/E_{e_{max}}$, $x_K = E_K/E_{K_{max}}$.

four D decay modes. These are summarized in Table XVIII.

$D^0 \to \mu^{\pm}e^{\mp}$

The lepton-number family violating decay $D^0 \to \mu^{\pm}e^{\mp}$ could be mediated by, for example, an SU(5) scalar leptoquark.[62] The stringent upper limit on the related decay $B(K_L^0 \to \mu^{\pm}e^{\mp}) < 6 \times 10^{-6}$ at 90% confidence level,[63] does not exclude a measurable rate for the D^0 into this channel. Five groups have now set limits on this decay which are summarized in Table XIX.

The rate for $D^0 \to \mu^{\pm}e^{\mp}$ is proportional to $f_D^2 M_{LQ}^{-4}$, where M_{LQ} is the mass of the leptoquark. With a value of f_D at the limit set by Mark III, the limits of Table XIX provide a lower limit of M_{LQ} which is greater than 1 TeV.

This decay would be helicity-suppressed in the case of vector exchange. We therefore note for completeness that E691 has also set a 90% confidence level limit $B(D^+ \to \mu^{\pm}e^{\mp}\pi^+) < 2 \times 10^{-4}$.

D_s^+ Hadronic Decays

Much less is known experimentally about the hadronic decays of the D_s^+, although this is an area in which rapid progress can be expected. Observations of the D_s^+ have had a checkered history, with several early experiments[64,65,66,67] finding signals in $\eta\pi^+$, $\eta\pi^+\pi^+\pi^-$, etc. at masses above 2 GeV, while current measurements clearly place the mass in the 1970 MeV region (see Table I and Figure 2).

Modern observations of D_s^+ decays are summarized in Table XX.[68] In the absence of a well-defined feature in the e^+e^- cross section analogous to the $\psi(3770)$, extraction of D_s^+ branching ratios requires a number of assumptions. The clearest set of assumptions applies to production in high energy e^+e^- annihilation far above threshold. A typical set of such assumptions is:

a) charmed quark production accounts for 4/10 of the total hadronic cross section below $b\bar{b}$ threshold and 4/11 above.

b) the probability of producing an $s\bar{s}$ pair from the vacuum is 15%,

TABLE XVIII

Limits on Charm-Changing Neutral Currents at 90% Confidence Level

Mode	Experiment				
	CLEO[54]	ARGUS[55]	E691[56]	EMC[57]	E615[58]
$D^0 \to \mu^+\mu^-$	$< 2.0 \times 10^{-4}$	$< 8 \times 10^{-5}$	$< 1 \times 10^{-4}$	$< 3.4 \times 10^{-4}$	$< 1.1 \times 10^{-5}$
$D^0 \to e^+e^-$	$< 2.4 \times 10^{-4}$	$< 1.3 \times 10^{-4}$	$< 8 \times 10^{-5}$		
$D^+ \to \pi^+\mu^+\mu^-$	$< 8.6 \times 10^{-4}$				
$D^+ \to \pi^+e^+e^-$	$< 1.2 \times 10^{-3}$				

TABLE XIX

Limits on $D^0 \to \mu^\pm e^\mp$

Experiment	Limit at 90% Confidence Level
Mark II[59]	$< 2.1 \times 10^{-3}$
ACCMOR[60]	$< 9 \times 10^{-4}$
Mark III[61]	$< 1.2 \times 10^{-4}$
ARGUS[55]	$< 9 \times 10^{-5}$
E691[56]	$< 8 \times 10^{-5}$

TABLE XX

D_s^+ Hadronic Branching Ratio Measurements

Decay Mode	Experiment	Measurement	$\sqrt{s}$	Branching Ratio (%)	Comments
$D_s^+ \to \phi\pi^+$	CLEO	$R(D_s^+)B(D_s^+ \to \phi\pi^+) = 2.2 \pm 0.5$	10	4.4 ± 1.1	a
	TASSO		35-44	$3.3 \pm 1.6 \pm 0.4$	a,b
	ARGUS	$R(D_s^+)B(D_s^+ \to \phi\pi^+) = 1.47 \pm 0.32 \pm 0.20$	10	$3.2 \pm 0.7 \pm 0.5$	a
	HRS	$R(D_s^+)B(D_s^+ \to \phi\pi^+) = 1.35 \pm 0.34$	29	3.3 ± 1.0	a
	MARK III	$\sigma \cdot B = (30 \pm 6 \pm 11)$pb	4.14	—	c,d
	E691	$\sigma(\gamma + Be \to D_s^+ X) \cdot B(D_s^+ \to \phi\pi^+)$	~ 16	3.5 ± 0.8	
$\bar{K}^{*0}K^+$	ARGUS	$\frac{B(D_s^+ \to K^{*0}K^+)}{B(D_s^+ \to \phi\pi^+)} = 1.44 \pm 0.37$	10	5.0 ± 1.3	e
	MARK III	$\sigma \cdot B = (31 \pm 6 \pm 11)$ pb	4.14	3.6 ± 2.1	c,d,e
	E691	$\frac{B(D_s^+ \to K^{*0}K^+)}{B(D_s^+ \to \phi\pi^+)} = 0.74 \pm 0.12 \pm 0.06$	~ 16	2.6 ± 0.5	e
$(K^-K^+\pi^+)_{\text{nonres}}$	TPC	$\frac{\sigma_{D_s^+}(z > 0.4)B(D_s^+ \to \text{“}K^+K^-\pi^+\text{”})}{\sigma_h} = (3.34 \pm 0.95 \pm 1.11) \times 10^{-2}$	29		
	ACCMOR	5 events hadronically produced			
	E691	$\frac{B(D_s^+ \to K^+K^-\pi^+)}{B(D_s^+ \to \phi\pi^+)} = 0.23 \pm 0.07 \pm 0.07$	~ 16	0.8 ± 0.35	e
$\phi\pi^+\pi^-\pi^+$	ARGUS	$\frac{B(D_s^+ \to \phi\pi^+\pi^-\pi^+)}{B(D_s^+ \to \phi\pi^+)} = 1.11 \pm 0.37 \pm 0.28$	10	3.9 ± 1.6	e
	E691	$\frac{B(D_s^+ \to \phi\pi^+\pi^-\pi^+)}{B(D_s^+ \to \phi\pi^+)} = 0.77 \pm 0.31 \pm 0.15$	~ 16	2.7 ± 1.2	e
$\bar{K}^0K^+$	MARK III	$\sigma \cdot B = (32 \pm 6 \pm 10)$ pb	4.14	3.7 ± 2.0	c,d,e
$\rho^0\pi^+$	ARGUS	$\frac{B(D_s^+ \to \rho^0\pi^+)}{B(D_s^+ \to \phi\pi^+)} < 0.77$ @ 90% CL	10	< 0.77 @ 90% CL	e
	E691	$\frac{B(D_s^+ \to \rho^0\pi^+)}{B(D_s^+ \to \phi\pi^+)} < 0.08$ @ 90% CL	~ 16	< 0.28 @ 90% CL	c,e
$(\pi^-\pi^+\pi^+)_{\text{nonres}}$	E691	$\frac{B(D_s^+ \to \pi^-\pi^+\pi^+)}{B(D_s^+ \to \phi\pi^+)} = 0.29 \pm 0.07 \pm 0.05$	~ 16	1.0 ± 0.4	c,e

a) See text

b) Assumes s/d quark suppression factor of 0.33 ± 0.02.

c) Preliminary

d) $\sigma \cdot B \equiv \sigma(e^+e^- \to D_s^\pm D_s^{\mp *}) \cdot B(D_s^+ \to \text{mode})$

e) Branching ratios derived using $B(D_s^+ \to \phi\pi^+) = (3.5 \pm 0.5)\%$, the average of the four high energy e^+e^- measurements. The error on the $\phi\pi$ branching ratio is not propagated.

f) Measurement scaled to $B(D_s^+ \to \phi\pi^+) = 3.5\%$ with statistical error of E691 measurement.

which thereby specifies the fraction of D_s^+ production compared to total charm production.

c) the D_s^+ production cross section is typically measured only above a given D_s^+ momentum. Thus, the measured cross section $\frac{s}{\beta}\frac{d\sigma}{dx_{D_s^+}}$ must be extra extrapolated to $x_{D_s^+} = 0$.

The observations of $D_s^+ \to \phi\pi^+$ can be used with this *ansatz* to extract the $\phi\pi^+$ branching ratio. The ratio of other D_s^+ branching fractions, which of course can be obtained indirectly, are also shown in Table XX. The uncertainties quoted are statistical only. Figure 41 shows three D_s^+ hadronic decays observed by Mark III.

There is, as yet, no experimental information on D_s^+ semileptonic decay. Such an inclusive measurement would provide an interesting comparison with D^0 and D^+ rates: a small branching ratio would support the speculation that D^0 and D_s^+ lifetimes are shorter than that of the D^+ due to non-leptonic enhancement associated with non-spectator diagrams. As Cabibbo-allowed D_s^+ semileptonic decay can proceed via $c\bar{s}$ annihilation as well as by a spectator process, a measurement of exclusive semileptonic modes would be particularly interesting.

Charmed Baryon Hadronic Decays

Four charmed baryons (Λ_c^+, Ξ_c^+, Ξ_c^0 and Ω_c^0) decay weakly. The others will decay to these states via radiative or strong decays. In contrast to the case of charmed mesons, there is no particular inhibition of non-spectator amplitudes in charmed baryon decay. We might, therefore, expect the charmed baryons which decay weakly to have shorter lifetimes than charmed mesons. This is indeed the case for the Λ_c^+ (see Table V). The small number of events seen in Ξ_c^+ and Ω_c^0 decay indicate that these states have somewhat longer lifetimes than the Λ_c^+.

Seven Λ_c^+ hadronic decay modes have been observed. These are summarized in Table XXI.[69] A typical Λ_c^+ invariant mass distribution is shown in Figure 42. The normalization of these branching ratios depends on measurements from Mark II,[70] together with a string of assumptions. The first hint of charmed baryon production in e^+e^-

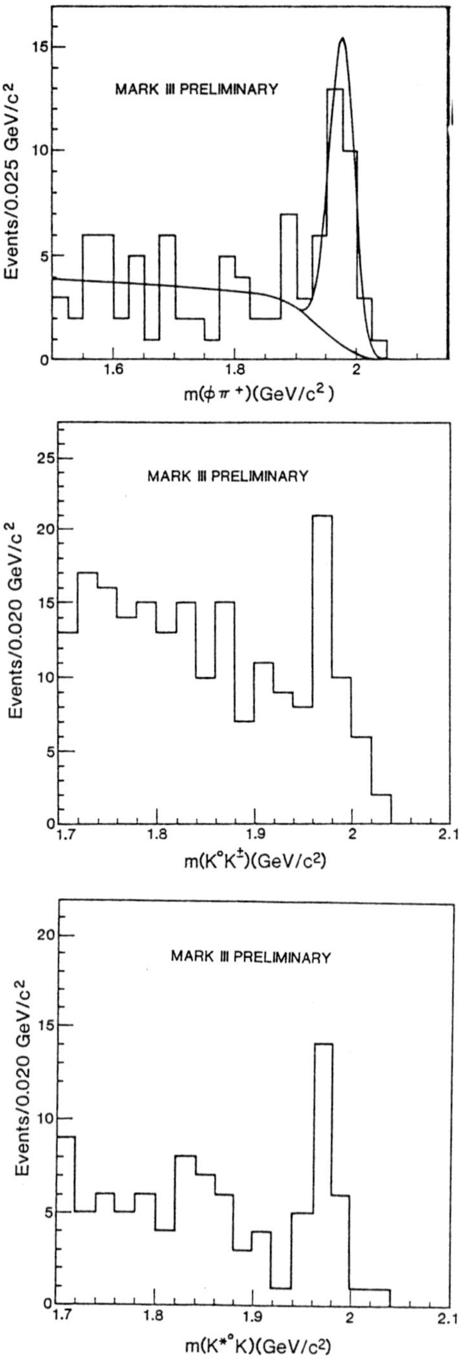

Fig. 41 Three Mark III D_s^+ hadronic decays:

a) $D_s^+ \to \phi\pi^+$

b) $D_s^+ \to K^0 K^+$

c) $D_s^+ \to K^{*0} K^+$.

TABLE XXI

Λ_c^+ Hadronic Branching Ratio Measurements

Decay Mode	Experiment	Measurement	Branching Ratio (%)	Comments
$\Lambda_c^+ \to pK^-\pi^+$	Mark II	$\sigma(\Lambda_c + \bar{\Lambda}_c)B(\Lambda_c^+ \to pK^-\pi^+) = (34 \pm 10)$pb	2.0 ± 0.8	a
	ARGUS	$\sigma \cdot B = (10.8 \pm 1.4 \pm 1.2)$pb	—	
$p\bar{K}^0$	Mark II	$\frac{B(\Lambda_c^+ \to p\bar{K}^0)}{B(\Lambda_c^+ \to pK^-\pi^+)} = 0.72 \pm 0.32$	1.4 ± 0.9	a
	ARGUS	$\frac{B(\Lambda_c^+ \to p\bar{K}^0)}{B(\Lambda_c^+ \to pK^-\pi^+)} = 0.62 \pm 0.15 \pm 0.03$	1.2 ± 0.6	b
	CLEO	$\frac{B(\Lambda_c^+ \to p\bar{K}^0)}{B(\Lambda_c^+ \to pK^-\pi^+)} = 0.66 \pm 0.20 \pm 0.19$	1.3 ± 0.8	b
$p\bar{K}^{*0}$	Mark II	$\frac{B(\Lambda_c^+ \to p\bar{K}^{*0})}{B(\Lambda_c^+ \to pK^-\pi^+)} = 0.18 \pm 0.10$	0.36 ± 0.25	a
	Basile *et al.*	$\frac{B(\Lambda_c^+ \to p\bar{K}^{*0})}{B(\Lambda_c^+ \to pK^-\pi^+)} = 0.42 \pm 0.24$	0.84 ± 0.59	b
	ARGUS	$\frac{B(\Lambda_c^+ \to p\bar{K}^{*0})}{B(\Lambda_c^+ \to pK^-\pi^+)} = 0.42 \pm 0.21$	0.84 ± 0.54	b
	NA-32	$\frac{B(\Lambda_c^+ \to p\bar{K}^{*0})}{B(\Lambda_c^+ \to pK^-\pi^+)} = 0.28 \pm 0.07$	0.56 ± 0.26	b
$\Delta^{++}K^-$	Mark II	$\frac{B(\Lambda_c^+ \to \Delta^{++}K^-)}{B(\Lambda_c^+ \to pK^-\pi^+)} = 0.17 \pm 0.07$	0.34 ± 0.20	a
	Basile *et al.*	$\frac{B(\Lambda_c^+ \to \Delta^{++}K^-)}{B(\Lambda_c^+ \to pK^-\pi^+)} = 0.40 \pm 0.17$	0.80 ± 0.47	b
	ARGUS	$\frac{B(\Lambda_c^+ \to \Delta^{++}K^-)}{B(\Lambda_c^+ \to pK^-\pi^+)} < 0.33$ @ 90% CL	< 0.7 @ 90% CL	b
$p\bar{K}_0\pi^+\pi^-$	ARGUS	$\frac{B(\Lambda_c^+ \to p\bar{K}^0\pi^+\pi^-)}{B(\Lambda_c^+ \to pK^-\pi^+)} < 0.91$ @ 90% CL	$< 1.8 \pm$ @ 90% *CL*	b
	BIS-2	$\frac{B(\Lambda_c^+ \to p\bar{K}^0\pi^+\pi^-)}{B(\Lambda_c^+ \to \Lambda\pi^+\pi^-\pi^+)} = 4.3 \pm 1.2$	3.9 ± 1.1	c
	CLEO	$\frac{B(\Lambda_c^+ \to p\bar{K}^0\pi^+\pi^-)}{B(\Lambda_c^+ \to pK^-\pi^+)} = 0.67 \pm 0.28 \pm 0.29$	1.3 ± 1.0	b
$\Lambda\pi^+$	Mark II	$\frac{B(\Lambda_c^+ \to \Lambda\pi^+)}{B(\Lambda_c^+ \to pK^-\pi^+)} < 0.25$ @ 90% CL	< 0.5 @ 90% CL	b
	Baltay *et al.*	$\frac{B(\Lambda_c^+ \to \Lambda\pi^+)}{B(\Lambda_c^+ \to p\bar{K}^0)} = 0.67 \pm 0.78 \pm 0.35$	1.3 ± 1.8	d
	Kitagaki *et al.*	$\frac{B(\Lambda_c^+ \to \Lambda\pi^+)}{B(\Lambda_c^+ \to p\bar{K}^0)} = 0.51 \pm 0.62 \pm 0.27$	1.0 ± 1.4	d
	ARGUS	$\frac{B(\Lambda_c^+ \to \Lambda\pi^+)}{B(\Lambda_c^+ \to p\bar{K}^0)} < 0.26$ @ 90% CL	< 0.5 @ 90% CL	d
	CLEO	$\frac{B(\Lambda_c^+ \to \Lambda\pi^+)}{B(\Lambda_c^+ \to pK^-\pi^+)} = 0.22 \pm 0.07 \pm 0.29$	0.4 ± 0.6	b
$\Lambda\pi^+\pi^+\pi^-$	ARGUS	$\frac{B(\Lambda_c^+ \to \Lambda\pi^+\pi^-\pi^+)}{B(\Lambda_c^+ \to pK^-\pi^+)} = 0.61 \pm 0.16 \pm 0.04$	1.2 ± 0.6	b
	CLEO	$\frac{B(\Lambda_c^+ \to \Lambda\pi^+\pi^-\pi^+)}{B(\Lambda_c^+ \to pK^-\pi^+)} = 0.80 \pm 0.19 \pm 0.22$	1.6 ± 0.9	b

a) Revised from original publication. See text.
b) Normalized to Mark II $B(\Lambda_c^+ \to pK^-\pi^+)$
c) Normalized to ARGUS $\sigma \cdot B(\Lambda_c^+ \to \Lambda\pi^+\pi^+\pi^-)/\sigma \cdot B(\Lambda_c^+ \to pK^-\pi^+)$
d) Normalized to Mark II $B(\Lambda_c^+ \to p\bar{K}^0)$

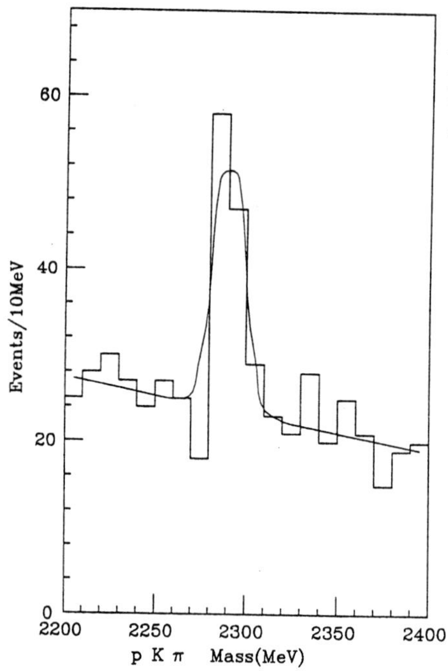

Fig. 42 Invariant mass of $pK\pi$ from experiment E691, showing Λ_c^+ photoproduction.

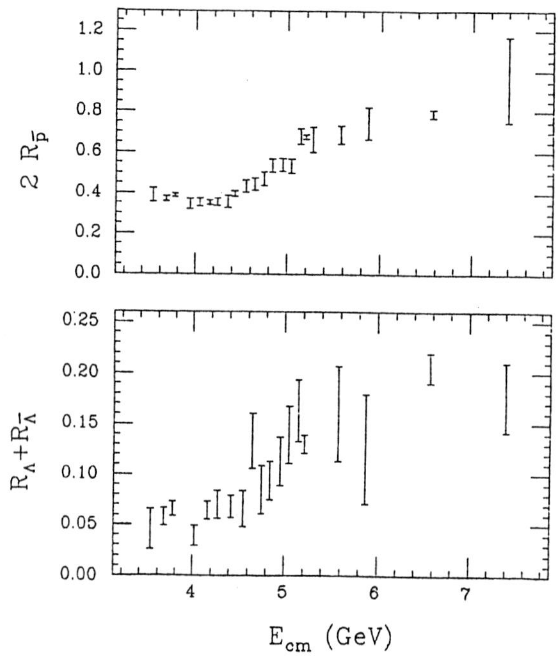

Fig. 43 Step in the $\sigma(P+\bar{P})$ and $\sigma(\Lambda+\bar{\Lambda})$ observed by the Mark II at SPEAR. Note that the values quoted in the text, as well as these figures are corrected from the original publication.

annihilation came from the observation of a step in $p+\bar{p}$ production at SPEAR energies.[71] The Mark II then remeasured this with increased precision, and also found a step in $\Lambda + \bar{\Lambda}$ production ($\Delta R(\Delta + \bar{\Lambda}) = 0.10 \pm 0.03$) (see Fig. 43). Steps in the normalization procedure are:

1) the step $\Delta R(p + \bar{p}) = 0.31 \pm 0.06$ is entirely due to the onset of charmed baryon production
2) all charmed baryons so produced cascade to the Λ_c^+
3) $\dfrac{\Lambda_c^+ \to p}{\Lambda_c^+ \to \text{all}} = 0.6 \pm 0.1.$

Thus,
$$\sigma(\Lambda_c^+ + \bar{\Lambda_c^+}) = \frac{\Delta R(p+p) \times \sigma_{\mu\mu}}{\sigma(\Lambda_c^+ \to p)/\sigma(\Lambda_c^+ \to \text{all})}$$
$$= 1.7 \pm 0.4 \text{ nb} \quad \text{at} \quad \sqrt{s} = 5.2 \text{ GeV}\,.$$

When the charmed baryon weak decay normalization described above is applied to the various individual branching ratios, we see that we can account thus far for only a bit more than 10% of total decays. This is surprising for several reasons:

1) As, by analogy with charmed meson decays, we might expect a dominance of two body decays, it would be difficult to find a sufficient number of additional low multiplicity modes to account for a major fraction of total decays.

2) Using the current normalization, the cross sections derived for hadron production, photoproduction, and e^+e^- annihilation cross sections for charmed baryons are surprisingly large. In particular, both photoproduction and e^+e^- cross sections for charmed baryons would appear to be comparable to D production cross sections, an unlikely conclusion.

Further investigation of the assumptions involved in Λ_c^+ branching ratio normalization, as well as new measurements of the $p + \bar{p}$ and $\Lambda + \bar{\Lambda}$ inclusive cross sections in the $\sqrt{s} = 5$ GeV region, are clearly warranted.

The Ξ_c^+ has been observed in the charged hyperon beam[72] at CERN in the channel $\Lambda K^-\pi^+\pi^+$ at $m(\Xi_c^+) = 2460 \pm 15$ MeV. The observed width is compatible with the experimental resolution.

The Ξ_c^+ has also recently been seen in 800 GeV neutron beam production at FNAL.[73] Two peaks are seen in the $\Lambda K\pi\pi$ invariant mass spectrum, which are interpreted as

$$\Xi_c^+ \to \Lambda K^-\pi^+\pi^+ \quad \text{and} \quad \Xi_c^+ \to \Sigma^0 K^-\pi^+\pi^+$$
$$\hookrightarrow \Lambda\gamma .$$

The mass is $m(\Xi_c^+) = 2448 \pm 5 \pm 30$ MeV, with $\sigma \cdot B(\Sigma^0 K\pi\pi)/\sigma \cdot B(\Lambda K\pi\pi) = 0.83 \pm 0.4$.

The two observations are not obviously compatible, as one would expect both experiments to show the two peak structure in $m(\Lambda K\pi\pi)$.

Three events interpreted at $\Omega_c^0 \to \Xi^- K^-\pi^+\pi^+$ have also been reported,[74] with a mass $m(\Omega_c^0) = 2740$ MeV and a lifetime $\tau(\Omega_c^0) = (7.9 \pm 2.8 \pm 2.0) \times 10^{-13}$ sec.

The Σ_c^{++} was found in the first bubble chamber event at BNL,[75] which also found the Λ_c^+, via the cascade

$$\nu_\mu p \to \mu^-\Sigma_c^{++}(2426)$$
$$\hookrightarrow \Lambda_c^+\pi^+$$
$$\hookrightarrow \Lambda\pi^+\pi^-\pi^- .$$

Twenty events involving Σ_c^{++} cascades to Λ_c^+ in the $\Lambda\pi^+$, $\Lambda 3\pi$, $\bar{K}^0 p$ and $\bar{K}^0 p\pi^+\pi^-$ channels were found in the 15 foot bubble chamber at FNAL,[76] allowing a measurement of the $\Sigma_c^{++} - \Lambda_c^+$ mass difference: $m(\Sigma_c^{++} - m(\Lambda_c^+) = 168 \pm 3$ MeV. The Σ_c^+ was observed in BEBC[77] with a track-sensitive target in the cascade

$$\nu_\mu p \to \mu^-\Sigma_c^+\pi^+$$
$$\hookrightarrow \Lambda_c^+\pi^0$$
$$\hookrightarrow K^-p\pi^+ ,$$

resulting in $m(\Sigma_c^+) - m(\Lambda_c^+) = 168 \pm 3$ MeV.

ARGUS[78] has recently reported new measurements of the Σ_c^0 and Σ_c^{++}, finding $m(\Sigma_c^{++}) - m(\Lambda_c^+) = 168.7 \pm 0.5$ MeV, and $m(\Sigma_c^0) - m(\Lambda_c^+) = 166.9 \pm 0.6$ MeV. It thus appears that the $\Sigma_c^{++}, \Sigma_c^+, \Sigma_c^0$ multiplet is nearly degenerate, as expected.

Of the four charmed baryons which are expected to decay by weak interactions, three, the Λ_c^+, Ξ_c^+ and Ω_c^0 have thus been seen, the Ξ_c^0 has yet to be found. There are fundamental differences between the phenomenology of weak decays of charmed baryons and that of charmed mesons. We will use the Λ_c^+ as an example to explore the differences.

There has been a single measurement of the Λ_c^+ semileptonic branching ratio:[79]

$$B(\Lambda_c^+ \to e^+ X) = 0.045 \pm 0.017 \quad ,$$

which, together with the short Λ_c^+ lifetime, points up the importance of non-leptonic enhancements in charmed hadron decay.

CHARMED MESON HADRONIC DECAYS – THEORY

The wealth of precise data on charmed particle lifetimes and hadronic and semileptonic branching ratios has led to a number of detailed theoretical treatments. In order to orient ourselves, let us begin with a few simple observations

If we assume the validity of the QCD-motivated discussion above, we are justified in using the current-current Hamiltonian derived therein, with a range of c_-/c_+ values close to that calculated with the leading-log approximation and its extensions. For the case of Cabibbo-allowed charmed mesons, the amplitudes at the quark level for the light quark spectator Hamiltonian are shown in Fig. 44. The four quark operators generate two distinct amplitudes for each charmed meson. The first, with coefficient $c_1 = \frac{1}{2}(c_+ + c_-)$ is called "color-allowed". The second, with coefficient $c_2 = \frac{1}{2}(c_+ - c_-)$, is due to the effective neutral current generated by the Fierz transformation; it is usually referred to

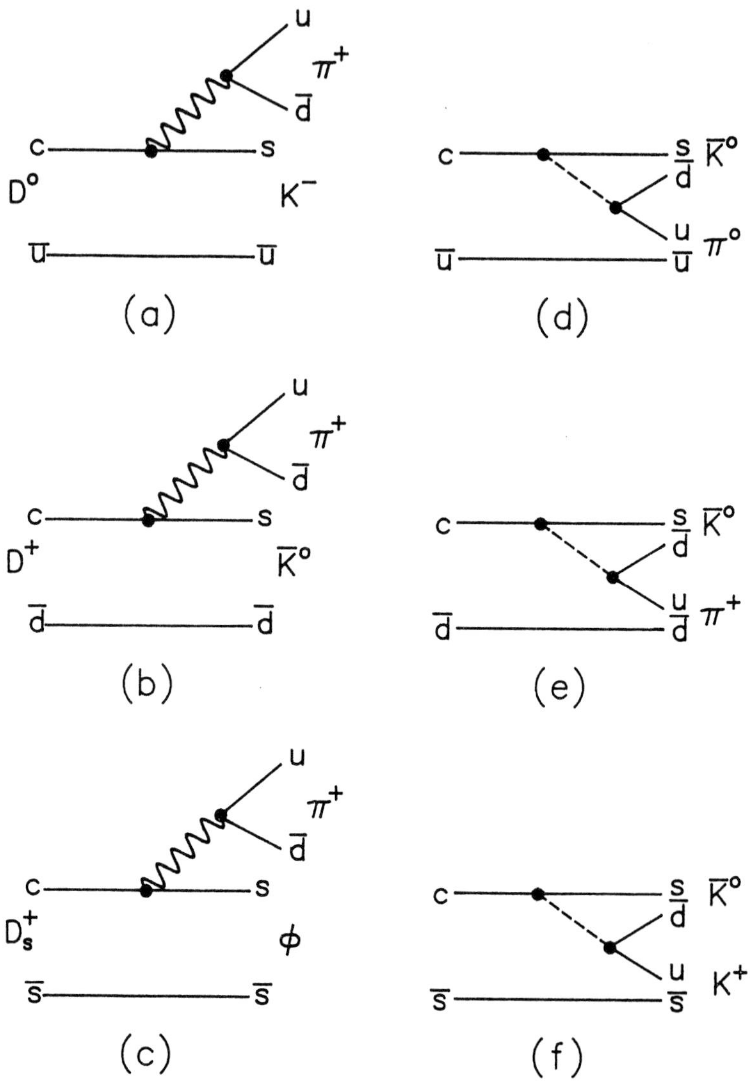

Fig. 44 Spectator amplitudes for charmed meson hadronic decay arising from QCD-motivated renormalization of operator coefficients. Diagrams a), b) and c) have the same operator structure as in the simple valence quark picture, although they now enter with coefficient $c_1 = (c_+ + c_-)/2$. Diagrams d), e) and f) are "effective neutral current" or "color suppressed" amplitudes, which have coefficient $c_2 = (c_+ - c_-)/2$. Note that in the D^+ case, the two amplitudes lead to the same final state, and may thus interfere.

as the "color-suppressed" amplitude due to the color mismatch which impedes final state hadron formation.

There are other amplitudes which should also be considered. Figure 9 shows these two "non-spectator" processes, which can, in principle, contribute to Cabibbo-allowed D^0 and D_s^+ decays. These amplitudes are helicity-suppressed at the light quark vertex: their neglect yields the light quark spectator model.

In treating exclusive hadronic decays, it is important to recognize from the outset the complex relation between the quark level operators and the actual hadrons; the explicit structure of hadrons is certain to play an important role in understanding exclusive decays. Final state interactions are also likely to be important, as charmed meson masses lie in a region where resonance rescattering effects cannot be neglected. We will return to this question presently.

Bauer, Stech and and Wirbel[80] have dealt with the hadron problem by appealing to factorization to replace the product of quark currents with products of effective hadron currents:

$$H_{\text{eff}}(\Delta c = -1) = \frac{G_F}{\sqrt{2}} V_{cs} V_{ud}^* \Big[a_1 (\bar{u} d')_H (\bar{s}' c)_H + a_2 (\bar{s}' d')_H (\bar{u} c)_H \Big] \ ,$$

where the subscript H now denotes hadron field operators. The QCD coefficients c_1 and c_2 are replaced by new effective coefficients a_1 and a_2:

$$a_1 = (c_1 + \xi c_2)_{\mu = m_c} \ , \ a_2 = (c_2 + \xi c_1)_{\mu = m_c} \ ,$$

where the color-mismatch factor ξ, which has the value $1/N$ (i.e., $1/3$) at the quark level, is now a free parameter. This is justified by the fact that the effect of color octet currents within explicit hadrons is not, *a priori*, understood.

The factorization hypothesis then allows us to write the amplitude for specific exclusive processes in terms of meson decay constants and hadron current matrix elements. For example:

$$A(D^0 \to K^-\pi^+) = \frac{G_F}{\sqrt{2}} V_{cs} V_{ud}^* a_1 (-i f_\pi) p_\lambda^\pi \langle K^- | (\bar{s}c)^\lambda | D^0 \rangle_{q^2=m_\pi^2} \quad ,$$

$$A(D^0 \to \bar{K^0}\pi^0) = \frac{G_F}{\sqrt{2}} V_{cs} V_{ud}^* a_2 \Big(\frac{-i f_\pi}{\sqrt{2}}\Big) p_\lambda^\pi \langle \bar{K^0} | (\bar{s}c)^\lambda | D^0 \rangle_{q^2=m_\pi^2} \quad .$$

The hadron matrix elements may be evaluated, using, for example, relativistic harmonic oscillator wave functions.

At the quark level, with $\xi = \frac{1}{3}$, we can see that the ratio of these two $D \to K\pi$ rates would be given by

$$R = \frac{\Gamma(D^0 \to \bar{K^0}\pi^0)}{\Gamma(D^0 \to K^-\pi^+)} = \frac{1}{2}\Big(\frac{2c_+ - c_-}{2c_+ + c_-}\Big)^2 \quad ,$$

for the nominal values of the QCD coefficients.

Table XXII shows the experimental value of this ratio: $R = 0.45 \pm 0.08 \pm 0.05$. This quark-level estimate cannot account for the data. Two D^0 decay modes into pseudoscalar-vector channels; $D^0 \to K^*\pi$ and $K\rho$ can also be estimated in the same way (see Table XXII). In the ratio $\Gamma(00)/\Gamma(-+)$ the hadron matrix elements, which are, in general, different for PP and PV channels, drop out; the dependence on the operator coefficients is identical. Clearly, this naïve estimate is incorrect. If we recognize that it is the *effective* coefficients which contribute to exclusive decays, however, and that the assumption $\xi = \frac{1}{3}$ is not justified in this picture, we can do better. If there are no color-suppressed amplitudes at all ($\xi = 0$), then we have

$$R = \frac{1}{2}\Big(\frac{c_+ - c_-}{c_+ + c_-}\Big)^2$$

$$\simeq .17 \quad for \quad c_-/c_+ = 2.4 \quad ,$$

with larger values possible for larger ratios of a_2/a_1. This is dramatic evidence for the lack of color-suppressed amplitudes in these decays.

Table XXII

Ratios of Charmed Meson Branching Fractions Involving Color-Suppressed Amplitudes

Ratio	Mark III Measurement
$\frac{B(D^0 \to \bar{K}^0 \pi^0)}{B(D^0 \to K^- \pi^+)}$	$0.45 \pm 0.08 \pm 0.05$
$\frac{B(D^0 \to \bar{K}^0 \rho^0)}{B(D^0 \to K^- \rho^+)}$	$0.11 \pm 0.04 \pm 0.02$
$\frac{B(D^0 \to \bar{K}^{*0} \pi^0)}{B(D^0 \to K^{*-} \pi^+)}$	$0.29 \pm 0.14 \pm 0.09$
$\frac{B(D^+ \to \phi \pi^+)}{B(D^+ \to K^- \pi^+ \pi^+)}$	$0.08 \pm 0.02 \pm 0.01$
$\frac{B(D_s^+ \to \bar{K}_s^0 K^+)}{B(D_s^+ \to \phi \pi^+)}$	$0.44 \pm 0.12 \pm 0.21$

Table XXIII

Isospin Amplitudes and Phase Shifts Derived from $D \to PP$ & $D \to PV$

Channel	$\mid A_{1/2}/A_{3/2} \mid$	$\delta_{1/2} - \delta_{3/2}$
$D \to \bar{K}\pi$	3.67 ± 0.27	$(77 \pm 11)^0$
$D \to \bar{K}^*\pi$	3.22 ± 0.97	$(84 \pm 13)^0$
$D \to \bar{K}\rho$	3.12 ± 0.40	$(0 \pm 26)^0$

Two other experimental ratios which involve color-mismatched decays, one for D^+ and one for D_s^+, also shown in Table XXII, support the idea that there is an inhibition of color-suppressed amplitudes.

Let us focus now on what we can learn about the weak hadronic decay mechanism by considering some individual decay rates. We must first observe that the naïve light quark spectator model, even with QCD-motivated corrections to the c_+ and c_- operator coefficients, fails not only to account for the lifetime differences among charmed mesons, but also cannot explain individual hadronic branching ratios. As these comparisons are treated in detail in Gilman's[81] and Bigi's[82] lectures, I will be brief, limiting the discussion to a few main points.

The first observation is that the large value for the ratio $\frac{B(D^0 \to \bar{K}^0\pi^0)}{B(D^0 \to K^-\pi^+)}$ cannot be accounted for in the naïve picture. In fact, the spectator model with QCD-corrected coefficients predicts an even smaller ratio than does the naïve model. Two effects are at work here. The first is that color-suppressed diagrams do not seem to contribute to D decays, as pointed out by Stech and co-workers, and motivated by the study of the $1/N$ expansion by Buras *et al.*[83] The absence of color-suppressed processes is not, however, sufficient to explain the large value of $B(D^0 \to \bar{K}^0\pi^0)/B(D^0 \to K^-\pi^+)$. Comparison of the three $D \to K\pi$ modes makes it clear that final state interactions play an important role, as the extracted $I = \frac{1}{2}$ and $\frac{3}{2}$ amplitudes are not relatively real. The recent Mark III analysis of $D \to K\pi\pi$ Dalitz plots has now allowed extension of these ideas to the $D \to K\rho$ and $D \to K^*\pi$ systems.

Additional information comes from the ratios $B(D^+ \to \phi\pi^+)/B(D^+ \to K^-\pi^+\pi^+)$ and $B(D_s^+ \to \bar{K}_s^0K^+)/B(D_s^+ \to \phi\pi^+)$, which also involve color-suppressed amplitudes.

Bauer *et al.* have extended the factorization idea to a large number of hadronic D (and B) decays. With appropriate pole masses parametrizing the q^2 dependence of the current matrix elements calculated at $q^2 = 0$ with oscillator wave functions, they are able to account for a large number of measurements (see Fig. 45). Their model includes

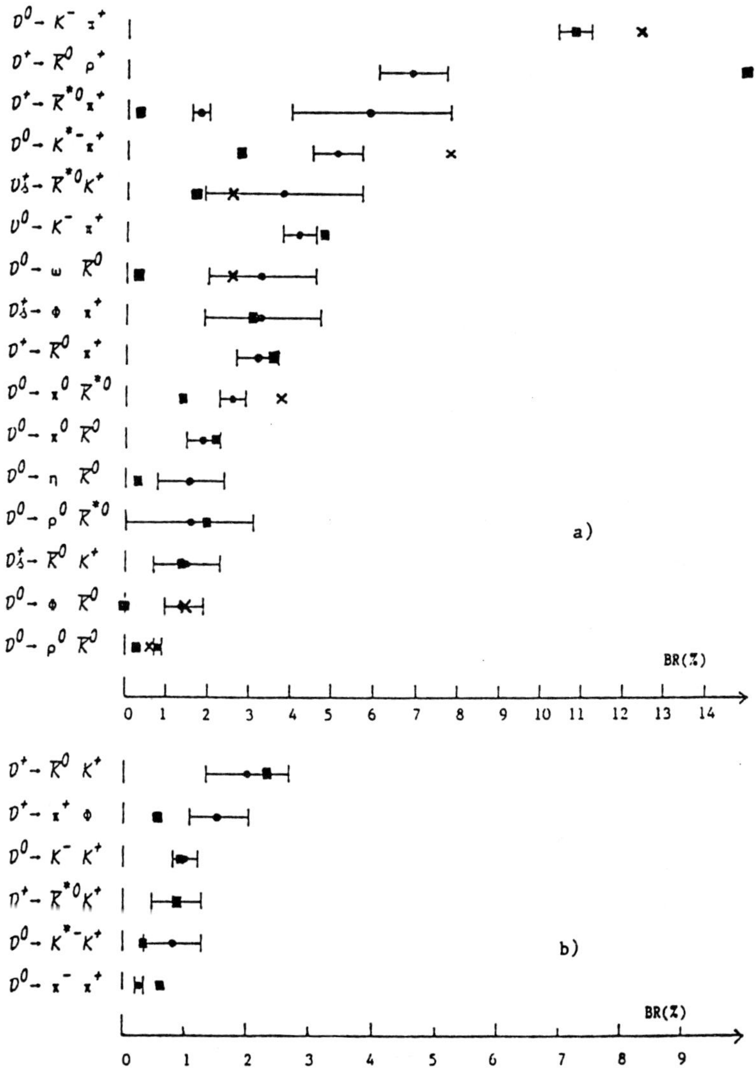

Fig. 45 Comparison of the experimental charmed meson branching ratios with the theoretical results of Bauer, Stech and Wirbel for a) Cabibbo-allowed decays and b) Cabibbo-suppressed decays. Only statistical errors are shown. Calculated values indicated by black squares are derived from factorized amplitudes with $a_1 \simeq 1.2$ and $a_2 \simeq -0.5$. Calculated values indicated with an "x" include annihilation contributions.

primarily spectator amplitudes. Annihilation amplitudes are included by fitting to the $D^0 \to \bar{K}^0\phi$ rate (see below). Final state interactions are also included.

A more detailed comparison of $D \to PP$ and $D \to PV$ decay modes related by isospin allows study of the role of final state interactions, which do turn out to play a role. Such an analysis has been done by Chau and Cheng,[84] and Kamal,[85] as well as by Bauer, Stech and Wirbel. An isospin decomposition of related modes yields:

$$\begin{aligned} A(D^0 \to K^-\pi^+) &= \frac{1}{\sqrt{3}}(\sqrt{2}A_{1/2} + A_{3/2}) \\ A(D^0 \to \bar{K}^0\pi^0) &= \frac{1}{\sqrt{3}}(-A_{1/2} + \sqrt{2}A_{3/2}) \\ A(D^+ \to \bar{K}^0\pi^+) &= \sqrt{3}A_{3/2} \quad , \end{aligned}$$

where the isospin $\frac{1}{2}$ and $\frac{3}{2}$ amplitudes are, in general, complex:

$$A_I = |A_I| \exp(i\delta_I) \quad .$$

These amplitudes obey the sum rule

$$A(D^0 \to K^-\pi^+) + \sqrt{2}A(D^0 \to \bar{K}^0\pi^0) = A(D^+ \to \bar{K}^0\pi^+) \quad .$$

Similar isospin relations hold for the $D \to K^*\pi$ and $D \to K\rho$ channels.

No solutions are possible for relatively real amplitudes. The Mark III data can, however, be fit with complex ratios of the isospin $\frac{1}{2}$ and $\frac{3}{2}$ amplitudes (see Table XXIII). It is clear that final state interactions play a role in these two-body decays. The $\bar{K}\rho$ and $\bar{K}^*\pi$ channels can themselves communicate via final state interactions, and can also mix with the (as yet unmeasured) $\bar{K}^*\rho$ channel and (for $I = \frac{1}{2}$ amplitudes) with $\bar{K}^0\phi$, $\bar{K}^0\omega$ and $\bar{K}^{*0}\omega$.

There has been much discussion in the literature of two additional effects which could, in principle, play a role in charmed meson lifetime differences. The first is Pauli interference,[92,93] which can occur only in D^+ decay. Note (see Fig. 44) that both the color-allowed ($\sim c_1$) and color-suppressed ($\sim c_2$) amplitudes produce the same final state in D^+ decay and that the final state has two identical fermions. It is, therefore, reasonable to expect interference between the two amplitudes. This interference can be seen to be destructive by solving the isospin analysis referred to above for the c_1 and c_2 coefficients. The D^0 lifetime is unaffected by this effect. The amount of interference can be accounted for by a parameter α (neglecting Cabibbo-suppressed decays):

$$\frac{\tau(D^+)}{\tau(D^0)} = \frac{2 + 2c_+^2 + c_-^2}{2 + 2(1+\alpha)c_+^2 + (1-\alpha)c_-^2} ,$$

where α can be evaluated as an overlap integral:

$$\alpha = 192\pi^2 \frac{|\psi(0)|^2}{M_D} .$$

In potential models

$$f_D^2 = \frac{12|\psi(0)|^2}{M_D} .$$

Thus,

$$\alpha = 16\pi^2 \left(\frac{f_D}{M_D}\right)^2 .$$

Maximal destructive interference can occur for $\alpha = 1$, which occurs for the reasonable value $f_D = 150$ MeV. With maximal interference, however, the ratio $\tau(D^+)/(\tau(D^0)$ does not exceed 1.5. If, however, we neglect color-suppressed terms, we have

$$\frac{\tau(D^+)}{\tau(D^0)} = \frac{2 + c_1^2 + c_2^2}{2 + c_1^2 + c_2^2 + 2\alpha c_1 c_2} ,$$

which admits larger values of the lifetime ratio.

Table XXIV compares D^0 and D^+ lifetimes and semileptonic branching ratios calculated by Buras *et al.* assuming large Pauli interference in D^+ decay with and without the inclusion of color-suppressed amplitudes. The neglect of color-suppressed amplitudes clearly improves the agreement with experiment.

The underlying reasons for the success of the phenomenological approach of Bauer, Stech and Wirbel can perhaps be found in a study of the relation of the $1/N$ expansion to the D hadronic decay matrix elements recently undertaken by Buras, Gérard and Rückl. We have already seen that neglect of color-suppressed amplitudes dramatically improves our understanding of the Mark III D branching ratio measurements. How may we justify this procedure? The work of Buras *et al.* provides a motivation. They first classify the amplitudes into three classes, then further investigate the order of the various contributions in a $1/N$ expansion. Buras *et al.* also define a diagrammatic language which facilitates such a classification. I will reproduce only sufficient detail here to make the basic points; the reader is referred to the original paper for more detail.

Consider the leading $(O(\sqrt{N}))$ contributions to the decays $D \rightarrow K\pi$. These are of three types:

$$\text{I:} \quad \Gamma(D^0 \rightarrow K^-\pi^+) \sim |a_1|^2$$

$$\text{II:} \quad \Gamma(D^0 \rightarrow \bar{K}^0\pi^0) \sim |a_2|^2$$

$$\text{III:} \quad \Gamma(D^+ \rightarrow \bar{K}^0\pi^+) \sim |a_1 + xa_2|^2 .$$

The factor x is process dependent; in the SU(3) limit, $x = 1$.

In addition to the leading terms, there are others. The effective neutral current or color-suppressed terms, produced by the Fierz transformation and color algebra, generate an amplitude proportional to $a_2/\sqrt{N}$ from the term $\sim a_1\sqrt{N}$. Annihilation amplitudes are also proportional to $a_1/\sqrt{N}$, as are final state interactions. Soft gluon interactions are responsible for non-planar diagrams $\sim a_1/N^{3/2}$. The terms

TABLE XXIV

Comparison of Inclusive D^0 and D^+ Measurements with Estimates Including Pauli Interference, with and without Inclusion of Color-Suppressed Amplitudes

Measurement	Color-suppressed terms neglected	Color-suppressed terms included	Experiment
$\tau(D^0)$ ($\times 10^{-13}$ sec)	5.5	7.0	4.27 ± 0.10
$B(D^0 \to e^+X)$ (%)	11.5	14.4	7.5 ± 1.2
$\tau(D^+)$ ($\times 10^{-13}$ sec)	9.5	9.8	$10.45 \pm^{0.31}_{0.29}$
$B(D^+ \to e^+X)$ (%)	19.7	20.4	17.0 ± 2.0
$\tau(D^+)/\tau(D^0)$	1.7	1.4	2.45 ± 0.09

proportional to $1/\sqrt{N}$ can be further divided into two classes. There is a factorizable part, which is retained in the standard formulation, and a non-factorizable part due to soft gluon exchange, which is dropped. The Buras *et al.* ansatz can thus be summarized as follows. The amplitude for the channel $D \to K\pi$ can be written as an expansion in $1/N$:

$$A(D \to K\pi) = \sqrt{N}\left[b_0 + \frac{1}{N}(b_1 + b'_1) + O\left(\frac{1}{N^2}\right) + \cdots\right].$$

The b_0 and b_1 terms are factorizable; both are retained in the standard approach. The b'_1 and higher order terms are non-factorizable, as they are due to multiple soft gluon exchange and final state interactions, and are usually not retained. Thus, the rate for $D^0 \to K^-\pi^+$ in the standard approach is:

$$\Gamma(D^0 \to K^-\pi^+) \sim |a_1 + \frac{1}{3}a_2|^2 .$$

Buras *et al.* point out that since the color-suppressed amplitude, b_1, is higher order in $1/N$, it, too, should be dropped. Thus,

$$\Gamma(D^0 \to K^-\pi^+) \sim |a_1|^2$$

to leading order in $1/N$. Thus, the phenomenological fit of Bauer *et al.*, which finds $\xi = 0$, is, in fact, just the neglect of factorizable terms which are higher order in $1/N$.

With this observation, one can just take over the fit to Mark III data of Bauer *et al.*, obtaining, thereby, both its successes and failures. Since annihilation diagrams and final state interactions are both higher order in $1/N$, this approach cannot cope with either the large $D^0 \to \bar{K}^0\phi$ rate nor with differences in the Cabibbo-suppressed regime such as $B(D^0 \to K^-K^+) \neq B(D^0 \to \pi^-\pi^+)$. The contribution of the annihilation process is, in fact, accounted for by fitting to the $\bar{K}^0\phi$ rate. Nonetheless, the $1/N$ expansion does serve to clarify major areas in the phenomenology of hadronic D decays.

Non-Spectator Decays

The lifetime differences among D^0, D^+ and D_s^+ mesons are due to a combination of effects in the hadronic decay sector. Final state interference effects are almost certainly significant in D^+ decay, while non-spectator diagrams could be important in D^0 and D_s^+ decay.

A search for the process $D^0 \to \bar{K}^0\phi$ was originally proposed by Bigi and Fukugita[88] as a "smoking gun" test of the existence of exclusive non-spectator processes (see Fig. 46). They predicted a rate of $.2-1\%$, for the non-spectator decay. Spectator processes can produce this mode only through OZI violation, leading to a much smaller predicted rate. Another non-spectator process, e.g., $D_s^+ \to \bar{K}^{*0}K^+$ can also arise from color-suppressed spectator processes (Fig. 47), so this mode does not rate smoking gun status. The $\bar{K}^0\phi$ mode was first found by ARGUS,[89] and has now also been seen by Mark III[90] and CLEO.[91] The branching ratios are shown in Table IV, normalized to the Mark III value for $B(D^0 \to \bar{K}^0\pi^+\pi^-)$. The rate is indeed large. There has been some debate over the extraction of the $\bar{K}^0\phi$ signal from background, but the three measurements are in good agreement.

Does the observation of $D^0 \to \bar{K}^0\phi$, then, provide the "smoking gun" for exclusive non-spectator processes? This has engendered a lively debate, which is treated in some detail in Bigi's lectures. The problem is that, considered purely as a weak decay process, it is difficult to understand how the effective value of f_D can be sufficiently large to produce a 1% branching ratio. Donoghue[92] has pointed out that rescattering effects required by unitarity can generate $D^0 \to \bar{K}^0\phi$ from spectator amplitudes via diagrams shown in Figure 48. Hussain and Kamal[93] contend, however, that the inelasticity required to generate $\bar{K}^0\phi$ at the 1% level would also produce a large rate for $D^0 \to \bar{K}^0\eta$, which is not observed. Chau[94] has emphasized the equivalence of the rescattering diagram and the annihilation diagram in her quark diagram classification scheme. Shifman[95] has pointed out that non-perturbative effects blur the distinction between the two types of language employed in the discussion. In the $1/N$ approach, all such processes are of higher order, and thus do not naturally appear to be large.

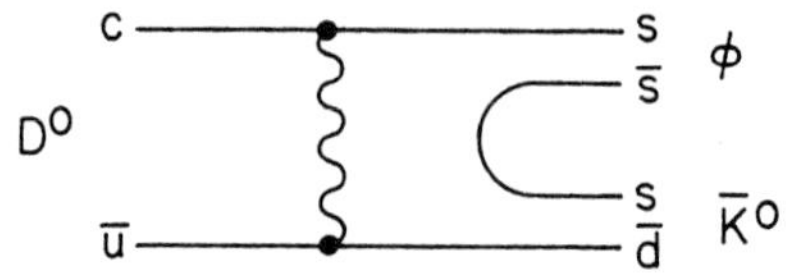

Fig. 46 Exchange amplitude for $D^0 \to \bar{K}^0\phi$ decay, the "smoking gun" test, as this decay cannot arise from a simple spectator diagram.

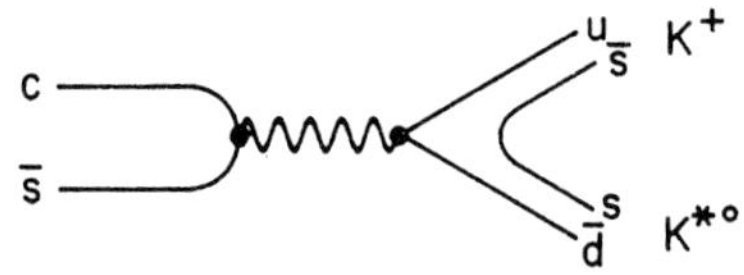

Fig. 47 Annihilation amplitude for $D_s^+ \to K^{*0}K^+$ decay. This decay can also arise from a color-suppressed spectator amplitude.

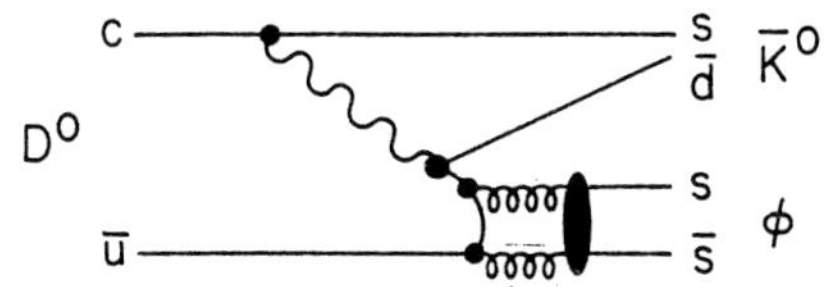

Fig. 48 Rescattering amplitude for $D^0 \to \bar{K}^0\phi$ decay which casts doubt on the simple interpretation of this mode as clear evidence for the existence of non-spectator processes.

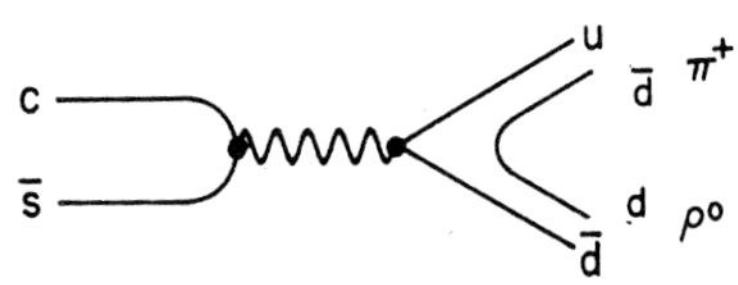

Fig. 49 Annihilation amplitude for $D_s^+ \to \rho\pi^+$ decay.

We may turn to D_s^+ hadronic decays for possible enlightenment. D_s^+ decays to all pion final states also occur via an exchange diagram (Fig. 49). As rescattering effects are unlikely to be similar to those in D^0 decay, the observation of a large rate for $D_s^+ \to \pi^+\pi^-\pi^+$ or $\rho\pi^+$ would provide evidence for weak exchange processes. ARGUS and E691 have recently placed stringent limits (see Table XX) on $D_s^+ \to \rho\pi^+$, and E691 has found the three pion mode at a small rate. These results tend to argue against prominent weak annihilation, although Chau has pointed out the presence of a possible cancellation in this mode within the quark diagram picture.

Cabibbo-Suppressed Decays

Hadronic D meson decays involving a single Cabibbo-suppressed vertex should naively occur at a rate of $tan^2\theta_c \sim 5.5\%$ compared to related allowed decays. Inspection of Tables VIII and IX, however, show that while this may be approximately the case in the large, their are several anomalies in detail.

In particular, the rates for $D^0 \to K^-K^+$ and $D^0 \to \pi^-\pi^+$ differ by a factor of 3.7. Even allowing for phase space differences, this result is surprising. It is not clear at this time whether this difference arises from final state interactions or from the contribution of Penguin diagrams[96] (see Figure 50). The very large ratio $B(D^+ \to \bar{K}^0K^+)/B(D^+ \to \bar{K}^0\pi^+)$ is likely to arise from the effect of destructive Pauli interference, which is operative in the case of $D^+ \to \bar{K}^0\pi^+$ but not for the Cabibbo-suppressed channel.

Further progress in understanding Cabibbo-suppressed channels likely awaits a measurement of $B(D^+ \to \pi^+\pi^0)$. This exotic final state should not be subject to final state effects and thus is fairly unambiguously predicted to have the value $\frac{1}{2}tan^2\theta_c \times B(D^+ \to \bar{K}^0\pi^+)$.

The $\Lambda_c^+(cdu)$ can decay weakly via a Cabibbo-allowed spectator process (Fig. 51a), or a W exchange process (Fig. 51b). The latter is not helicity-suppressed as in the charmed meson case, and, in fact, exchange diagrams are expected to dominate Λ_c^+ decay. In the naïve quark model, the (ud) quarks are in a $J = 0$, $I = 0$ state. Thus,

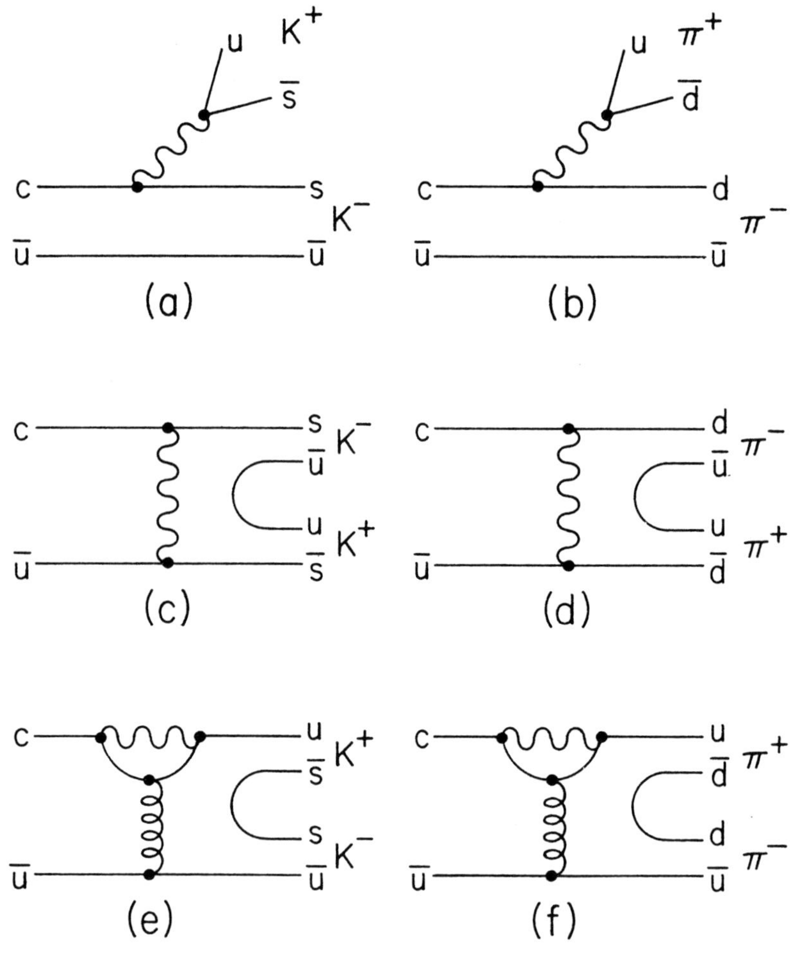

Fig. 50 Amplitudes contributing to the Cabibbo-suppressed decays, $D^0 \to K^- K^+$ and $D^0 \to \pi^+ \pi^-$. Diagrams a) and b) are the ordinary spectator amplitudes; c) and d) are exchange amplitudes; e) and f) are Penguin diagrams, which explicitly involve gluon exchange.

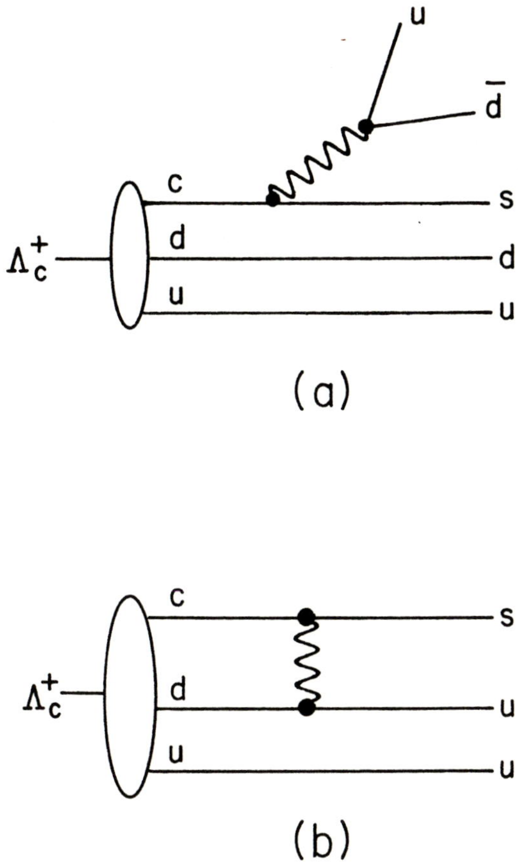

Fig. 51 Typical amplitudes for the weak decay of the Λ_c^+: a) Spectator diagram; b) exchange diagram, which in the case of baryons is not helicity-suppressed.

the (cd) quarks are in $J = 1$ or $J = 0$ configurations with statistical weights $3 : 1$. The decay of the spin 1 configuration to two fermions in an s state is helicity-suppressed, but such a decay of the spin 0 system is allowed. As the spectator diagram results in two u quarks in the final state, Pauli interference effects will also be important. Since the baryon wavefunction is antisymmetric in color, only matrix elements of O_- are non-zero. The most significant distinction between charmed baryon and meson decay is thus that non-spectator effects enter at the valence quark level in baryon decay,[18] and are therefore relatively more important. There have been a number of detailed calculations of the non-leptonic decay branching ratios of charmed baryons.[97–101]

NEUTRAL MESON STATE MIXING

The phenomenon of mixing observed three decades ago in the $K^0 - \bar{K}^0$ system can also occur in meson systems comprised of heavy quarks. This mixing is generated by second-order weak processes: to the extent that such diagrams are calculable, they provide another possible window into the regime of higher masses.

States containing heavy quarks in flavor isosinglet combinations, i.e., $(K^0, \bar{K}^0)$, $(D^0, \bar{D})$, $(B_d^0, \bar{B}_d^0)$, $(B_s^0, \bar{B}_s^0)$ or generically $(M^0, \bar{M}^0)$, which are eigenstates of the (flavor-conserving) strong and electromagnetic interactions, will be mixed by second order weak interactions. M^0 and $\bar{M}^0$ have opposite CP; we will use the convention $|\bar{M}^0\rangle = CP|\bar{M}^0\rangle$. The behavior of these states is easily described in the Wigner-Weisskopf formalism. The effective Hamiltonian H is a 2×2 matrix, with a hermitean part, and an anti-hermitean part which represents decay out of the $(M^0, \bar{M}^0)$ system:

$$H = M - \frac{i}{2}\Gamma .$$

CPT invariance implies $M_{11} = M_{12}$; $M_{21} = M_{12}^*$; $\Gamma_{11} = \Gamma_{22}$ and $\Gamma_{21} = \Gamma_{12}^*$. The physical eigenstates are obtained by diagonalizing H:

$$|M_1\rangle = \frac{1}{\sqrt{2(1+|\epsilon|^2)}}[(1+\epsilon)|M^0\rangle + (1-\epsilon)|\bar{M}^0\rangle]$$

$$|M_2\rangle = \frac{1}{\sqrt{2(1+|\epsilon|^2)}}[(1+\epsilon)|M^0\rangle - (1-\epsilon)|\bar{M^0}\rangle],$$

where

$$\left(\frac{1-\epsilon}{1+\epsilon}\right)^2 = \frac{M_{12}^* - \frac{i}{2}\Gamma_{12}^*}{M_{12} - \frac{i}{2}\Gamma_{12}}.$$

The masses m_i and widths γ_i of the physical states M_i are given by

$$m_i - \frac{i}{2}\gamma_i = M_{11} - \frac{i}{2}\Gamma_{11} \pm \left[\left(M_{12}^* - \frac{i}{2}\Gamma_{12}^*\right)\left(M_{12} - \frac{i}{2}\Gamma_{12}\right)\right]^{\frac{1}{2}}.$$

If CP is conserved ($\epsilon = 0$), then M_1 and M_2 are simultaneous eigenstates of CP, M and Γ, and the expressions simplify to

$$m_i = M_{11} \pm M_{12}$$

$$\gamma_i = \Gamma_{11} \pm \Gamma_{12}.$$

If CP is not conserved, then M_1 and M_2 are no longer orthogonal, by an amount

$$\langle M_1|M_2\rangle = \frac{2Re\epsilon}{1+|\epsilon|^2}.$$

As the observation of CP violating effects in the neutral D and B systems is not likely in the near future, let us assume that CP is conserved in these systems.

If we prepare a state $|M^0\rangle$ at $t = 0$ by strong or electromagnetic production, it will evolve as:

$$\begin{aligned}|M^0(t)\rangle = &\frac{1}{2}\left(e^{-im_1t-\frac{\gamma_1}{2}t} + e^{-im_2t-\frac{\gamma_2}{2}t}\right)|M^0\rangle \\ &+ \frac{1}{2}\left(e^{-im_1t-\frac{\gamma_1}{2}t} - e^{-im_2t-\frac{\gamma_2}{2}t}\right)|\bar{M^0}\rangle.\end{aligned}$$

Thus, in general, we expect $M^0, \bar{M^0}$ mixing to occur. The amount of mixing is seen to be governed by

$$\delta m \equiv m_2 - m_1$$

and

$$\Gamma_{\pm} \equiv \frac{\gamma_1 \pm \gamma_2}{2} .$$

In the K^0 system, $\gamma_2 \gg \gamma_1$, due to the large phase space difference for the dominant processes $K_1^0 \rightarrow 2\pi$ and $K_2^0 \rightarrow 3\pi$. In heavier systems, such differences do not amount to very much, so that $\gamma_1 \simeq \gamma_2$. Thus $\Gamma_- \simeq 0$ and there will be a single lifetime $\Gamma \cong \Gamma_+$. In general, the amount of mixing is governed by

$$\frac{\delta m}{\Gamma} \equiv x ,$$

and

$$\frac{\Delta\Gamma}{\Gamma} \equiv y (\simeq 0 \,\text{for}\, D^0, \bar{D} \,\text{and}\, B^0 \bar{B^0}) .$$

We can usefully define a mixing parameter, r, for a state prepared as $|M^0\rangle$ at $t = 0$, by:

$$r = \frac{\int_0^\infty I(\bar{M^0})dt}{\int_0^\infty (M^0)dt} = \frac{x^2 + y^2}{2 + x^2 - y^2} .$$

Evidently, $r = 0$ signifies no mixing, while $r = 1$ implies complete mixing.

In the $K^0, \bar{K^0}$ system, the fast component rapidly decays away, leaving only the slow component, the familiar $K^0, \bar{K^0}$ mixture. In heavier neutral systems, both components have similar lifetimes, so that significant mixing can occur through the $M^0, \bar{M^0}$ transition. We can get a good estimate of Γ and $\Delta\Gamma$ through our familiar spectator picture. In order to estimate δm, we must estimate the various possible contributions:

1) Charged current box diagrams of the type shown in Figure 52 may be calculated in the short distance limit. In the case of $D^0, \bar{D}$ mixing, each diagram is at least doubly Cabibbo-suppressed. The dominant contribution to Δm is:

$$\Delta m(D^0 \bar{D}) = \frac{G_F^2 B_D f_D^2 m_D}{4\pi^2}(m_s^2 - m_d^2) V_{cs}^2 V_{ud}^{*2} \quad .$$

The factor B_D is the so-called "Bag Constant", which enters through the definition

$$\langle D^0 | [(c\bar{s})_L]^2 | \bar{D} \rangle = \frac{4}{3} B_D f_D^2 m_D \quad .$$

In the vacuum insertion approximation, $B = 1$. There have been numerous estimates of B, particularly in view of the role analogous calculations play in estimates of CP violation in the K^0 system. Values of B obtained in the static quark model, harmonic oscillator model and MIT bag model, sum rules, chiral perturbation theory and lattice calculations range from -0.4 to 2.9, with the majority clustering around 0.4 or so. The dominant opinion seems to be that B is of order 1 multiplied by uncertain scale-dependent logarithmic corrections, which may be substantial.

With reasonable estimates for other parameters, the box diagrams lead to the estimate:

$$r_{D^0} \sim 10^{-7} \quad .$$

2) Long distance contributions may, in the case of the K^0 and D^0 systems, be as large as or larger than both diagrams, although they are almost certainly unimportant in the B^0 sector. Such diagrams, shown schematically in Figure 53, lead to terms in the case of D^0's, of the form

$$\Sigma_I \frac{\langle D^0 | H_w | I \rangle \langle I | H_w | \bar{D} \rangle}{m_D^2 - m_I^2 + i\epsilon} \, .$$

In the SU(3) limit, the GIM cancellation works as well as it does in the short distance approximation, but SU(3) breaking lifts this cancellation in the dispersive sector, leading to estimates of effects for D^0's

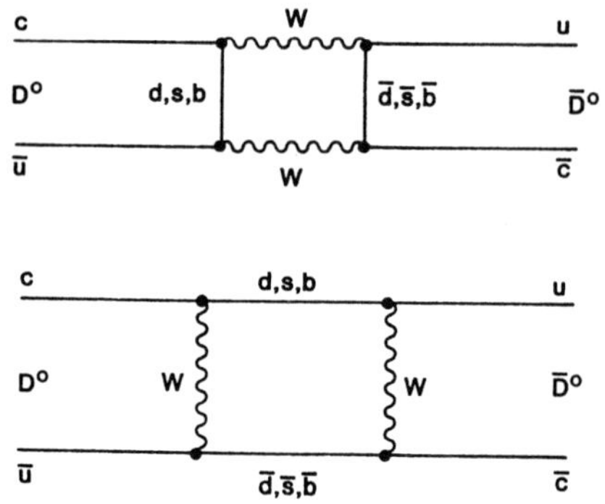

Fig. 52 Second order weak diagrams contributing to $D^0\bar{D}^0$ mixing

Fig. 53 Long distance contributions to $D^0 - \bar{D}^0$ mixing.

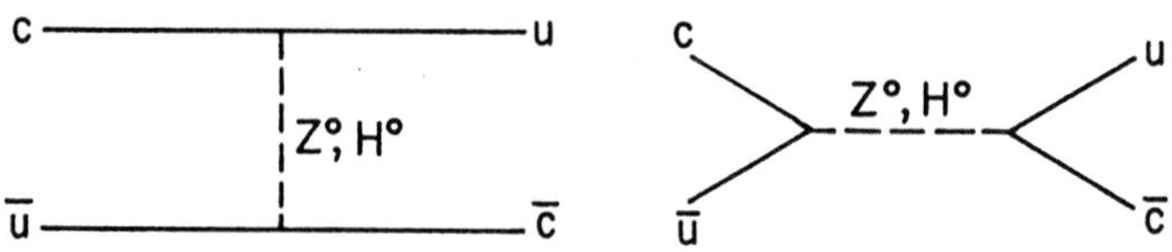

Fig. 54 Flavor-changing neutral current contributions to $D^0 - \bar{D}^0$ mixing.

which are one or two orders of magnitude larger than the box diagrams. Donoghue *et al.*[102] have parametrized the contribution to the D^0 sector as:

$$\begin{aligned}\Delta m_D^{\rm disp} &\simeq \frac{1}{2\pi}\ln\frac{m_d^2}{\mu^2}\Big[\Gamma(D^0\to K^+K^-)+\Gamma(D^0\to\pi^+\pi^-)\\ &\qquad -2\Big(\Gamma(D^0\to K^-\pi^+)\Gamma(D^0\to K^+\pi^-)\Big)^{\frac{1}{2}}\Big]\\ &\simeq 2\times 10^{-15}\ \text{GeV}\Big[1-0.65(f_{K^+\pi^-})^{\frac{1}{2}}\Big],\end{aligned}$$

which, for $\mu = 1$ GeV and $f_{K^+\pi^-} = 1$, is more than an order of magnitude larger than the box diagram contribution. Other intermediate states, involving 3 or more pseudoscalars, surely also must contribute.

3) Mechanisms not contained in the Standard Model can also generate mixing, either through direct first order flavor-changing neutral currents (Fig. 54) or, as in the case of right-left symmetric models, through additional box diagrams which can lead back to $r_D \sim O(.1)-O(1)$ with $g_L, g_R \sim 10^{-3}$.

Thus, both through the dependence of mixing on the t quark mass and couplings, and through the sensitivity to nonstandard mechanisms for the generation of flavor-changing neutral currents, the search for evidence of mixing in neutral heavy quark systems is well motivated. Should such evidence appear, however, particularly in the $D^0\bar{D}$ system, a more complete understanding of the long distance effects will be necessary before definite conclusions can be drawn.

Experimental Signatures for Mixing

Searches for mixing in the charm sector fall into three general classes: those which use 1) a lepton tag, 2) a daughter quark tag, through identification of a kaon, and 3) those which use full reconstruction of the complete final state. The first two methods depend on the GIM mechanism to ensure a unique signature.

The **lepton tag** method makes use of the fact that the sign of a lepton produced in D quark decay tags the sign of the $W^{\pm}$ and, therefore, the quark or antiquark current to which the W couples uniquely (Fig. 55). Thus, an l^+ arises from a c while an l^- signifies a $\bar{c}$. Three *caveats* are important:

1) It is necessary to know that the parent meson was a D. This is typically ensured by taking advantage of the quite differing regions of p_T populated by leptons from K, D and B decay.

2) It is necessary to know that the source of the lepton is the primary vertex.

3) It is necessary to have high quality lepton identification in order that misidentified hadrons not overwhelm the time lepton signal.

The **daughter quark tag** method makes use of the correlation of kaon charge with D flavor:

a) The GIM mechanism provides the primary correlation of, e.g., a K^- daughter with a D^0 parent (Fig. 56a) in Cabibbo-allowed decay.

b) Cabibbo-suppressed decays can, however, produce wrong sign daughters (Fig. 56b). Thus, inclusive K's are not an unambiguous tag of the parent D producing a background at the level of $\sim \tan^2\theta_c$. This problem can be evaded by reconstruction of exclusive D decay modes (see below).

c) There is, in addition, an irreducible physics background due to double-Cabibbo-suppressed decays (Fig. 56c), which produces a signal indistinguishable from true flavor oscillation at the level of $\sim \tan^4\theta_c$ i.e., at $10^{-3} - 10^{-2}$.

Thus, while the exclusive production of $D\bar{D}$ pairs at the $\psi(3770)$ resonance in e^+e^- annihilation would seem to provide an ideal place to search for mixing through complete reconstruction of both D's, one must bear in mind that observation of, for example, two $K^-\pi^+$ pairs

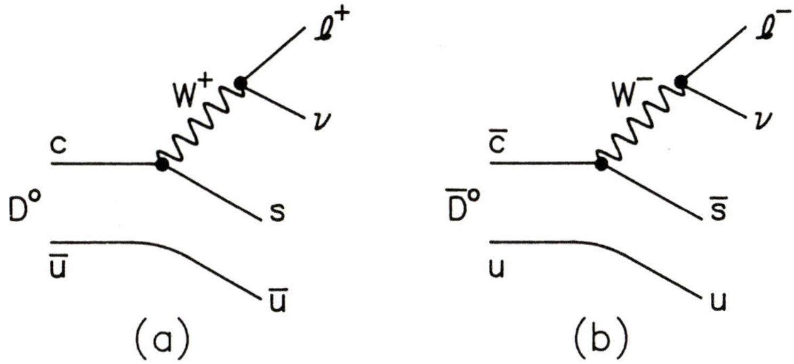

Fig. 55 Lepton tag signatures for $D^0 \bar{D}^0$ mixing.

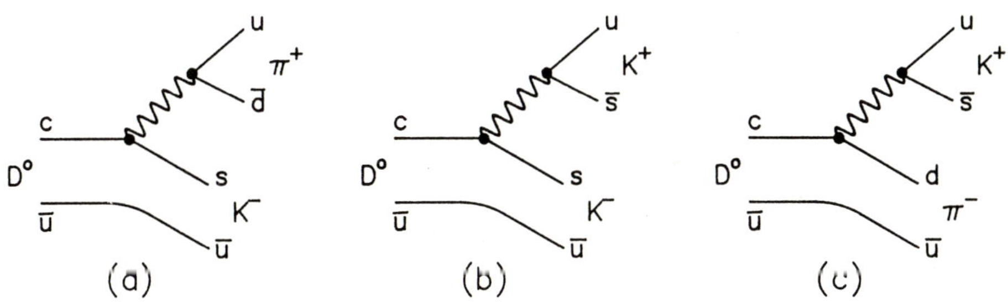

Fig. 56 Daughter quark tag signatures for $D^0 - \bar{D}^0$ mixing: a) normal Cabibbo-allowed amplitude; b) wrong-sign kaon originating from Cabibbo-suppressed amplitude. This background requires that the D^0 be reconstructed in an exclusive mode; c) wrong-sign kaon originating from doubly-Cabibbo-suppressed amplitude. This produces a signal identical to that from $\bar{D}^0$ decay, producing a background which can be isolated by study of the time dependence of the reconstructed events.

at m_{D^0} signifies **either** $D^0\bar{D}$ mixing **or** $\tan^4\theta_c$ decay.

There are two possible ways around this limitation. The first is to reconstruct exclusive final states in which, e.g., the D^0 decays hadronically while the $\bar{D}$ decays semileptonically, since the lepton in D decay unambiguously identifies the parent quark flavor. The second makes use of the detailed study of quantum statistics made by Bigi and Sanda.[103] We can distinguish two cases:

a) Both D^0 and $\bar{D}$ decay semileptonically. Here, the size of the effect depends on the relative angular momentum of the pair:

$$\left|\frac{N(l^{\pm}l^{\pm})}{N(l^{+}l^{-})}\right|_{D^0\bar{D}} = \begin{cases} \frac{x^2+y^2}{2} & \text{l odd} \\ \frac{3(x^2+y^2)}{2} & \text{l even} \end{cases} .$$

b) Both D's decay to the $K\pi$ final state, for example:

$$\left|\frac{N(K^-\pi^+,K^-\pi^+)}{N(K^-\pi^+,K^+\pi^-)}\right|_{D^0\bar{D}} = \begin{cases} \frac{x^2+y^2}{2} & \text{l odd} \\ \frac{3(x^2+y^2)}{2} + 4|\rho|^2 + 8y\rho & \text{l even} \end{cases} ,$$

where

$$|\rho|^2 = \left|\frac{A(\bar{D}\to K^-\pi^+)}{A(\bar{D}\to K^-\pi^+)}\right|^2 .$$

In the useful case of $e^+e^- \to D^0\bar{D}^0$, l must be odd. Therefore, only mixing and not $\tan^4\theta_c$ processes can produce $s=\pm 2$ final states, where the two final states are identical, e.g., $K\pi$, $K\rho$, $K^*\pi$.

Experimental Searches for $D^0\bar{D}^0$ Mixing

We will here briefly review the experimental situation in searches for $D^0-\bar{D}$mixing, as an illustration of some of the different techniques just described.

DELCO[104] has used its good particle identification capability to search its $\sqrt{s} = 29$ GeV e^+e^- sample for wrong sign $K\pi$, using tagged D^*'s. In the decay $D_s^+ \to D^0\pi^+$, the sign of the π tags the charm of the D. A subsequent $K^-\pi^+$ decay is then "normal", while a $K^+\pi^-$ signifies either mixing or doubly Cabibbo-suppressed decay. They observe 97 normal combinations and 15 of the wrong sign. The latter are consistent with combinatorial background and particle misidentification, so there is no evidence for mixing. A maximum likelihood fit yields $r_D < 8.3 \times 10^{-2}$ at 90% confidence level.

E615 at FNAL[58] has produced a limit for mixing by studying the reaction $\pi N \to \mu\mu X$ at 225 GeV. In this case, both muons would originate from semileptonic D decays. They observe 3973 same sign events with $m_{\mu\mu} > 2$ GeV. Study of the $|\cos\theta|$ distribution of the same sign pairs indicates that all events are consistent with uncorrelated pion interactions; less than 63 events are consistent with $D^0\bar{D}^0$ mixing. Assuming $\sigma(D^0\bar{D}^0) = (3.8 \pm 0.5)\mu b$/nucleon, this leads to a limit $r_D < 5.6 \times 10^{-3}$ at 90% confidence level.

Better limits have recently been set by ARGUS[105] ($r_D < 0.023$) and HRS[106] ($r_D < 0.040$), using the sign of the pion in, e.g., $D^{*+} \to \pi^+ D^0$ events to tag the neutral D species.

Mark III has provided the only example[107] of an attempt to fully reconstruct $D^0\bar{D}^0$ final states. Using the 9 pb^{-1} sample of $D\bar{D}$ events produced at the $\psi(3770)$, 162 $D^0\bar{D}^0$ events having total strangeness 0 have been fully reconstructed with very stringent particle identification requirements. These are shown in Figure 57, which also shows three events which have total strangeness ± 2. Events in which both a K and π are misidentified should yield a background of 0.4 ± 0.1 events. Less than one event is expected from $\tan^4\theta_c$ processes. *If* the events in excess of these background estimates are attributed to mixing, then $r_D \simeq 10^{-2}$, but the statistics are too meager to reach a definite conclusion.

E691[108] has employed its excellent vertex spatial resolution to study the time dependence of a sample of $D^0 \to K^-\pi^+$ decays tagged using pions from D^*'s. Figures 58 and 59 show these results. Correcting for

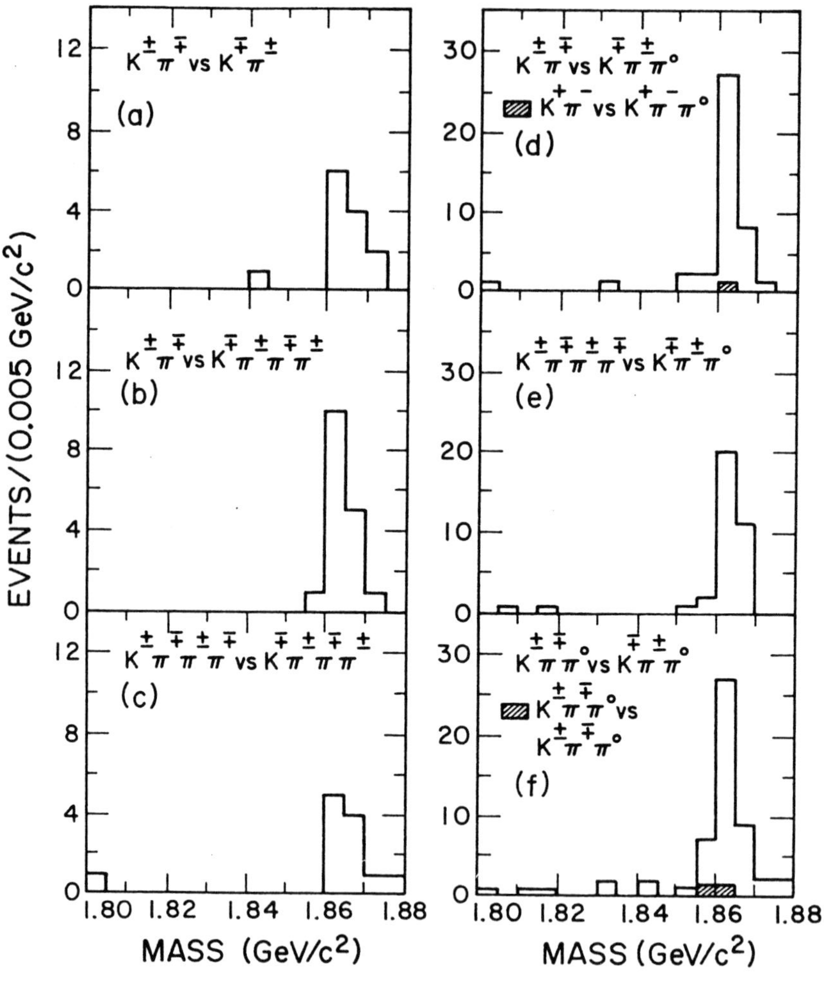

Fig. 57 Mark III double-tag exclusively reconstructed D^0 hadronic events at the $\psi(3770)$. The shaded regions show a single $K^+\pi^-$ *vs* $K^+\pi^-\pi^0$ event and two $K^\pm\pi^\mp\pi^0$ *vs* $K^\pm\pi^\mp\pi^0$ events which could be evidence either for $D^0\bar{D}^0$ mixing or large doubly-Cabibbo-suppressed decays.

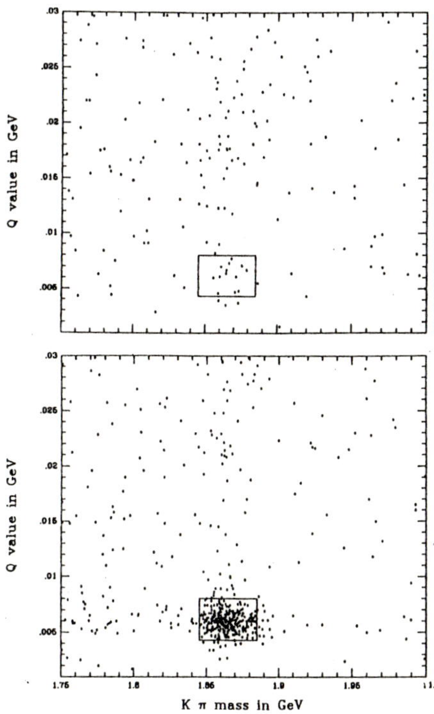

Fig. 58 Scatter plots of $K\pi$ mass *vs* $Q[= M_{K\pi\pi} - M_{K\pi} - M_{\pi}]$ for E691 $K\pi\pi$ events with $t >$ 0.22 psec. The upper plot shows $K^-\pi^+\pi^-$ events; the lower plot shows $K^-\pi^+\pi^+$ events. Charge conjugate modes are also shown. The box contains approximately 90% of $D^* \to \pi D$, $D \to K\pi$ events.

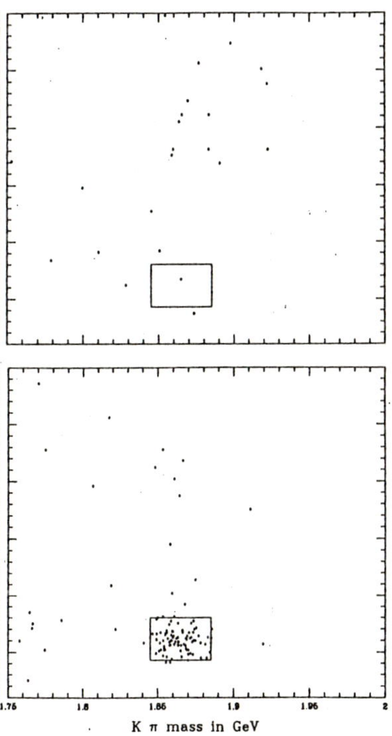

Fig. 59 Scatter plots of $K\pi$ mass *vs* Q for E691 $K\pi\pi$ events with $t >$ 0.88 psec.

events missed at short decay times, there are a total of 597 ± 38 right sign D^* events. For the sample cut at $t > 0.22$ psec, there are 9.6 ± 4.9 events of the wrong sign. For the sample cut at $t > 0.88$ psec, there is one event. *If interference effects are neglected*, then mixing events would have a time distribution $\sim t^2 exp(-\Gamma t)$ while $tan^4\theta_c$ events would follow the usual $exp(-\Gamma t)$ distribution. Fitting to all events, E691 is able to set a limit of $r_D < 5 \times 10^{-3}$ at 90% confidence level. On the assumption of no interference, the Mark III result would have implied ~ 7 events. Analysis of this sample allowing for the effect of interference is clearly warranted. In particular, the sign of the ρ parameter, which is unknown, can dramatically alter this conclusion.

REFERENCES

1. H. J. Schnitzer, *Phys. Lett.* **134B**, 253 (1984); M. Frank and P. J. O'Donnell, *Phys. Lett.* **157B**, 174 (1985).

2. M. Aguilar-Benitez *et al.* (Particle Data Group) *Phys. Lett.* **170B**, 1 (1986).

3. G. T. Blaylock *et al.* (Mark III Collaboration) *Phys. Rev. Lett.* **58**, 2171 (1987).

4. J. McKenna *et al.* (ARGUS Collaboration) Contribution to the 1987 International Symposium on Lepton and Photon Interactions at High Energies, Hamburg (1987).

5. A. DeRujula, H. Georgi and G. Glashow, *Phys. Rev.* **D12**, 147 (1975); D. B. Lichtenberg, *Nuovo Cim.* **28A**, 563 (1985); J. M. Richard, *Ann. Phys.* **150**, 267 (1983); E. Eich *et al., Z. Phys.* **C28**, 225 (1985); D. Izatt, C. Detar and M. Stephenson, *Nucl. Phys.* **B199**, 269 (1982).

6. References for Table I – Measurements of the D_s^+ Mass:

   CLEO : A. Chen *et al., Phys. Rev. Lett.* **51**, 634 (1983)

   TASSO : M. Althoff *et al., Phys. Lett.* **136B**, 130 (1984)

   ARGUS : H. Albrecht *et al., Phys. Lett.* **179B**, 1398 (1986)

   ARGUS : H. Albrecht *et al., Phys. Lett.* **153B**, 343 (1985)

   NA–11/32: H. Becker *et al., Phys. Lett.* **184B**, 277 (1987)

   NA–32ccd: S. Barlag *et al.*, Contribution to the 1987 International Symposium on Lepton and Photon Interactions at High Energies, Hamburg (1987)

   TPC : H. Aihara *et al., Phys. Rev. Lett.* **53**, 2465 (1984)

   HRS : M. Derrick *et al., Phys. Rev. Lett.* **54**, 2568 (1985)

   Mark III : G. T. Blaylock *et al., Phys. Rev. Lett.* **58**, 2171 (1987)

   E691 : J. C. Anjos *et al.*, Contribution to the 1987 International

Symposium on Lepton and Photon Interactions at High Energies, Hamburg (1987).

7. References for Table II – Measurements of the Λ_c^+ Mass:

Cazzoli : E. G. Cazzoli *et al., Phys. Rev. Lett.* **34**, 1125 (1975)

Knapp : B. Knapp *et al., Phys. Rev. Lett.* **37**, 882 (1976)

Baltay : C. Baltay *et al., Phys. Rev. Lett.* **42**, 1721 (1979)

Cnops : A. M. Cnops *et al., Phys. Rev. Lett.* **42**, 197 (1979)

Giboni : K. L. Giboni *et al., Phys. Rev. Lett.* **85B**, 437 (1979); A. Kerman, *Proceedings of the 1979 International Symposium on Lepton and Photon Interactions at High Energies*, Batavia (1980)

Mark II : G. S. Abrams *et al., Phys. Rev. Lett.* **44**, 10 (1980)

Allasia : D. Allasia *et al., Nucl. Phys.* **B176**, 13 (1980)

Calicchio : M. Calicchio *et al., Phys. Lett.* **93B**, 521 (1980)

Kitagaki I : T. Kitagaki *et al., Phys. Rev. Lett.* **45**, 955 (1980)

Russell : J. J. Russell *et al., Phys. Rev. Lett.* **46**, 799 (1981)

Bosetti : P. C. Bosetti *et al., Phys. Lett.* **109B**, 234 (1981)

Kitagaki II : T. Kitagaki *et al., Phys. Rev. Lett.* **48**, 299 (1982)

WA–58: M. Adamovich *et al.*, CERN-EP/86-76 (1986), submitted to *Yad. Fiz.*

Aleev : A. N. Aleev *et al., Z. Phys.* **C23**, 233 (1984)

CLEO : T. Bowcock *et al., Phys. Rev. Lett.* **55**, 923 (1985)

ACCMOR : S. Barlag *et al.*, Contribution to the 1987 International Symposium on Lepton and Photon Interactions at High Energies, Hamburg (1987)

LEBC–EHS: M. Aguilar-Benitez *et al., Phys. Lett.* **189B**, 254 (1987)

ARGUS : H. Albrecht *et al.*, Contribution to the 1987 International Symposium on Lepton and Photon Interactions at High Energies, Hamburg (1987)

E691 : M. Witherell, these Proceedings

R–608: P. Chauvat *et al.*, Contribution to the 1987 International Symposium on Lepton and Photon Interactions at High Energies, Hamburg (1987).

8. M. Kobayashi and T. Maskawa, *Prog. Theor. Phys.* **49**, 652 (1973).

9. L. Maiani, in *Proceedings of the 1977 International Symposium on Lepton and Photon Interactions at High Energies*, Hamburg, 1977, p. 867; L.-L. Chau, *Phys. Rep.* **95**, 1 (1983).

10. S. K. Bose and E. A. Paschos, *Nucl. Phys.* **B169**, 384 (1980); V. Barger, K. Whimsat and R. Phillips, *Phys. Rev.* **D23**, 2773 (1981); M. Gronau and J. Schechter, *Phys. Rev.* **D31**, 1668 (1985); F. J. Botella and L.-L. Chau, *Phys. Lett.* **168B**, 97 (1986).

11. K. Kleinknecht and B. Renk, *Z. Phys.* **C34**, 209 (1987); see also F. J. Gilman and K. Kleinknecht, in *Review of Particle Properties* (Particle Data Group), *Phys. Lett.* **170B**, 74 (1986).

12. D. M. Coffman, Ph.D. Thesis, California Institute of Technology (1986).

13. W. Schmidt-Parzefall, Contribution to the 1987 International Symposium on Lepton and Photon Interactions at High Energies, Hamburg (1987).

14. V. A. Novikov *et al., Phys. Rev. Lett.* **38**, 626 (1977)

E. Golowich *et al., Phys. Lett.* **91B**, 271 (1980)

H. Krasemann, *Phys. Lett.* **96B**, 397 (1980)

V. S. Mathur *et al., Phys. Lett.* **107B**, 127 (1981)

E. Shuryak, *Nucl. Phys.* **B198**, 83 (1982)

B. F. L. Ward, *Phys. Rev.* **D28**, 1215 (1983)

L. Reinders, *Phys. Rep.* **127**, 1 (1985)

M. Suzuki, UCB-PTH-85/32 (1985)

S. Godfrey and N. Isgur, *Phys. Rev.* **D32**, 189 (1985)

P. J. O'Donnell, CERN-TH/4419-86 (1986)

R. M. Woloshyn *et al.*, TRI-PP-87-62 (1987).

15. J. Adler *et al.* (Mark III Collaboration), SLAC-PUB-4441 (1987), submitted to *Phys. Rev. Lett.*

16. References for Table IV – Measurements of Lifetimes of Charmed Mesons:

E531 : N. Ushida *et al., Phys. Rev. Lett.* **56**, 1767 (1986); N. Ushida *et al., Phys. Rev. Lett.* **56**, 1771 (1986)

WA–58: M. Adamovich *et al.* CERN-EP/86-77 (1986), submitted to *Phys. Lett. B*; V. Castillo-Gimenez, Thesis Doctoral, Universidad de Valencia (1986)

SHF : K. Abe *et al., Phys. Rev.* **D33**, 1 (1985)

NA–16: M. Aguilar-Benitez *et al., Phys. Lett.* **122B**, 312 (1983)

NA–18: A. Badertscher *et al., Phys. Lett.* **123B**, 471 (1982)

NA–27: M. Aguilar-Benitez *et al.* (LEBC-EHS Collaboration), CERN-EP/87-38 (1987), submitted to *Phys. Lett. B*

NA–1: E. Albini *et al., Phys. Lett.* **B110**, 339 (1982); S. R. Amendolia *et al.*, CERN-EP/87-20 (1987), submitted to *Phys. Lett. B*

NA–11: R. Bailey *et al.* (ACCMOR Collaboration), *Z. Phys.* **C28**, 357 (1985); H. Becker *et al.* (ACCMOR Collaboration), *Phys. Lett.* **B184**, 277 (1987); H. Palka *et al.* (ACCMOR Collaboration), *Z. Phys.* **C35**, 151 (1987).

NA–32: H. Becker *et al.* (ACCMOR Collaboration), *op. cit.*; S. Barlag *et al.* (ACCMOR Collaboration), Contribution to the 1987 International Symposium on Lepton and Photon Interactions at High Energies, Hamburg (1987)

NA–14: M. Cattaneo, Ph.D. Thesis, RAL T048 (1987).

NA–14′: P. Roudeau, LAL 87/39, preliminary result presented at the Second Topical Conference on Heavy Flavors, San Miniato (1987)

E691 J. R. Raab *et al.* (Tagged Photon Spectrometer Collaboration), submitted to *Phys. Rev. Lett.*

DELCO : H. Yamamoto *et al., Phys. Rev.* **D22**, 2901 (1985)

Mark II : L. Gladney *et al., Phys. Rev.* **D34**, 2601 (1986)

Mark II Upgrade : S. R. Wagner *et al.*, SLAC-PUB-4304, submitted to *Phys. Rev. D*

HRS : S. Abachi *et al.*, ANL-HEP-CP-86-73;S. Abachi *et al.*, ANL-HEP-CP-86-62;S. Abachi *et al.*, ANL-HEP-CP-86-60; S. Abachi *et al., Phys. Rev. Lett.* **56**, 1775 (1986)

TASSO : W. Braunschweig *et al.*, DESY 86-159 (1986); W. Braunschweig *et al.* Contribution to the 1987 International Symposium on Lepton and Photon Interactions at High Energies, Hamburg (1987).

CLEO : S. C. Csorna *et al.*, CLNS 86/747 (1986)

ARGUS : H. Albrecht *et al.*, DESY Preprint (1987).

17. References for Table V – Measurements of Lifetimes of Charmed Baryons:

E531 : N. Ushida *et al, Phys. Rev. Lett.* **56**, 1767 (1986)

WA–58: M. Adamovich *et al.* CERN-EP/86-77, submitted to *Phys. Lett. B*; V. Castillo-Gimenez, Thesis Doctoral, Universidad de Valencia (1986)

NA–27: M. Aguilar-Benitez *et al.* (LEBC-EHS Collaboration), *Phys. Lett.* **189B**, 254 (1987).

NA–1: S. R. Amendolia *et al.*, CERN-EP/87-139, submitted to *Z. Phys. C*

NA–32: S. Barlag *et al.* (ACCMOR Collaboration) Contribution to the 1987 International Symposium on Lepton and Photon Interactions at High Energies, Hamburg (1987)

E691 : J. R. Raab *et al.* (Tagged Photon Spectrometer Collaboration), submitted to *Phys. Rev. Lett.*

E400 : J. C. Anjos *et al.*, UCSB-HEP-87-7 (1987)

WA–62: S. F. Biagi *et al., Phys. Lett.* **122B**, 455 (1983); S. F. Biagi *et al., Phys. Lett.* **150B**, 230 (1985).

18. For a detailed exposition, see R. Rückl, Habilitationsschrift, University of Munich (1983).

19. M. K. Gaillard and B. W. Lee, *Phys. Rev. Lett.* **33**, 108 (1974); G. Altarelli and L. Maiani, *Phys. Lett.* **52B**, 351 (1974).

20. F. J. Gilman and M. G. Wise, *Phys. Rev.* **D20**, 2392 (1979).

21. G. Altarelli, G. Curci, G. Martinelli and R. Petrarca, *Phys. Lett.* **99B**, 141 (1981) and *Nucl. Phys.* **B187**, 461 (1981).

22. R. M. Baltrusaitis *et al.* (Mark III Collaboration) *Phys. Rev. Lett.* **54**, 1976 (1985).

23. J. C. Anjos *et al., Phys. Rev. Lett.* **58**, 311 (1987).

24. R. H. Schindler *et al., Phys. Rev.* **D21**, 2716 (1980).

25. J. M. Feller, Ph.D. Thesis, University of California, Berkeley, LBL-9017 (1979).

26. Y. Zhu, private communication.

27. H. J. Lipkin, *Phys. Lett.* **179B**, 278 (1986).

28. A. J. Hauser, Ph.D. Thesis, California Institute of Technology (1984); R. M. Baltrusaitis *et al., Phys. Rev. Lett.* **55**, 150 (1985).

29. References for Tables VI and VII – Cabibbo-Allowed D^0 and D^+ Hadronic Branching Ratio Measurements:

LGW : I. Peruzzi *et al., Phys. Rev. Lett.* **39**, 1301 (1977); D. Scharre *et al., Phys. Rev. Lett.* **40**, 74 (1978)

Mark II : R. H. Schindler *et al., Phys. Rev.* **D24**, 78 (1981)

Mark III : A. J. Hauser, Ph.D. Thesis, California Institute of Technology (1984); R. M. Baltrusaitis *et al., Phys. Rev. Lett.* **55**, 150 (1985); R. M. Baltrusaitis *et al., Phys. Rev. Lett.* **56** 2140 (1986); R. M. Baltrusaitis *et al., Phys. Rev. Lett.* **55**, 150 (1985); J. Adler *et al., Phys. Lett.* **B196**, 107 (1987); J. Adler *et al.*, SLAC-PUB-4291 (1987), submitted to *Phys. Rev. Lett.*

LEBC–EHS: M. Aguilar-Benitez *et al.*, CERN-EP/87-136 (1987), submitted to *Z. Phys. C*

ARGUS : H. Albrecht *et al., Phys. Lett.* **158B**, 525 (1985); H. Albrecht *et al., Z. Phys.* **C35**, 359 (1987); H. Albrecht *et al., Phys. Lett.* **153B**, 343 (1985)

CLEO : C. Bebek *et al., Phys. Rev. Lett.* **56**, 1983 (1986)

Avery : P. Avery *et al., Phys. Rev. Lett.* **51**, 1139 (1983)

Bailey : R. Bailey *et al., Phys. Lett.* **132B**, 237 (1983)

ACCMOR : S. Barlag *et al., op. cit.*

30. References for Tables VIII and IX – Cabibbo-Suppressed D^0 and D^+ Hadronic Branching Ratio Measurements:

Mark II : G. Abrams *et al., Phys. Rev. Lett.* **43**, 481 (1979); R. H. Schindler *et al., Phys. Rev.* **D24**, 78 (1981)

Mark III : R. M. Baltrusaitis *et al. Phys. Rev. Lett.* **54**, 1976 (1985); A. J. Hauser, *op. cit.*

LEBC–EHS: M. Aguilar-Benitez *et al.*, CERN-EP/87-136 (1987), submitted to *Z. Phys. C*

Crystal Ball : R. A. Partridge, Ph.D. Thesis, California Institute of Technology (1984)

E691 : J. C. Anjos *et al.*, Contribution to the 1987 International Symposium on Lepton and Photon Interactions at High Energies, Hamburg (1987).

31. P. Rapidis *et al., Phys. Rev. Lett.* **39**, 526 (1977).

32. R. H. Schindler *et al., Phys. Rev.* **D21**, 2716 (1980).

33. H. Sadrozinski, in *High Energy Physics – 1980*, ed. L. Durand and L. Pondrom, AIP Conference Proceedings No. 68 (AIP, New York, 1981), p. 681.

34. G. T. Blaylock, Ph.D. Thesis, University of Illinois (1986).

35. R. M. Baltrusaitis *et al., Phys. Rev. Lett.* **56** 2140 (1986).

36. J. Adler *et al.*, SLAC-PUB-4291 (1987), submitted to *Phys. Rev. Lett.*

37. See, for example, R. H. Schindler, *Proceedings of the International Europhysics Conference on High Energy Physics*, Uppsala (1987).

38. V. Vuillemin *et al., Phys. Rev. Lett.* **41**, 1149 (1978)

39. R. H. Schindler *et al., Phys. Rev.* **D24**, 78 (1981).

40. M. Aguilar-Benitez *et al.*, CERN-EP/87-135 (1987), submitted to *Z. Phys. C.*

41. H. Albrecht *et al.* (ARGUS Collaboration), *Phys. Lett.* **150B**, 235 (1985).

42. D. Bortoletto *et al., Phys. Rev.* **D35**, 19 (1987).

43. M. Derrick *et al., Phys. Rev. Lett.* **53**, 1971 (1984).

44. R. Ammar, *Proceedings of the International Europhysics Conference on High Energy Physics*, Uppsala (1987).

45. T. Bowcock, *Proceedings of the International Europhysics Conference on High Energy Physics*, Uppsala (1987).

46. J. Adler *et al., Phys. Lett.* **B196**, 107 (1987).

47. R. M. Baltrusaitis *et al.* (Mark III Collaboration) *Phys. Rev. Lett.* **54**, 1976 (1985).

48. R. H. Schindler *et al., Phys. Rev.* **D24**, 78 (1981).

49. W. Bacino *et al., Phys. Rev. Lett.* **45**, 329 (1980).

50. J. M. Feller, *op. cit.*

51. J. C. Anjos *et al.*, Contribution to the 1987 International Symposium on Lepton and Photon Interactions at High Energies, Hamburg (1987).

52. B. Grinstein, M. B. Wise and N. Isgur, *Phys. Rev. Lett.* **56**, 298 (1986).

53. M. Bander, D. Silverman and A. Soni, *Phys. Rev. Lett.* **44**, 7 (1980).

54. P. Haas *et al.*, Contribution to the 1987 International Symposium on Lepton and Photon Interactions at High Energies, Hamburg (1987).

55. K. Shubert, T. Bowcock, *Proceedings of the International Europhysics Conference on High Energy Physics*, Uppsala (1987).

56. J. C. Anjos *et al.*, Contribution to the 1987 International Symposium on Lepton and Photon Interactions at High Energies, Hamburg (1987).

57. J. J. Aubert *et al., Phys. Lett.* **155B**, 461 (1985).

58. W. C. Louis *et al., Phys. Rev. Lett.* **56**, 1027 (1986).

59. K. Riles *et al., Phys. Rev.* **D35**, 2914 (1987).

60. H. Palka *et al.* (ACCMOR Collaboration) *Phys. Lett.* **189B**, 238 (1987).

61. J. Becker *et al.* (Mark III Collaboration) *Phys. Lett.* **193B**, 147 (1987).

62. W. Buchmüller and D. Wyler, *Phys. Lett.* **177B**, 377 (1986).

63. A. Clark *et al., Phys. Rev. Lett.* **26**, 1667 (1971).

64. R. Brandelik *et al., Phys. Lett.* **80B**, 412 (1979).

65. R. Ammar *et al., Phys. Lett.* **94B**, 118 (1980).

66. N. Ushida *et al., Phys. Rev. Lett.* **45**, 1053 (1980).

67. D. Aston *et al., Phys. Lett.* **100B**, 91 (1981).

68. References for Table XX – D_s^+ Hadronic Decay Branching Ratio Measurements:

CLEO : C. Chen *et al., Phys. Rev. Lett.* **51**, 634 (1983)

TASSO : W. Braunschweig *et al.*, DESY 86-159 (1986)

ARGUS : H. Albrecht *et al., Phys. Lett.* **153B**, 343 (1985); H. Albrecht *et al., Phys. Lett.* **179B**, 398 (1986); H. Albrecht *et al.*, DESY 87-045 (1987)

HRS : M. Derrick *et al., Phys. Rev. Lett.* **54**, 2568 (1985)

Mark III : G. T. Blaylock *et al., Phys. Rev. Lett.* **58**, 2171 (1987)

E691 : J. C. Anjos *et al.*, Contribution to the 1987 International Symposium on Lepton and Photon Interactions at High Energies, Hamburg (1987)

TPC : H. Aihara *et al., Phys. Rev. Lett.* **53**, 2465 (1984)

ACCMOR : R. Bailey *et al., Phys. Lett.* **139B**, 320 (1984).

69. References for Table XXI – Λ_c^+ Hadronic Branching Ratio Measurements:

Mark II : G. S. Abrams *et al., Phys. Rev. Lett.* **44**, 10 (1980); J. M. Weiss, *Proceedings of Baryon 80*, Toronto (1980); E. Vella, Ph.D. Thesis, University of California, Berkeley, LBL-13845 (1981). Note that this thesis contains revised values of $\sigma(\Lambda + \bar{\Lambda})$, which have been used to derive corrected values of the Mark II Λ_c^+ branching ratios.

ARGUS : H. Albrecht *et al.*, Contribution to the 1987 International Symposium on Lepton and Photon Interactions at High Energies, Hamburg (1987)

CLEO : T. Bowcock *et al., Phys. Rev. Lett.* **55**, 923 (1985)

Basile : M. Basile *et al., Nuovo Cim.* **62A**, 14 (1981)

NA–32: S. Barlag *et al., Phys. Lett.* **184B**, 283 (1987)

BIS–2: A. N. Aleev *et al., Z. Phys.* **C23**, 233 (1984)

Baltay : C. Baltay *et al., Phys. Rev. Lett.* **42**, 1721 (1979)

Kitagaki : T. Kitagaki *et al., Phys. Rev. Lett.* **45**, 955 (1980).

70. G. S. Abrams *et al., Phys. Rev. Lett.* **44**, 10 (1980); E. Vella *op. cit.*

71. M. Piccolo *et al., Phys. Rev. Lett.* **39**, 1503 (1977).

72. S. F. Biagi *et al., Phys. Lett.* **122B**, 455 (1983).

73. P. Coteus *et al.*, COLO-HEP-140 (1986).

74. S. F. Biagi *et al., Z. Phys.* **C28**, 175 (1985).

75. E. G. Cazzoli *et al., Phys. Rev. Lett.* **34**, 1125 (1975).

76. C. Baltay *et al., Phys. Rev. Lett.* **42**, 1721 (1979).

77. H. Grassler *et al., Phys. Lett.* **99B**,159 (1981).

78. H. Seywerd (ARGUS Collaboration), *Proceedings of the 1987 Rencontres de Moriond*, ed. J. Tram Thanh Van, Editions Frontières, Gif-sur-Yvette, France (1987).

79. E. N. Vella *et al., Phys. Rev. Lett.* **48**, 1515 (1982).

80. M. Bauer, B. Stech and M. Wirbel, *Z. Phys.* **C34**, 103 (1987); B. Stech, *Proceedings of the Second Topical Conference on Heavy Flavors*, San Miniato (1987).

81. F. Gilman, these Proceedings.

82. I. Bigi, these Proceedings.

83. A. J. Buras, J. M. Gérard and R. Rückl, *Nucl. Phys.* **B268**, 16 (1986).

84. L.-L. Chau and H.-Y. Cheng, *Phys. Rev.* **D36**, 137 (1987).

85. A. N. Kamal, *Phys. Rev.* **D33**, 1344 (1986).

86. R. D. Peccei and R. Rückl, *Proceedings of the Ahrenhoop Symposium* (1981).

87. M. B. Voloshin and M. A. Shifman, *Yad. Fiz.* **41**, 187 (1985).

88. I. I. Bigi and M. Fukugita, *Phys. Lett.* **91B**, 121 (1980).

89. H. Albrecht *et al., Phys. Lett.* **158B**, 525 (1985).

90. R. M. Baltrusaitis *et al., Phys. Rev. Lett.* **56**, 2136 (1986).

91. C. Bebek *et al., Phys. Rev. Lett.* **56**, 1983 (1986).

92. J. Donoghue, *Phys. Rev.* **D33**, 1516 (1986).

93. F. Hussain and A. N. Kamal, ALBERTA-THY-1-86 (1986).

94. L. L. Chau and H.-Y. Cheng, *op. cit.*

95. M. A. Shifman, *Proceedings of the 1987 International Symposium on Lepton and Photon Interactions at High Energies*, Hamburg (1987).

96. M. A. Shifman *et al., JEPT Lett.* **22**, 55 (1975); M. A. Shifman *et al., Nucl. Phys.* **B120**, 316 (1977); J. Finjord, *Nucl. Phys.* **B181**, 74 (1980); B. Guberina, D. Tadic and J. Trampetic, *Nucl. Phys.* **B202**, 317 (1982); M. A. Shifman, *op. cit.*

97. J. G. Körner, G. Kramer and J. Willrodt, *Phys. Lett.* **78B**, 492 (1978) (Erratum *ibid.* **81B**, 419 (1979)); *Z. Phys.* **C2**, 117 (1979).

98. V. Barger, J. P. Leveille and P. M. Stevenson, *Phys. Rev. Lett.* **44**, 226 (1980).

99. D. Ebert and W. Kallies, *Phys. Lett.* **131B**, 183 (1983) (Erratum, *ibid.* **148B**, 502 (1984)); *Z. Phys.* **C29**, 643 (1985).

100. H.-Y. Cheng, *Z. Phys.* **C29**, 453 (1985).

101. J. G. Körner, MZ-TH/86-11 (1986).

102. J. F. Donoghue, E. Golowich and B. R. Holstein, *Phys. Rev.* **D33**, 179 (1986).

103. I. I. Bigi and A. I. Sanda, *Phys. Lett.* **B171**, 320 (1986).

104. H. Yamamoto *et al., Phys. Rev. Lett.* **54**, 522 (1985).

105. H. Albrecht *et al.* (ARGUS Collaboration), DESY 87-085 (1987).

106. S. Abachi *et al., Phys. Lett.* **182B**, 101 (1986).

107. G. Gladding (Mark III Collaboration), *Proceedings of Physics in Collision V*, Autun (1985).

108. J. C. Anjos *et al.*, Contribution to the 1987 International Symposium on Lepton and Photon Interactions at High Energies, Hamburg (1987).

ON CHARM DECAYS —
PRESENT STATUS AND FUTURE GOALS*

I. I. BIGI†

Stanford Linear Accelerator Center

Stanford University, Stanford, California, 94305

ABSTRACT

After a qualitative introduction into the dynamics underlying charm decays I describe in some detail three different theoretical treatments: the Stech et al. description based on factorization, the 1/N approach and an ansatz employing QCD sum rules. The overall agreement of the emerging theoretical picture with the data is rather encouraging and indicates that the effects of hadronization on these decays are under reasonable control. Yet more and more detailled data are needed to confirm (hopefully) this simple picture. I list the processes most relevant in this respect and emphasize the need for increasing our theoretical sophistication. Once this is achieved we have on one hand acquired the theoretical tools to deal with B physics; on the other hand we will then be ready to exploit charm physics to the fullest in searching for exotic D decays, $D^0 - \bar{D}^0$ mixing and CP violation.

* Work supported by the Department of Energy, contract DE–AC03–76SF00515.

† Heisenberg fellow.

1. Introduction: Motivation

The decays of hadrons carrying the quantum number charm have been studied for quite some time now, both on the experimental as well as the theoretical side. Large amounts of data have been accumulated in the last few years and their analysis has reached the maturity level. It is then tempting to argue that after these *existing* data have been analyzed one should move on to newer, greener pastures, i.e. topics of study.

I believe such a conclusion would be quite erroneous – there is actually still a strong motivation for further dedicated charm studies. The answer to the question, "What is the physics interest in charm decays?", consists of two parts and goes quite clearly beyond the mountain climber's reply who when asked, "Why do you climb mountains?", might just state, "Because they are there!".

A. The first part of the answer notes that according to the Standard Model the electroweak forces in charm decays are of a rather uninspiring, if not even dull, kind:

(*i*) The charged current couplings of charm quarks—$V(cs)$ and $V(cd)$—are known to be given by the Cabibbo angle θ_c to a high accuracy (the main ingredient here is the "long" bottom life time of roughly 1 psec).

(*ii*) Very tiny $D^0 - \overline{D}^0$ mixing, no observable CP violation and other flavor changing neutral currents are expected (as explained in more detail later on).

This apparent vice can however be swiftly turned into a virtue by noting that charm decays therefore offer a "clean" laboratory to study hadronization effects in quark transitions. The benefits we can derive from it are two-fold: (a) we will improve our understanding of strong interactions at the interface between the perturbative and non-perturbative regime. (b) We can apply these lessons to other dynamical systems where the underlying forces are much less known. The most urgent need for these lessons arises in bottom decays where the charged current couplings $V(ub)$ and $V(cb)$ are not or are fairly poorly known; where $B^0 - \overline{B}^0$ mixing which apparently has been observed has to be interpreted properly; where Penguins could make a small, yet significant, impact and where—most importantly and most ambitiously—CP violation might be observed.

B. There is no guarantee that the Standard Model represents the complete cast of forces and states. On the contrary there is widespread suspicion in the community that it is incomplete. One can also draw from historical precedent, namely, K decays: the observation of the $\theta - \tau$ puzzle, of $K^0 - \overline{K}^0$ mixing and of $K_L \to \pi\pi$ lead to the realization that "New Physics" was present; namely, parity violation, the charm quark and CP violation. Something like this might happen again in charm decays: the observation of $D^0 - \overline{D}^0$ mixing, CP violation, etc., would—apart from their intrinsic fascination—represent almost zero-background signals for New Physics.

These lecture notes will be organized as follows: in Section 2 we discuss the qualitative picture that can be obtained from a simple approach based on quark diagrams; Section 3 contains a more quantitative description of exclusive charm decays based on a largely phenomenological approach; its success will prompt us to consider more formal approaches in Section 4; namely, a $1/N$ treatment and QCD sum rules. In Section 5 we analyze how further lessons can be obtained from F, $\Lambda_c, \ldots$ decays and from once and twice Cabibbo suppressed decays and how they can profitably be applied in B decays. Section 6 is devoted mainly to heresies, i.e. New Physics in D decays producing flavor changing neutral currents, $D^0 - \overline{D}^0$ mixing and CP violation. The whole discussion is summed up in Section 7 and an outlook on future developments is presented.

2. A Qualitative Picture from Quark Diagrams

The effective Lagrangian for Cabibbo allowed charm decays is obtained by dressing the bare current-current coupling that includes the intermediate W boson with gluon lines and then integrating out the W boson.The result reads[1]

$$\mathcal{L}(\Delta C = 1) \propto \frac{c_+ + c_-}{2}\; \bar{s}_L \gamma_\mu c_L \, \bar{u}_L \gamma_\mu d_L + \frac{c_+ - c_-}{2}\; \bar{u}_L \gamma_\mu c_L \bar{s}_L \gamma_\mu d_L$$

$$c_\pm = \left[\frac{\alpha_S(M_W^2)}{\alpha_S(m_c^2)}\right]^{\gamma_\pm}, \qquad \gamma_+ = \frac{2}{b} = -\frac{1}{2}\gamma_-, \qquad b = 11 - \frac{2}{3} n_F \ . \tag{1}$$

For D^0 decays there are two quark diagrams with different topologies, as shown in Fig. 1a,b, while for D^+ there is only one—Fig. 1a. The open circle in the decay vertex denotes both the couplings given in Eq. (1). The diagrams in Fig. 1a represent the spectator mechanism since the anti-quarks do not participate directly in the c quark decays. On dimensional grounds one obtains for the corresponding decay width[2]

$$\Gamma_{\text{Spect}}(D^+) = \Gamma_{\text{Spect}}(D^0) \propto G_F^2 m_c^5 \ . \tag{2}$$

The diagram in Fig. 1b represents the "W-exchange" or "weak annihilation" process (hereafter summarily referred to as WA). On

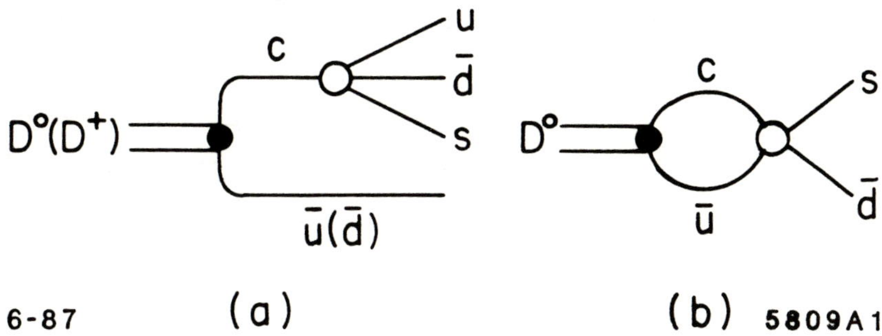

Fig. 1

Fig.1a: Spectator diagram for D^0,D^+ decays.

Fig.1b: WA diagram for D^0 decays.

general grounds one finds:

$$\Gamma_{\rm WA}(D^0) \propto G_F^2 |f_D|^2 m_s^2 m_D \ . \tag{3}$$

f_D—the decay constant of D mesons (the precise definition will be given later)—depends on $|\phi(0)|$, the D meson wavefunction at the origin. Its presence reflects the pointlike nature of weak forces at this scale; m_s—the strange quark mass—represents the helicity suppression for spin one currents, in complete analogy to $\pi \to \ell\nu_\ell$ decays. Since

$$f_D \sim 0.2\ GeV, \qquad m_s \lesssim 0.5\ GeV \ll m_c \sim 1.5\ GeV \tag{4}$$

one predicts $\Gamma_{\rm Spect} \gg \Gamma_{\rm WA}$ and expects therefore a universal charm lifetime which is very simply related to the muon lifetime:

$$\tau_D \simeq \tau(\text{charm}) \simeq \Gamma^{-1}_{\rm Spect} \simeq \frac{1}{2+3}\,\tau_\mu \left(\frac{m_c}{m_\mu}\right)^5 \sim 7\times 10^{-13}\,\text{sec} \tag{5}$$

where the last relation holds for $m_c \simeq 1.5\ GeV$.

Comparing (5) with the experimental findings[3]

$$\tau(D^0) \simeq 4.3\times 10^{-13}\,\text{sec} \qquad \tau(D^+) \simeq 10\times 10^{-13}\,\text{sec} \tag{6}$$

i.e. $\langle\tau_D\rangle \simeq 7\times 10^{-13}$ sec one should be duly impressed—after all, Eq. (5) represents an extrapolation over more than six orders of magnitude.

After having savored this moment of triumph it is however appropriate to note that the agreement between expectation and the data is certainly not perfect since experimentally

$$\frac{\tau(D^+)}{\tau(D^0)} \sim 2.5 \tag{7}$$

holds—in contrast to the expectation of a universal lifetime. The situation is clearly not as dramatic as in K decays—
where $\tau(K^+)/\tau(K_s) \sim 135$ holds—but there is a definite need for refining our theoretical picture.

One word of caution is in order here: drawing cute diagrams is one thing, deriving reliable predictions is another. After all the D mass is such that charm hadrons are placed into an environment of clear resonances. Therefore final state interactions a priori cannot be neglected and one has to allow for a non-trivial hadronization process.

Some examples can illustrate the meaning of this caveat:

(1) *Helicity suppression:*
As already mentioned the amplitude for $D^0 \to s\bar{d}$ vanishes like m_s as $m_s \to 0$. Alternatively one can look at a WA transition between hadrons, for example $D^0 \to P_1P_2$ where P_1, P_2 denote pseudoscalar mesons; its amplitude is proportional to $m^2_{P_1}, m^2_{P_2}$. Thus it vanishes as $m_s > m_{u,d} \to 0$ (for Goldstone bosons P_1, P_2)—as anticipated. For $D^0 \to PV$ decays on the other hand where V denotes a vector meson the situation is quite different: the amplitude does not go to zero anymore for vanishing quark masses. (Technically this comes about due to the presence of a new variable, the spin vector for V; dynamically it is effected by a nearby 0^- resonance with the quantum numbers of $\overline{K}^0$ which will contribute to $D^0 \to PV$, but not to $D^0 \to PP$.)

Thus we see that considerable care has to be employed when one appeals to the concept of duality for equating quark transitions with transitions between hadrons.

(2) $D^0 \to K^-\pi^+$ *vs.* $\overline{K}^0\pi^0$:

For $c_+ = 1 = c_-$ there is only one spectator diagram each for $D^0 \to K^-\pi^+$ and $D^0 \to \overline{K}^0\pi^0$, Fig. 2 , from where one reads off

$$\frac{\Gamma(D^0 \to \overline{K}^0\pi^0)}{\Gamma(D^0 \to K^-\pi^+)} \simeq \frac{1}{2}\frac{1}{N^2} = \frac{1}{18} \tag{8}$$

where N denotes the number of colors.

For $c_- > 1 > c_+$ there are two more contributions due to the induced charm changing neutral current and one finds

$$\frac{\Gamma(D^0 \to \overline{K}^0\pi^0)}{\Gamma(D^0 \to K^-\pi^+)} \simeq \frac{1}{2}\left(\frac{2c_+ - c_-}{2c_+ + c_-}\right)^2 \sim \frac{1}{50} \,. \tag{9}$$

The superficial lesson of Eq. (9) would be to expect a tiny branching ratio for $D^0 \to \overline{K}^0\pi^0$; the correct lesson is to realize that this prediction is highly unstable under a change in parameters. Therefore one cannot ignore the impact of final state interactions (hereafter referred to as FSI): looking at the isospin decomposition of $K\pi$

$$\begin{aligned} |K^-\pi^+\rangle &= \frac{1}{\sqrt{3}}\left(\frac{3}{2},\frac{1}{2}\right)_I + \sqrt{\frac{2}{3}}\left(\frac{1}{2},\frac{1}{2}\right)_I \\ |\overline{K}^0\pi^0\rangle &= \sqrt{\frac{2}{3}}\left(\frac{3}{2},\frac{1}{2}\right)_I - \frac{1}{\sqrt{3}}\left(\frac{1}{2},\frac{1}{2}\right)_I \end{aligned} \tag{10}$$

illustrates this point when one keeps in mind that the phase shifts are not expected to be even approximately the same for the $I = 3/2$ and $I = 1/2$ channels.

(3) *Color Coherence in D^+ Decays:*
Considering $D^+ = (c\bar{d}) \to (s\bar{d})_{\overline{K}^0}(u\bar{d})_{\pi^+}$ one notices that there are two $\bar{d}$ quarks in the final state[2] . Thus one has to address the issue of coherence already on the quark level. It is sometimes argued that the ensuing interference has to be negative since the identical states —$\bar{d}$—are fermions. This conclusion—that there is destructive interference—is correct, the reasoning is however misleading. A cleaner argument is based on V spin: V spin groups u and s quarks into a doublet, while d and c are singlets. D^+ is thus a V spin singlet whereas D^0 and F^+ form a doublet; secondly, Bose statistics requires $\overline{K}^0$ and π^+ which are V spin partners to form a $V = 1$ state; thirdly, in $\mathcal{L}_{\text{eff}}(\Delta C = 1)$ there is a $\Delta V = 0$ term with coefficient c_- and a $\Delta V = 1$ term with coefficient c_+. Combining these three observations one notes that $\Gamma(D^+ \to \overline{K}^0\pi^+)$ depends on c_+^2 whereas $\Gamma(D^0 \to K\pi)$ contains also c_-^2. Therefore $D^+ \to \overline{K}^0\pi^+$ is *suppressed* relative to $D^0 \to K\pi$ by $(c_+/c_-)^2$.

The conclusions I want to draw at this point are:

(*a*) the simplest picture—the pure spectator mechanism—already gives the correct order of magnitude for charm lifetimes.

(*b*) Even so there is a definite need for refining the theoretical description since the measurements yield[3] :

$$-\frac{\tau(D^\pm)}{\tau(D^0)} \sim 2.5 > 1 \tag{11a}$$

$$-\frac{\Gamma(D^0 \to \overline{K}^0\pi^0)}{\Gamma(D^0 \to K^-\pi^+)} \sim 0.4 \gg \frac{1}{18} \tag{11b}$$

$$-\frac{\Gamma(D^0 \to K^+K^-)}{\Gamma(D^0 \to \pi^+\pi^-)} \sim 3 - 4 > 1 \ . \tag{11c}$$

(*c*) To achieve this necessary improvement one has to go beyond the purely probabilistic treatment of a parton model approach, coherence effects have to be included as well.

First we will address the challenge posed by $\tau(D^+)/\tau(D^0) > 1$. One reply to this challenge consists of including WA diagrams which—on the Cabibbo allowed level—contribute to D^0, but not to D^+ decays. This diagram is added incoherently since WA generates only a $q_1\overline{q}_2$ final state in the first step while the spectator process leads directly to a $q_1\overline{q}_2 q_3\overline{q}_4$ final state. (In passing it should be noted that WA can therefore proceed only if the quantum numbers of the final state are such that they can be carried by a $q\overline{q}$ pair; it is exactly those "non-exotic" channels that are affected by FSI.) Therefore one obtains quite naturally:

$$\frac{\tau(D^+)}{\tau(D^0)} \simeq 1 + \frac{\Gamma_{\rm WA}(D^0)}{\Gamma_{\rm Spect}(D)} > 1 \ . \tag{12}$$

To reproduce the observed lifetime ratio (see Eq. (11a)) from this mechanism alone one has to require

$$\frac{\Gamma_{\rm WA}(D^0)}{\Gamma_{\rm Spect}(D)} \sim 1 \ . \tag{13}$$

Thus one has to find a way around the arguments given above for $\Gamma_{\rm WA} \ll \Gamma_{\rm Spect}$. This could conceivably be done in three steps:

(*i*) On the quark level the spectator process $c \to s\, q_1\overline{q}_2$ leads to a three-body final state while WA $c\overline{u} \to s\overline{d}$ yields a two-body final state; phase space thus favors WA.

(*ii*) The helicity suppression can be circumvented in two ways. Close to the D meson there could be a resonance with the quantum numbers of the K which could overcome the helicity suppression[4]. It should be noted that such a resonance would produce only PV, but not PP final states where $P[V]$ stands for pseudoscalar [vector] meson. Alternatively, as a more intriguing possibility[5,6], the D meson wave function might contain a non-vanishing component of $c\bar{q}g$ where g denotes one (or more gluons). In that case $c\bar{q}$ can form quite naturally a spin-one configuration—hence no helicity suppression for WA!

(*iii*) The weak forces are still pointlike, WA therefore depends, as before, on the $c\bar{q}$ overlap wavefunction at zero distance. It should be noted however that this is not the same quantity which is measured in $D^+ \to \ell^+\nu_\ell$ and which is referred to as f_D. Either there are strong resonance effects which make these arguments somewhat academic or—as argued above—WA proceeds predominantly in the presence of gluons. In that case one probes a part of the hadronic wavefunction, namely, $c\bar{q}+$ glue, that is not accessible in purely leptonic D decays. We will refer to the $c\bar{q}$ overlap in the presence of gluons as $\tilde{f}_D$ There is actually a qualitative argument for

$$|\tilde{f}_D| > |f_D| \,. \tag{14}$$

The D meson is described by a $c\bar{q}$ pair connected via a color flux tube made up by gluons. Such a system can produce a purely leptonic final state – as in D^+, $F^+ \to \ell^+\nu_\ell$ – only if the $c\bar{q}$ pair reabsorbs the flux tube before it annihilates. For this to happen the flux tube has to occupy the straight line of approach of c and $\overline{q}$. This should however be a rather unlikely configuration; it appears much more likely—due to the independent motion of the flux tube—that substantial amounts of it and thus of hadronic energy survive the $c\overline{q}$ annihilation. WA therefore contributes almost exclusively to nonleptonic and semi-leptonic decays—the latter for $F^+[D^+]$ mesons on the Cabibbo allowed [disallowed] level. Continuing this argument leads furthermore to the expectation that WA will contribute mainly to genuine multi-body final states and not to two-body modes.

There are apparently two ways to interpret $\tilde{f}_D$:

(A) It reflects a $(c\overline{q}g)$ or $(c\overline{q}gg)$, etc. component in the meson wavefunction; thus it evidently reflects non-perturbative dynamics.

(B) One invokes bremsstrahlung to procure the required gluon background. This appears to be a perturbative phenomenon which could be treated quantitatively. That impression is however misleading as can be seen in the following way: a non-relativistic model calculation yields for D^0 transitions[6]

$$\tilde{f}_D^2 \propto f_D^2 \left(\frac{m_c}{m_u}\right)^2 . \tag{15}$$

From it one reads off that the integration implicit in this ansatz receives most of its contributions from momenta p around $m_u \sim$ 300 MeV, say $m_u \lesssim p \lesssim 2m_u$. This is however the realm of "soft" dynamics which cannot be treated perturbatively. Alternatively one can note that the gluon bremsstrahlung of D mesons which are color singlets has an intrinsic infrared cut-off which is provided by the internal structure of the meson; its scale is given by the inverse Compton wavelength of the light quark; i.e. its mass m_u. Thus one is still very sensitive to soft gluons and a perturbative treatment cannot be trusted. In heavy quarkonia $(Q\overline{Q})$, on the other hand, the situation is quite different since there is only one scale $\sim m_Q$.

In conclusion, WA can be enhanced to a significant level only by invoking some non-perturbative phenomena (though the pattern exhibited in Eq. (15) might still be true for some other reason). Thus at present one cannot predict the strength of WA.

Instead of giving up on WA one can adopt the following strategy: (a) postulate some strength for WA like $\Gamma_{\text{WA}}(D^0) = \Gamma_{\text{Spect}}$ (charm); (b) then predict some other rates. Such a program can indeed be pursued on the semi-quantitative or at least qualitative level. It leads to predictions on global rates like lifetimes, on Cabibbo suppressed decays and on some special decay modes:

(i) WA, unless it is driven by the accidental presence of an appropriate resonance, leads quite unambiguously to

$$\tau_{\Lambda_c} < \tau_{D^0} < \tau_{D^+} \tag{16}$$

since in Λ_c decays WA does not have to contend with helicity suppression and its amplitude is proportional to the enhanced coefficient c_- (the quarks form color antisymmetric combinations in baryons).

For F decays no such clear statement can be made and two scenarios are conceivable

$$\tau_F < \tau_{D^0} < \tau_{D^+} \tag{17a}$$

$$\tau_{D^0} < \tau_F < \tau_{D^+} \ . \tag{17b}$$

The first case, (17a), appears somewhat more probable, yet the second one, (17b), could be realized in a scenario as described by Eq. (15) which leads to

$$\tilde{f}^2_{D^0} \sim \left(\frac{m_s}{m_u}\right)^2 \tilde{f}^2_F \sim (2-3) \times \tilde{f}^2_F \ .$$

Isospin arguments yield

$$\Gamma(D^+ \to \ell\nu X) = \Gamma(D^0 \to \ell\nu X) + \mathcal{O}(\sin^2\theta_c) \tag{18}$$

and therefore

$$\frac{\tau(D^+)}{\tau(D^0)} \simeq \frac{b_{\rm SL}(D^+)}{b_{\rm SL}(D^0)} \tag{19}$$

up to corrections of order θ_c^2, θ_c = Cabibbo angle. $\Gamma(D^0 \to \ell\nu X)$ and $\Gamma(F^+ \to \ell\nu X)$ are related by V spin and not by I spin; since WA is very likely to break V spin invariance one arrives at the general expectation:

$$\frac{b_{\rm SL}(F^+)}{b_{\rm SL}(D^0)} > \frac{\tau(F^+)}{\tau(D^0)} \ . \tag{20}$$

(ii) WA contributes directly to those Cabibbo suppressed D^+ modes that evolve from a ($u\bar{d}$+ glue) system; those should exhibit an enhancement over their naively expected level:

$$\frac{\Gamma(D^+ \to \pi\text{'s})}{\Gamma(D^+ \to K + \pi\text{'s})} > tg^2\theta_c \ . \tag{21}$$

(iii) There are those special decay modes that—it seems—can be generated only via WA on an appreciable level. Some prominent

examples of this family:

$$BR(D^0 \to \overline{K}^0 \phi) \underset{\text{WA}}{\sim} 0.5 - 1\% \tag{22}$$

$$BR(F^+ \to \pi'\text{s}) \underset{\text{WA}}{\sim} 3 - 5\% \tag{23}$$

$$BR(\Lambda_c^+ \to \Delta^{++} K^-) \underset{\text{WA}}{\sim} \text{few}\% \,. \tag{24}$$

The valence quark structure of these final states cannot be produced via spectator decays in a direct way—a contention we return to in the next chapter.

The second answer to the challenge provided by $\tau(D^+)/\tau(D^0) \sim 2.5$ consists of generalizing color coherence as introduced above for $D^+ \to \overline{K}^0 \pi^+$ to all D^+ decays. From V spin arguments one concludes

$$\tau(D^0) \simeq \tau(F^+) \,, \qquad b_{\text{SL}}(D^0) \simeq b_{\text{SL}}(F^+) \tag{25}$$

(in Section 5 we will give some estimates on the size of V spin breaking in this scheme). For Λ_c decays no identical quanta appear in the final state and thus

$$\tau(\Lambda_c) \simeq \tau(D^0) \,, \qquad b_{\text{SL}}(\Lambda_c) \sim b_{\text{SL}}(D^0) \,. \tag{26}$$

Coherence effects in the Cabibbo suppressed $D^+ \to \pi$'s transition should play the same role as in $D^+ \to K + \pi$'s since on the quark level they read $(c\bar{d}) \to u\bar{d}d\bar{d}$ and $(c\bar{d}) \to u\bar{d}s\bar{d}$. For $D^+ \to K\overline{K} + \pi$'s no such coherence emerges: $(c\bar{d}) \to u\bar{s}s\bar{d}$ and therefore[8]

$$\frac{BR(D^+ \to K\overline{K} + \pi\text{'s})}{BR(D^+ \to K + \pi\text{'s})} > tg^2\theta_c \simeq 0.05 \ . \qquad (27)$$

To summarize Chapter 2:

(*i*) $\tau(D^+)/\tau(D^0) > 1$ can be accommodated (though not really explained) by WA or Color Coherence.

(*ii*) Both schemes make new genuine predictions beyond $\tau(D^+)/\tau(D^0) > 1$: for example Cabibbo suppressed D^+ decays should be enhanced relative to the $tg^2\theta_c$ level. According to Color Coherence [WA] this enhancement should occur in the $D^+ \to K\overline{K} + \pi$'s [$D^+ \to \pi$'s] modes.

(*iii*) For pure Color Coherence one expects

$$\tau(\Lambda_c^+) \sim \tau(F) \simeq \tau(D^0) < \tau(D^+)$$

these relations should be reflected also in the corresponding relations between the semi-leptonic branching ratios.

In WA one obtains instead

$$\tau(\Lambda_c) < \tau(D^0) < \tau(D^+) \qquad \frac{b_{SL}(F)}{b_{SL}(D^0)} > \frac{\tau(F)}{\tau(D^0)} \ .$$

(*iv*) WA leads to some special decay modes.

Two important warnings should be added at this point:

(*a*) So far we have completely ignored non-trivial hadronization effects as expressed by final state interactions, resonances, etc.

(*b*) It is very hard on the level of reasoning pursued so far to make the statements listed above under (i)-(iv) more quantitative.

3. The Stech *et al.* Approach to Charm Decays or "Modesty Rewarded"

It is important to note that by now almost all theoretical treatments of charm decays employ the same effective Lagrangian with coefficients $c_\pm$ of more or less the same numerical size. These schemes—the Spectator Ansatz, WA, etc.—differ only in the way they deal with the hadronic matrix elements that enter. There is one class of decay modes where these matrix elements appear to be simpler and therefore more tractable: these are the two-body decays $D, F \to PP, PV, VV$. At first there arises the concern that one might embark on a somewhat academic exercise since these decay channels might make up only a minor part of all decays. Nevertheless one proceeds in the spirit of "modesty."

The approach by Stech and co-workers[9] (Fakirov-Stech, Bauer-Stech, Bauer-Stech-Wirbel) is based on five ingredients:

(i) use the appropriate $\mathcal{L}_{\text{eff}}$ with QCD coefficients $c_\pm$.

(ii) Ignore WA contributions.

(iii) Draw diagrams for all the different quark decay topologies; two typical examples are shown in Fig. 2a,b. The $u\bar{d}$ pair in Fig. 2a forms a color singlet and thus carries all the quantum numbers of a pion; the quark current is then simply replaced by the hadronic current with the quantum numbers of the pion. The situation is less straightforward in Fig. 2b: the $s\bar{d}$ pair does not necessarily form a color singlet. When one replaces the quark current by its hadronic counterpart one has to include a new parameter ξ of *a priori* unknown size. Naive counting of the color degrees of freedom would lead to $\xi \sim 1/N_c = 1/3$; yet ξ is used here as a free parameter:

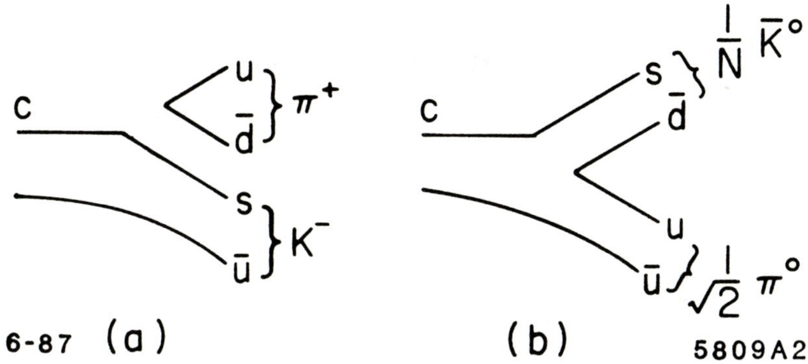

Fig. 2

Fig.2a: Spectator diagram for $D^0 \to K^-\pi^+$.

Fig.2b: Spectator diagram for $D^0 \to K^0\pi^0$.

$\xi = 0$ means color matching is necessary for forming a hadron, $\xi = 1$ it is not.

One then writes down for the transition *amplitude*

$$\begin{aligned} T(D \to f) \propto\ & a_1 \left\langle f \middle| (\bar{s}_L \gamma_\mu c_L)_H (\bar{u}_L \gamma_\mu d_L)_H \middle| D \right\rangle \\ & + a_2 \left\langle f \middle| (\bar{u}_L \gamma_\mu c_L)_H (\bar{s}_L \gamma_\mu d_L)_H \middle| D \right\rangle \end{aligned} \tag{28}$$

with

$$\begin{aligned} a_1 &= \frac{1}{2}\,(c_+ + c_-) + \frac{\xi}{2}\,(c_+ - c_-) \\ a_2 &= \frac{1}{2}\,(c_+ - c_-) + \frac{\xi}{2}\,(c_+ + c_-) \end{aligned} \tag{29}$$

where the subscript H in Eq. (28) refers to the use of *hadronic* currents carrying the quantum number of the quarks shown.

(iv) For the final states $f = PP, PV, VV$ one employs—and this is a crucial assumption—a factorization ansatz

$$\langle f \,|\, J_\mu J_\mu \,|\, D\rangle \simeq \langle P, V \,|\, J_\mu \,|\, 0\rangle \ \langle P, V \,|\, J_\mu \,|\, D\rangle \ . \tag{30}$$

Although the number of hadronic matrix elements is thus doubled, they become much more tractable:

$$\langle P \,|\, J_\mu \,|\, 0\rangle = i f_P k_\mu \tag{31}$$

$$\langle V \,|\, J_\mu \,|\, 0\rangle = i f_V m_V \epsilon^V_\mu \tag{32}$$

$$\langle P \mid J_\mu \mid D\rangle = \left(k^D + k^P - \frac{m_D^2 - m_P^2}{q^2}\, q\right)_\mu F_1(q^2) +$$

$$+ \frac{m_D^2 - m_P^2}{q^2}\, q_\mu F_0(q^2)$$

$$q = k^D - k^P\,, \quad F_1(0) = F_0(0) \tag{33}$$

$$\langle V \mid J_\mu \mid D\rangle = \frac{2}{m_D + m - V}\, \epsilon_{\mu\alpha\beta\gamma}\, \epsilon_\alpha^V k_\beta^D k_\gamma^V V(q^2)$$

$$+ i\Big\{ \epsilon_\mu^V (m_D + m_V) A_1(q^2)$$

$$- \frac{\epsilon^V \cdot q}{m_D + m_V}\, (k^D + k_V)_\mu A_2(q^2)$$

$$- 2\, \frac{\epsilon^V \cdot q}{q^2}\, m_V q_\mu A(q^2) \Big\} + 2i\, \frac{\epsilon^V \cdot q}{q^2}\, m_V q_\mu A_0(q^2)$$

$$A_3(q^2) = \frac{m_D + m_V}{2m_V}\, A_1(q^2) - \frac{m_D - m_V}{2m_V}\, A_2(q^2)$$

$$A_3(0) = A_0(0) \tag{34}$$

ϵ^V denotes the spin vector of V.

The q^2 dependence of the various formfactors $F_i, A_i(q^2)$ is assumed to be dominated by the nearest pole. E.g. for $D \to K$

transitions one uses

$$F_1(q^2) \simeq \frac{h_1}{1 - q^2/m_{F^*}^2} \ . \tag{35}$$

The residues of the pole term, e.g. h_1, are determined by the overlap integral of the appropriate hadronic wave functions as obtained in a relativistic harmonic oscillator model.

(v) Strong final state interactions (= FSI)—phase shifts, absorption, etc.—are included as best as possible.

A few comments might help in elucidating this lengthy recipe.

- Ingredients (ii)-(v) concern the treatment of matrix elements.
- In principle there are only two free parameters: a_1, a_2. In practice however there exists some considerable leeway since the final state interactions involved are not well known or determined.
- The parameters $c_\pm$ and ξ are *quite different in origin*: $c_\pm$ originates in the renormalization of the effective operator $\mathcal{L}_{\text{eff}}(\Delta C = 1)$ whereas ξ is connected with hadronic matrix elements. Off-setting changes in ξ by varying $c_\pm$ to maintain the same a_1, a_2 runs the risk of being little more than numerology.
- On the justification for ingredient (iv) the following can be said at this point: concerning factorization, try it! On f_π, f_K—take them from experiment, i.e. $f_\pi \simeq 133$ MeV, $f_K \sim 160$-170 MeV. On the residues h_i—they contain the biggest model uncertainties; at least make sure that your model reproduces f_π, f_K correctly.

Having prepared the machinery one proceeds to apply it: there are three types of transitions:

- class I transitions where the amplitude is proportional to a_1 only. these are the D^0 decays leading to two charged mesons: $D^0 \to M_1^{\pm} M_2^{\mp}$.

- class II transitions controlled by a_2: $D^0 \to M_1^0 M_2^0$.

- class III transitions: $D^+ \to M_1^+ M_2^0$. Here the amplitude depends on a linear combination: $a_1 + (1 + SB)a_2$. The quantity SB represents the amount of $SU(3)_{\text{FL}}$ breaking; in its absence one finds $a_1 + a_2 = (1 + \xi)c_+$ - as inferred from the V spin considerations presented in Section 2.

Fitting the (slightly revised) MARK III data on D^0, $D^{\pm}$ branching ratios one finds

$$a_1^{\text{exp}} \simeq 1.2 \pm 0.1 \qquad a_2^{\text{exp}} \simeq -0.5 \pm 0.1 \ . \tag{36}$$

Comparing these numbers with the expressions obtained when inserting typical QCD values for $c_{\pm}$:

$$a_1^{\text{QCD}} \simeq 1.3 - 0.6\,\xi \qquad a_2^{\text{QCD}} \simeq -0.6 + 1.3\,\xi \tag{37}$$

one sees that the experimental numbers are well reproduced for $\xi \sim 0$! In particular class II transitions depend very sensitively on the value of ξ. For example for $\xi = 1/3$ one gets $a_2^{\text{QCD}} \simeq -0.2 \neq a_2^{\text{exp}}$. Another useful relation is provided by

$$(a_1 + a_2)^2(a_1 - a_2) = (1 - \xi^2)(1 + \xi)\, c_+^2 c_- \simeq (1 - \xi^2)(1 + \xi) \tag{38}$$

where we have used $c_+^2 c_- = 1$ which is an identity on the leading log level[1] and still holds within a few percent on the two-loop level.

Before commenting on the quality of the fit to the data, one apparently technical, though important comment, has to be made. Considerable care has to be applied when extracting the coefficient a_1, a_2 from the data. For naively one would conclude if $\xi = 0$

$$\frac{\Gamma(D^0 \to \overline{K}^0 \pi^0)}{\Gamma(D^0 \to K^- \pi^+)} \simeq \frac{1}{2} \left(\frac{c_+ - c_-}{c_+ + c_-} \right)^2 \sim 0.1$$

which is by a factor of ~ 4 too small compared to the data. This embarrassment is avoided mainly by the intervention of FSI. The three $D \to K\pi$ amplitudes can be expressed in terms of two isospin amplitudes with isospin 1/2 and 3/2 in the final state:

$$\begin{aligned} A(D^0 \to K^- \pi^+) &= \frac{1}{\sqrt{3}} \left(\sqrt{2}\, A_{1/2} + A_{3/2} \right) \\ A(D^0 \to \overline{K}^0 \pi^0) &= \frac{1}{\sqrt{3}} \left(-A_{1/2} + \sqrt{2}\, A_{3/2} \right) \\ A(D^+ \to \overline{K}^0 \pi^+) &= \sqrt{3}\, A_{3/2} \ . \end{aligned} \tag{39}$$

MARK III measurements yield for the phase shift between the two isospin amplitudes

$$\delta_{1/2} - \delta_{3/2} \simeq 77^\circ \ .$$

This is a very reasonable value for FSI.[12] Including this effect (see prescription (v) above) allows one to increase $\Gamma(D^0 \to \overline{K}^0 \pi^0)/\Gamma(D^0 \to K^- \pi^+)$ from 0.1-0.2 to the experimental number while still using values for a_1, a_2 as given in Eq. (36).

A detailed representation of the data is given in D. Hitlin's lectures and does not have to be repeated here. Instead we will elaborate on the quality of the Bauer-Stech fit:

(i) Their description of some 20-odd D^0, D^+ decay modes with just two fit parameters a_1, a_2 reproduces the data rather well, though not perfectly. This is certainly impressive even when one keeps in mind that some "poetic license" is created by the way in which one deals with FSI.

(ii) There are some decay modes—in particular $D^0 \to \overline{K}^0\phi$, $\overline{K}^0\omega$, $\overline{K}^0\eta$—where this approach seems to seriously underestimate the experimental values. One can point out that the experimental findings on $D^0 \to \overline{K}^0\omega$, $\overline{K}^0\eta$ are far from settled and contain large uncertainties. Yet more importantly one notes that these three "problematic" modes are all class II transitions proportional to $(a_2)^2$; i.e. smallish; $BR(D^0 \to \overline{K}^0\phi) = 0$ is even expected for a one-step decay here since the spectator picture gives $(c\overline{u}) \to s\overline{d}\,u\overline{u} \neq s\overline{d}\,s\overline{s}$. Having realized that it is then plausible to consider two-step processes involving rescattering leading to the same final state. E.g.

$$D^0 \to \text{“}K^-\rho^+\text{”} \to \overline{K}^0\phi \ . \tag{40}$$

The important point is that the first step in the reaction chain in (40) is a class I transition, i.e. sizeable, which in this case actually commands a branching ratio of more than 10%. Therefore even a small rescattering probability can lead to an overall branching ratio of 1-2% in these cases. Thus these two-step processes might even be the dominant source for $D^0 \to \overline{K}^0\phi$, $\overline{K}^0\eta$, $\overline{K}^0\omega$. In particular the mode $D^0 \to \overline{K}^0\phi$ has attracted considerable attentions since WA can generate it in a one-step process[14]

$$D^0 = (c\overline{u}g) \to s\overline{d}g \to s\overline{s}\,s\overline{d} \ . \tag{41}$$

It might appear academic at this point yet it should be noted nevertheless that the WA process in (41), shown by Fig. 3a, describes a very different dynamical scenario than the rescattering reaction of (40), shown diagrammatically in Fig. 3b which could be called strong annihilation. In WA the $c\overline{u}$ quarks annihilate due to the weak forces. The latter are however pointlike, the $c\overline{u}$ quarks are therefore required to overlap; a measure of the probability for this to occur is provided by f_D. The WA rate is thus—as stated before—suppressed by $(f_D/m_D)^2$. In strong annihilation on the other hand it is the strong forces that drive rescattering or produce the $u\overline{u}$ annihilation. Their range is of order one Fermi—precisely like the typical $c\overline{u}$ separation. Therefore no suppression due to the finite range of the forces ensues—in contrast to WA. The two processes are thus clearly different in principle although in practice it represents a difficult task to disentangle the two.

Semi-leptonic $D \to \ell\nu K, K^*$ decays are treated in the same way and actually with more ease, since only one hadronic matrix element enters:
$\langle K, K^* \,|\, (\overline{c}_L\gamma_\mu s_L)_H \,|\, D^0\rangle$. One then finds:[10]

$$\Gamma(D \to \ell\nu\, K) \simeq 8 \times 10^{10}\,\mathrm{sec}^{-1} \tag{42}$$

$$\Gamma(D \to \ell\nu\, K^*) \simeq 9.5 \times 10^{10}\,\mathrm{sec}^{-1}\ . \tag{43}$$

Let us summarize the Stech *et al.* approach:

(i) The theoretical prediction for $D \to \ell\nu\, K, K^*$ nearly saturates the observed semi-leptonic width:[3]

$$\begin{gathered}\Gamma(D \to \ell\nu\, K, K^*)_{\text{theoret}} \sim (17-18) \times 10^{10}\,\mathrm{sec}^{-1} \\ \Gamma(D \to \ell\nu\, X)_{\text{exp}} \sim (18-19) \times 10^{10}\,\mathrm{sec}^{-1}\ .\end{gathered} \tag{44}$$

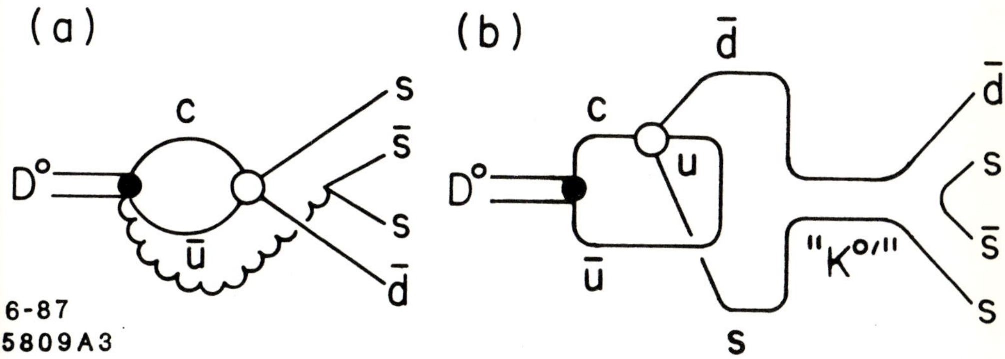

Fig. 3

Fig.3a: WA for $D^0 \to \bar{K}^0\phi$.

Fig.3b: Strong annihilation for $D^0 \to \bar{K}^{0\prime} \to \bar{K}^0\phi$

There are MARK III data[3] on $\Gamma(D \to \ell\nu K\pi)$ which case some doubt on the success of prediction (43); we will come back to this point in Section 5.

(ii) As explained in more detail in D. Hitlin's lecture the modes $D \to PP$, PV make up a large fraction of all non-leptonic D decays. Most of the remainder could be due to $D \to VV$. Thus the two-body modes seem to dominate non-leptonic D^0, D^+ decays—a surprise, yet a pleasant one: "modesty rewarded".

(iii)

$$\frac{\Gamma(D^0 \to \ell\nu K, \ell\nu K^*, PP, PV, VV)}{\Gamma(D^+ \to \ell\nu K, \ell\nu K^*, PP, PV, VV)} \sim 2-3 \tag{45}$$

i.e. the observed lifetime ratio is reproduced due to an interference in two-body non-leptonic D^+ decays that is destructive since $a_1 \cdot a_2 < 0$, see (36).

(iv) The semi-leptonic D^0 branching ratio is reproduced correctly, i.e.

$$\frac{\Gamma(D^0 \to \ell\nu K, K^*)}{\Gamma(D^0 \to \ell\nu K, \ell\nu K^*, PP, PV, VV)} \sim 8\% . \tag{46}$$

It should be noted that (iii) and (iv) are achieved without any WA! The "low" D^0 semi-leptonic branching ratio of around 7.5% is thus no unambiguous evidence for substantial WA. Hadronization effects, formfactors, etc.—as exemplified by (46)—lower the semi-leptonic branching ratio from its naive value of 12 [15]% obtained when using the quark-level expression

$$b_{\mathrm{SL(charm)}} \sim \frac{1}{2 + \frac{3}{2}\left(c_+^2 + c_-^2\right) + \frac{3}{2}\,\xi(c_+^2 - c_-^2)} \tag{47}$$

with $\xi = 0[1/3]$.

Therefore it is fair to state that so far no phenomenological need for WA contributions has been demonstrated in D^0, D^+ decays. FSI on the other hand do play a major role as seen most clearly when analyzing the isospin amplitudes in $D \to K\pi$, $K\rho$, $K^*\pi$: the MARK III data yield for the ratios of isospin 1/2 and 3/2 amplitudes and their phase shifts[3]

$$
\begin{aligned}
|A_{3/2}/A_{3/2}| &= 3.67 \pm 0.27, \quad 3.22 \pm 0.97; \quad 3.12 \pm 0.40 \\
\delta_{1/2} - \delta_{3/2} &= (77 \pm 11)^\circ, \quad (84 \pm 13)^0, \quad (0 \pm 26)^0
\end{aligned}
\tag{48}
$$

for the three modes $D \to K\pi$, $K^*\pi$, $K\rho$ respectively. Another observation can be made here: if WA played the major role in explaining the $D^+ - D^0$ lifetime difference due to the presence of a K' resonance one would not expect a nearly universal value for $|A_{1/2}/A_{3/2}|$ as indicated by the data in (48). For

$$D^0 \to \text{“}K'\text{”} \to PV$$

can proceed whereas

$$D^0 \to \text{“}K'\text{”} \to PP$$

cannot. Secondly, WA contributes to $A_{1/2}$ only. Yet keeping the error bars in mind one cannot draw really quantitative conclusions from this at present. More precise data would be quite desirable.

Since the data on D^+, D^0 decays do not exhibit any clear need for WA and since WA does not represent an intellectually lucid concept one might be tempted to forget about it all together. This would however be ill-advised.

(i) The fact that WA has not been established as a major source of D decays does not imply that it is insignificant there. If WA were the dominant generator for $\tau(D^+)/\tau(D^0) > 1$ it could conceivably—though personally I am rather skeptical about it—lead to $\tau(B^+)/\tau(B^0) \sim 1.5$ and could well open up new avenues for CP studies in B decays.

(ii) WA though not dominant could still be significant, say on the 20% level. If this were not due to the lucky presence of an appropriate resonance, it would reveal—as discussed in Section 2—the presence of gluons in the D meson wavefunction thus bringing them one step closer to a role in spectroscopy. Although this would not be a truly surprising result, it would be a nice one nevertheless.

(iii) Even if WA were rather unimportant for the overall D decay rates it could have a significant impact on some rare modes, like once and twice Cabibbo suppressed modes. This would affect our conclusion on the role of Penguins (see Section 5) and $D^0 - \overline{D}^0$ mixing (see Section 6).

(iv) WA could be much more significant in charm baryon than in charm meson decays.

In summary: the Bauer-Stech ansatz represents a considerable step forward in understanding charm decays. It is rather successful and its very success shows that there is a fairly simple dynamical pattern underlying most of charm decays. This realization should however not lead to complacency. The ansatz is based on *ad hoc* assumptions like factorization and valence quark description and on a rather *ad hoc* treatment of FSI. On a more technical level the ansatz is not very successful in Cabibbo suppressed D^0 decays:[2,3]

$$\frac{\Gamma(D^0 \to K^+K^-)}{\Gamma(D^0 \to \pi^+\pi^-)} \sim \begin{cases} 1.44 & \text{“theoret”} \\ 3-4 & \text{experim.} \end{cases} \tag{49}$$

A more detailed discussion of this point will be given in Section 5.

There are two ways to progress from here:

(A) One generalizes the Stech *et al.* approach by including WA and Penguin diagrams and determines their relative weight phenomenologically from a comprehensive data analysis.[11]

(B) One starts from a more theoretical treatment of hadron dynamics.

I will choose the second avenue in Section 4—partly because I believe that it will offer more theoretical insight—partly because I am quite skeptical at present that approach (A) allows a satisfactory treatment of FSI.

4. Creeping Towards a Truly Theoretical Description of Charm Decays—The $1/N$ Approach and QCD Sum Rules

That D^0, D^+ decays are rather well described if $\xi \simeq 0$ is used appears somewhat miraculous at this point: if one thinks of ξ as representing the impact of low energy strong interactions—say, soft gluons, etc.—one would hardly expect to find a universal value of ξ; after all the kinematics, etc., are different for the various two–body modes. And a priori ξ could have taken any value of order $\pm$ 1.

The same surprise emerges in $B \to \psi X$ decays which in the language of Bauer–Stech are Class II transitions and thus very sensitive to the value of ξ. Theoretically one predicts[15,16,9] :

$$\mathrm{BR}(B \to \psi X) \simeq 1 - 2\% \cdot (3a_2)^2 \tag{50}$$

$$\mathrm{BR}(B \to \psi K^*) \simeq 4 \cdot 10^{-3} \cdot (3a_2)^2 \tag{51}$$

Comparing Eq. (50, 51) with the data[17]

$$\mathrm{BR}(B \to \psi + X) \sim 1.2\% \tag{52}$$

$$\mathrm{BR}(B \to \psi K^*) \sim (4 \pm 2) \cdot 10^{-3} \tag{53}$$

leads to

$$|a_2^{\mathrm{exp}}| \sim \frac{1}{3} \tag{54}$$

Using short–distance coefficients $c_\pm$ appropriate for $\Delta B = 1$ transitions—$c_+ \sim 0.82, c_- \sim 1.5$—one obtains

$$a_2^{\mathrm{QCD}} \simeq -0.34 \pm 1.7\xi \tag{55}$$

and $\xi \simeq 0$ again emerges as a natural solution—this time for a much more massive hadronic system, the B meson:

$$\xi \text{ (charm) } \simeq \xi \text{ (bottom)} \tag{56}$$

An ocurrence like that begs for an explanation and it was Buras et coworkers who first gave one within the framework of the $1/N$ approach where they extended considerably the earlier work by other authors[18].

A. The $1/N$ Approach

The usual effective Lagrangian as given in Eq. (1) is employed. The matrix element is expanded into a power series in $1/N$, N being the number of colors:

$$< M_1, M_2 | \mathcal{L}_{\text{eff}}(\Delta C = 1) | D > = \sqrt{N} \left\{ b_0 + \frac{b_1}{N} + O\left(\frac{1}{N^2}\right) \right\} \tag{57}$$

The leading term (in $1/N$) has coefficient b_0, the next–to–leading b_1. We do not concern ourselves with mathematical niceties like whether such a power series is well–defined. The main assumption implicit in Eq. (57) is then the hope that the first term or at worst the first two terms of this expansion already represent a decent approximation to the full (yet unknown) result.

Very simple and almost self–evident rules are established for computing the coefficient in the $1/N$ expansion:

1. Draw all quark level diagrams with two weak vertices that contribute to a decay process; mesons are represented by their valence quarks.

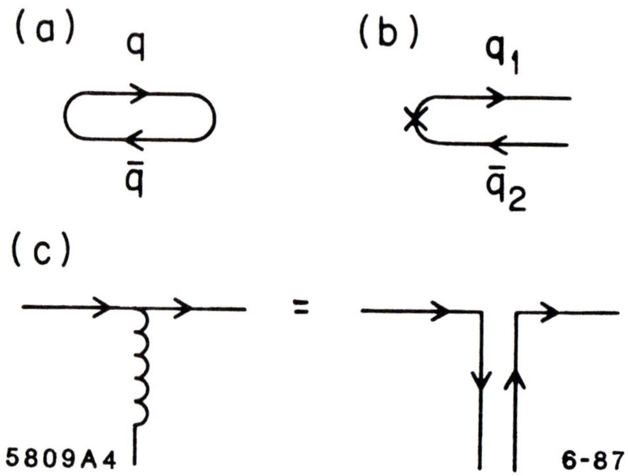

Fig. 4

Fig.4a,b,c: 1/N counting rules.

2. Assign a weight N to each closed quark loop, Fig. 4a.

3. Each meson wavefunction, denoted by a cross in Fig. 4b, introduces a normalization factor $1/\sqrt{N}$.

4. Gluon lines are replaced by $q\bar{q}$ lines, see Fig. 4c; the quark–gluon coupling carries a weight $1/\sqrt{N}$.

It is very easy to reproduce the following basic results:

1. There are two types of diagrams that contribute on the leading $1/N$ level, i.e., to b_0, namely one of the spectator- and one of the WA-type, see Fig. 5a,b.

2. There are three types of diagrams on the next–to–leading level contribution to b_1: a spectator diagram, Fig. 6a, a WA (or W–exchange) diagram, Fig. 6b, and diagrams involving FSI, Fig. 6c.

3. For later reference it should be noted that "hair–pin" diagrams[13] give suppressed contributions—not surprisingly, since they represent OZI-forbidden processes. For example the hair–pin diagram in Fig. 7 contributes only on the $1/(N\sqrt{N})$ level, i.e., is suppressed by N^2 in amplitude relative to the leading term.

Two pleasant surprises are immediately read off from these and the corresponding D^+ diagrams:

1. The leading $1/N$ diagrams factorize into the same kind of matrix elements as appear in the Stech et al. ansatz.

2. It is quite evident that the interference in $D^+ \to \bar{K}^0\pi^+$ decays has to be destructive since it is proportional to $(c_+ + c_-)(c_+ - c_-) = c_+^2 - c_-^2 < 0$!

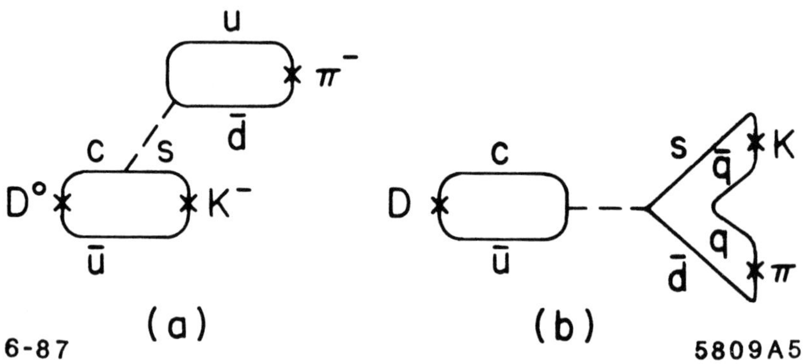

Fig. 5

Fig. 5a,b: Leading 1/N diagrams for $D^0 \to K\pi$; the broken line represents a weak current, either a charged current or a QCD induced neutral current.

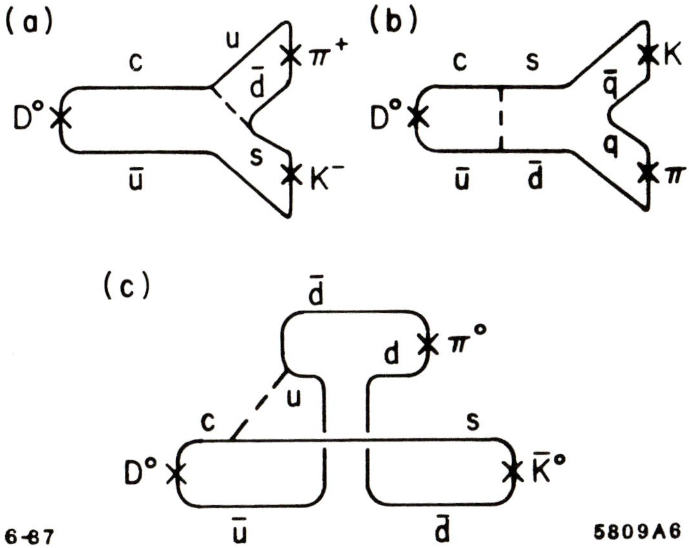

Fig. 6

Fig.6: Next-to-leading diagrams.

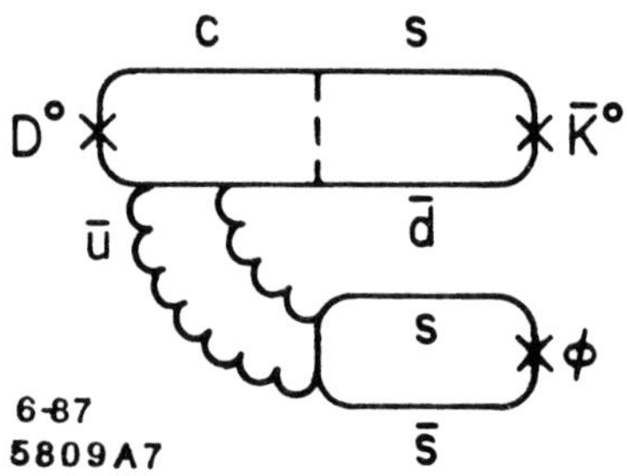

Fig. 7

Fig.7: A hairpin diagram.

A few typical predictions are listed below and compared with the data on one hand and with the (naive) Standard Approach (i.e., $\xi = \frac{1}{3}$) on the other hand (for a more complete listing see ref. 3,18):

Decay Mode	Standard Approach	Large N Approach	Data
	[Decay widths in units of $10^{10}\,\text{sec}^{-1}$]		
$D^0 \to K^-\pi^+$	14.6	20.1	9.8±1.5
$D^0 \to \bar{K}^0\pi^0$	0.26	4.1	4.4±1.0
$D^0 \to K^-\rho^+$	26.1	36.0	25±3.5
$D^0 \to \bar{K}^0\rho^0$	0.20	3.3	2.3±0.7
$D^0 \to \bar{K}^0\eta$	0.13	2.1	3.5±1.6
$D^0 \to \bar{K}^0\omega$	0.20	3.2	7.4±3.4
$D^0 \to \bar{K}^0\phi$	$\simeq 0$	0.2-0.5	2±1.2
$D^0 \to K^+K^-$	1.4	1.9	1.2±0.3
$D^0 \to \pi^+\pi^-$	1.0	1.4	0.33±0.12

Such a comparison leads to the following conclusions:

1. A leading $1/N$ treatment is not only less disastrous then the naive Standard Approach, but it produces also a decent representation of the data.

2. It is more transparent and thus more "user–friendly" than the Stech et al. ansatz.

3. Yet its fit to the data suffers from some evident deficiencies:

- The ratio $\frac{\Gamma(D^0 \to \bar{K}^0\pi^0)}{\Gamma(D^0 \to \bar{K}^-\pi^+)} \simeq 0.2$ is too small by a factor of roughly two relative to the data.

- $\mathrm{BR}(D^0 \to \bar{K}^0\phi) \sim 0.1 - 0.2\%$ appears too small by a factor of at least five relative to the measured numbers. WA contains a leading $1/N$ contribution and in $D \to$ PV transitions there is no automatic helicity suppression (in contrast to $D \to PP$ modes) yet it is a class II transition, i.e., proportional to $\frac{1}{4}(c_+ - c_-)^2$ and suffers from a hadronic formfactor suppression. (One should keep in mind that this last point does not represent an ironclad argument).

- $\frac{\mathrm{BR}(D^0 \to K^+K^-)}{\mathrm{BR}(D^0 \to \pi^+\pi^-)} \simeq 1.4$, i.e., apparently too small again by a factor of two or more.

- The predictions on class I transitions over-shoot the experimental numbers, in particular after the recent MARK III re-calibration. Re-scattering processes which have not been included here (see comment 2. below) will decrease this excess; this problem therefore appears more as a technical shortcoming than a fundamental defect.

The $1/N$ approach, as practiced so far, is still not fully satisfactory as a theoretical description:

1. The terms non-leading in $1/N$ are dropped by fiat, not by theoretical reasoning. The $1/N$ approach developed so far represents a nice way to memorize our knowledge on charm decays, but by itself it does not advance our understanding of these phenomena.

2. Self-consistency requires us to ignore FSI; yet there is clear evidence for their significance in the data.

3. Non–leading terms, in particular non-factorizable contributions like FSI cannot be calculated in such a simple way.

Yet the relative phenomenological success of this easy and transparent ansatz clearly suggests that the various non–leading terms have - by and large - a strong tendency to cancel each other.

B. QCD Sum Rules

(a) General Procedure:

It has become customary to state that all problems of hadronization will be solved some day once and for all by Monte Carlo simulations on a sufficiently large lattice. I am not so sure however that this will happen before this millennium is over. People who share my relative lack of patience have to employ other, maybe less reliable vehicles to approach hadronization. At present the most promising theoretical technology, in my judgment, is based on the judicious use of sum rules, as pioneered by the ITEP group. We will give here only an outline of the general strategy since a more detailed description goes well beyond the scope of these lectures and can readily be found in the original papers[19] or in the thorough review by Reinders, et al.[20] .

One always starts from a correlation function between two or more (generalized) currents; they are chosen such that they carry the same quantum numbers as the hadrons one wants to study. A well–known example is provided by the reaction $e^+e^- \rightarrow$ charm (at energies well below the Z^0 mass). The appropriate current is a Lorentz vector $j_\mu = \bar{c}\gamma_\mu c$ and one considers the time–ordered two–

point function

$$i\int d^4x e^{iq\cdot x} < 0|T\{j_\mu(x)j_\nu(0)\}10 > \equiv (g_{\mu\nu}q^2 - q_\mu q_\nu)\Pi^V(q^2) \tag{58}$$

Next one invokes analyticity to write a dispersion relation for the polarization function $\Pi(q^2)$. (In general it will contain subtraction terms, which can be removed by applying the operator $\frac{\partial}{\partial q^2}$ a sufficient number of times. Since we will do that anyway for different reasons, we will ignore this complication in the following.)

$$\Pi(q^2) = \frac{q^2}{\pi}\int \frac{\text{Im }\Pi(s)}{s(s-q^2)}ds \tag{59}$$

For $e^+e^- \to$ charm there is a very simple representation for Im $\Pi^V(q^2)$ due to the optical theorem:

$$\text{Im }\Pi^V(q^2) = \frac{9}{64\pi^2\alpha^2}\,\sigma(e^+e^- \to \text{had.})$$

$$\cong \frac{\pi}{e_q^2}\sum_R \frac{m_R^2}{g_R^2}\,\delta(q^2 - m_R^2) +$$

$$+\frac{1}{4\pi}\left(1+\frac{\alpha_s}{\pi}\right)\,\theta(q^2 - s_{\text{thresh.}}) \tag{60}$$

We have made two approximations in the last line where we have expressed the total cross section as a sum over a discrete spectrum of (say charmonium) resonances R with mass [width] m_R $[g_R]$ and a continuum of open flavour production above the threshold (say $s_{\text{thresh.}} > (2m_D)^2$) : firstly we have used a narrow resonance width ansatz and secondly we have inserted an expression for the continuum cross section that goes to first order in α_s only.

The first big theoretical assumption consists in employing an operator product expansion (=OPE) for the time ordered current–current product:

$$i \int d^4 x e^{q \cdot x} T\{j_\mu(x) j_\nu(0)\} = (g_{\mu\nu} q^2 - q_\mu q_\nu) \cdot$$

$$\cdot \left[c_0(q^2) I + \sum_k c_k(q^2) \mathcal{O}_k \right] \tag{61}$$

where the c_i are the Wilson coefficients, I denotes the identity operator and $\mathcal{O}_i$ the other local, gauge invariant operators. Analyses so far have included five such operators, namely quark and gluon "bilinears" $\mathcal{O}_m = m\bar{q}q$, $\mathcal{O}_G = G^a_{\mu\nu} G^a_{\mu\nu}$ of dimension four, and three manifestly non-bilinear operators of dimension six, namely

$$\mathcal{O}_\Gamma = \bar{q}\gamma_\mu q \bar{q}\gamma_\mu q,$$

$$\mathcal{O}_\sigma = \tilde{m}\bar{q}\sigma_{\mu\nu} \frac{\lambda_a}{2} q G^a_{\mu\nu},$$

$$\mathcal{O}_f = f_{abc} G^a_{\mu\nu} G^b_{\nu\gamma} G^c_{\gamma\mu},$$

The next step is equally important and unique to this approach: a perturbative treatment gives for all VEV's $< 0|\mathcal{O}_k|0 >= 0$ (apart from $< 0|I|0 >$ of course). Allowing for

$$< 0|\mathcal{O}_k|0 > = f_k \neq 0 \tag{62}$$

therefore incorporates some - and hopefully the most significant - non–perturbative aspects. Thus new unknowns f_k are introduced, like the quark and gluon condensates $< 0|\bar{q}q|0 >$, $< 0|G \cdot G|0 >$; yet they are process–independent quantities that can - at least in principle - be fitted in one reaction and then used in many other processes.

Putting everything together one arrives at

$$\text{LHS} \simeq \text{M}_n(q_0^2) \equiv \frac{1}{n!}\left(\frac{d}{dq^2}\right)^n \Pi(q^2)\Bigg|_{q=q_0^2} = \frac{1}{\pi}\int ds \frac{\text{Im }\Pi(s)}{(s+q_0^2)^{n+1}} =$$

$$\simeq \text{RHS} \tag{63}$$

$$\text{RHS} \simeq \frac{1}{e_q^2}\,\frac{m_R^2}{g_R^2}\,\frac{1}{(m_R^2+q_0^2)^{n+1}}\,\left[1+\Delta_n(q_0^2)\right] \tag{64}$$

$$\text{LHS} \simeq \frac{1}{n!}\left(\frac{d}{dq^2}\right)^n \left\{c_0(q^2) + \sum_k c_k(q^2) f_k\right\} \tag{65}$$

where the following comments have to be made:

(a) Eq. (63) represents the duality concept: one equates (or at least approximates) the right-hand side RHS which is expressed in terms of hadronic parameters m_R,g_R, etc., at a somewhat unphysical momentum scale q_0 with the left-hand side LHS calculated on the quark–gluon level where non–perturbative effects are included via the VEV's f_k.

(b) In Eq. (64) we have factored out the lowest lying resonance R whereas the higher lying resonances and the continuum are lumped together into $\Delta_n(q_0^2)$; its leading term is given by

$$\Delta_n(q_0^2) \sim \left(\frac{m_R^2 + q_0^2}{m_{R'}^2 + q_0^2}\right)^{n+1}, \quad m_{R'} > m_R \qquad (66)$$

(c) $< 0|I|0 >= 1$ was used in Eq. (65).

Applying Eq. (63-65) truly resembles a voyage between Scylla and Charybdis: the higher n is chosen, the more sensitive RHS becomes for the resonance parameters M_R, g_R since $\Delta_n(q_0^2) \to 0$ as $n \to \infty$; however at the same time the use of OPE in LHS with a *finite* number of operators $\mathcal{O}_k$ becomes more problematic. No clearcut criterion can be given for dealing with this road hazard. One assumes that it is treated in a satisfactory way numerically if one can find a "stability region" in (n, q_0^2) for LHS $\simeq$ RHS; i.e., if LHS and RHS do not vary too dramatically under changes of n and q_0^2. Therefore one has to apply these sum rules *judiciously*. An improvement is often, though not always achieved if a Borel transformation is applied to Eq. (63).

$$\begin{aligned} L_M\Pi(q^2) &\equiv \lim \frac{1}{(n-1)!}(q^2)^n \left(\frac{d}{dq^2}\right)^n \Pi(q^2) \\ &= \frac{1}{\pi M^2} \int ds e^{-s/M^2} Im \ \ \Pi(s) \end{aligned} \qquad (63a)$$

as $n, q^2 \to \infty$ with $\frac{q^2}{n} = \mathrm{M}^2$ kept fixed.

The general procedure can thus be summarized briefly:

1. Pick the appropriate current for your problem; for pseudoscalar Goldstone bosons one even has a choice, namely to use either $\bar{q}\gamma_5 q$ a là Reinders et al. or $\bar{q}\gamma_\mu\gamma_5 q$ a là Shifman et al.

2. Write down the relevant n-point function in these currents.

3. Compute the Wilson coefficients (as many as possible/necessary) in the resulting OPE.

4. Obtain the significant VEV's from previous fits to other measured quantities.

5. Find the "stability regime" in LHS $\simeq$ RHS and extract numbers.

(b) Applications, part I: Some Couplings and Masses

Two point functions of currents carrying open charm (like $\bar{c}\gamma_5 q$, $\bar{c}\gamma_\mu\gamma_5 q$ or $\bar{c}\gamma_\mu q$) allow us to obtain an estimate of f_D once the experimental m_D is used. The authors of Ref. 21 find

$$f_D(\text{QCD SR}) \sim 150 - 180\ MeV \tag{67}$$

in rough agreement with other theoretical estimates[22,23]

$$f_D \sim \begin{cases} 150 - 230\ MeV & \text{potential models} \\ 100 - 170\ MeV & \text{MIT bag models}\ . \end{cases} \tag{68}$$

Similarly one finds for the mass splittings between s-wave, D, and p-wave, D^{**}, states[24]

$$\langle m_{D^{**}} - m_D \rangle \sim 400 - 500\ MeV \tag{69}$$

as in potential models.

(c) Applications, part II: Two-body Decays of Charm mesons

Very recently Blok and Shifman[24,25] have developed a treatment of

$$D \to PP,\ PV \tag{70}$$

decays that is based on QCD sum rules. They analyzed a four-point correlation function between three currents—one carrying the quantum number of D, the other two those of the final state mesons generically referred to as A and B—and the weak Lagrangian $\mathcal{L}_W$:

$$\begin{aligned}\Pi_{\mu\nu}(Q_1, Q_2, q) = \int d^4x \int d^4y \int d^4z\, e^{iQ_2\cdot x + iq\cdot y}\cdot \\ \cdot \langle 0 | \{ j_A(y) j_B(x) j_D(0) \mathcal{L}_W(z) \} | 0 \rangle \ .\end{aligned} \tag{71}$$

This relation can be visualized as follows: one starts from three types of skeleton (or bare) diagrams as shown in fig. 8; fig. 8a represents WA, fig. 8b,c, Spectator processes.

These bare diagrams are then dressed by gluon exchanges, both soft and hard. If these gluon lines stay inside the triangle or the external loop, then the factorization property of the amplitude is not changed. If on the other hand the gluons communicate between the triangle and the external loop the diagram becomes definitely non-factorizable. In addition light quark (u, d, s) lines and soft gluon lines are cut to include the non-perturbative VEV's in the OPE treatment of LHS.

In the OPE Blok and Shifman use the five operators $\mathcal{O}_k$ (in addition to the identity I) that were introduced above after eq. (61). The Wilson coefficients are calculated at $Q_1^2 \sim Q_2^2 \sim q^2 \sim -(1.5\ GeV)^2$, i.e. in the Euclidean region. When extrapolating to the (physical)

Minkowski region to make contact with RHS one has to concern oneself with three types of singularities:

(*i*) a pole at the D mass—the objective of the analysis.

(*ii*) Cuts at higher masses corresponding to excited D mesons and to the $D+\pi\pi$, etc., continuum. Their importance is suppressed by performing a Borel summations a la eq. (63a).

(*iii*) Physical cuts, etc., corresponding to resonances, FSI that affect the final state AB. It is argued—and this might be the gravest theoretical uncertainty—that this duality ansatz when initiated in the Euclidean region possesses no sensitivity to such singularities.

I am not convinced that this rather complex analysis has already reached maturity; it is very likely indeed—judging from past experience—that at least some numbers will change. Nevertheless I want to present some of their numbers, be it only to strongly encourage and stimulate further work in this direction:

Decay Mode	Theoret. BR
$D^0 \to K^-\pi^+$	6.4%
$\overline{K}^0\pi^0$	1.5%
$K^-\rho^+$	15%
$\overline{K}^0\rho^0$	0.8%
$K^{-*}\pi^+$	9 %
$\overline{K}^{0*}\pi^0$	3 %
$\overline{K}^0\omega$	1.5%
$\overline{K}^0\phi$	1.3%

The following points should be especially noted:

(*i*) The agreement with the "old" MARK III branching ratios is quite good! These "predictions" cannot be confronted immediately with the recalibrated branching ratios: for a few of the experimental data act as input to determine the relative weight of the three diagrams in fig. 8—as is typical for the duality ansatz underlying QCD sum rules.

(*ii*) Most of the numbers are actually very similar to those obtained in the $1/N$ ansatz. Thus it lends theoretical justification for dropping non-leading $1/N$ terms in most cases.

(*iii*) This treatment represents a theoretical improvement also in the sense that it does include non-leading terms. For example

$$BR(D^0 \to \overline{K}^0 \phi) \sim 1.3\% \tag{72}$$

is generated purely from WA—without help from FSI! This implies furthermore that $\tilde{f}_D > f_D$ (see the discussion in Section 2) since $f_D \leq 0.2\ GeV$ in this treatment, Eq. (67).

(*iv*) WA emerges as a significant though not dominant process contributing roughly 20% to $\Gamma(D^0)$.

After listing all these successes some of the short-comings have to be mentioned as well:

(*a*) The present treatment does not allow for taking FSI properly into account.

(*b*) It is then not surprising that the ratio

$$\frac{BR(D^0 \to \overline{K}^0 \pi^0)}{BR(D^0 \to K^0 \pi^+)} \sim 0.2$$

turns out to be rather small.

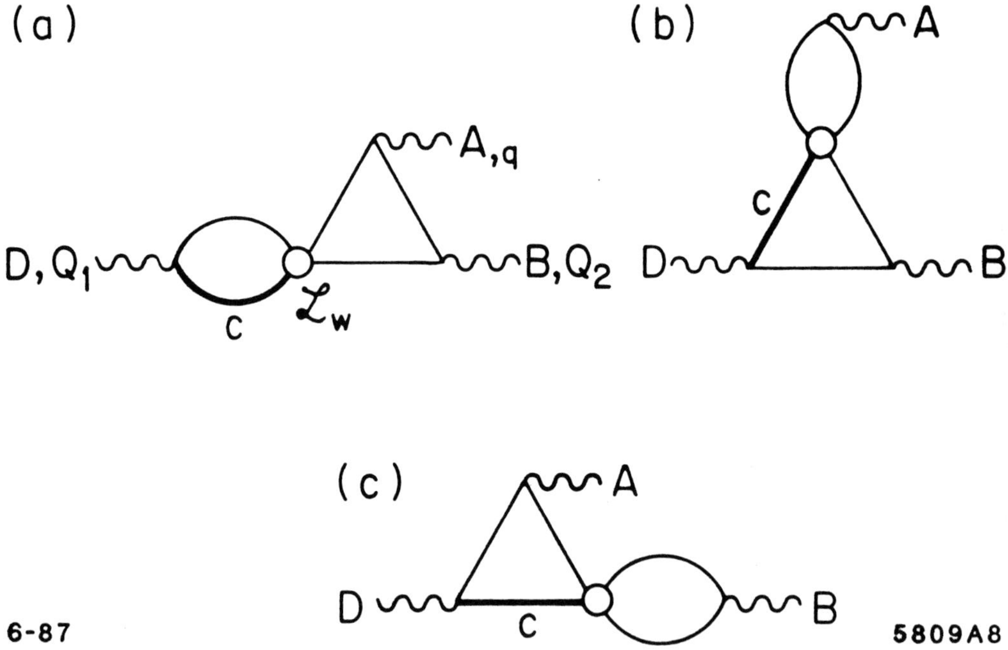

Fig. 8

Fig.8a,b,c: The three skeleton diagrams for $D \to AB$ decays.

(c) $\Gamma(D^0 \to K^+K^-) \simeq \Gamma(D^0 \to \pi^+\pi^-)$ is found in clear conflict to the data. Yet this equality is largely built into the ansatz from the beginning since $SU(3)_{\rm FL}$ breaking effects are ignored (like $(\langle \bar{s}s \rangle \neq \langle \bar{d}d \rangle)$.

Interlude: An Assessment of Our Theoretical Understanding of Charm Decays

The Stech *et al.* treatment revealed that—by and large—a fairly simple dynamical scenario underlies D^0, D^+ decays.

The $1/N$ approach allows us to express this overall simplicity in a concise manner. Yet—like Feynman diagrams and personal computers before—it does not necessarily deepen our understanding although it certainly facilitates most computations. Still it wets the appetite for a deeper understanding.

It is certainly not clear whether treatments that invoke QCD sum rules can take us to this goal; but I believe that eventually they will take us a good distance along this road.

While the final verdict on the theoretical treatments is therefore still not in, it is nevertheless fair to say that the state of the art in theoretical reasoning has improved quite considerably in the last two years. Looking at simple quark-level diagrams is still useful (and fun)—but only as a starting point for a dynamical analysis, not as the end point!

After having made these statements which might be perceived as a clear symptom of overconfidence, I will attempt to redeem myself and some of my colleagues by adding a few words of caution:

(*i*) The success of our theoretical description is still somewhat unstable. More data and more precise data can jeopardize our contentment.

(*ii*) Success is actually quite elusive so far in some rare D decay modes.

These issues will be addressed in the next section.

5. Present and Future Cross Checks and Lessons

A. F Decays

(i) The first question obviously refers to the F total lifetime: do $\tau(D^0)$ and $\tau(F^+)$ agree to within, say, 10% as suggested by V spin symmetry with moderate breaking? WA on the other hand allows for very substantial V spin breaking. The most recent data give an affirmative answer[3] :

$$\frac{\tau(F^+)}{\tau(D^0)} = 1.0 \pm 0.1 \ . \tag{73}$$

(ii) Semi-leptonic branching ratios are the next most interesting global quantities. V spin symmetry gives

$$\frac{b_{\rm SL}(F^+)}{b_{\rm SL}(D^0)} \simeq \frac{\tau(F^+)}{\tau(D^0)} \simeq 1 \tag{74}$$

whereas WA quite naturally leads to

$$\frac{b_{\rm SL}(F^+)}{b_{\rm SL}(D^0)} \gtrsim \frac{\tau(F^+)}{\tau(F^0)} \ . \tag{75}$$

(iii) There are special decay modes—$F \to \pi$'s—that at first sight appear to be produced by WA only, like $D^0 \to \overline{K}^0 \phi$. Yet the same caveat applies in both cases: FSI can in principle generate such final states via strong annihilation and thus fake WA. Nevertheless the following argument can be made: (a) If WA is assumed to dominate $D^0 \to \overline{K}^0 \phi$, and (b) $\tau(F^+) \simeq \tau(D^0) < \tau(D^+)$ is largely attributed to WA, then one concludes

$$BR(F^+ \to \pi\rho) \, , \quad BR(F^+ \to \pi\omega) \sim 3 - 5\% \tag{76}$$

has to hold. Such branching ratios are then quite similar or even larger than $BR(F^+ \to \pi\phi)$. Very recent E691 and ARGUS data however reveal[3]

$$\frac{BR(F^+ \to \pi^+ \rho^0)}{BR(F^+ \to \pi^+ \phi)} < 0.08 \qquad (90\%\ C.L.) \tag{77}$$

thus ruling WA out as the dominant motor behind $D^0 \to \overline{K}^0 \phi$ or F decays, or both.

There are two loop-holes in this argument. FSI could conceivably (though not very likely) conspire with WA in such a way that the transition $F^+ \to \pi^+ \rho^0$ is interfered away. There is another more intriguing and more interesting scenario[13] : if hairpin diagrams were important one would expect to see the mode $F^+ \to \pi^+ \omega$ as well as $F^+ \to \pi^+ \phi$ (and $F^+ \to \pi^+ G$ if a glueball state G were accessible), but not $F^+ \to \pi^+ \rho^0$ since the ρ^0 has no isoscalar component! Yet two notes of caution can be added here:

(α) In the $1/N$ approach $F^+ \to \pi^+ \omega$ is clearly suppressed relative to $F^+ \to \pi^+ \phi$ as explained in Section 4. Furthermore fairly little FSI is expected to intervene in either the $\pi\omega$ or $\pi\phi$ final state. Finding comparable rates, i.e.

$$BR(F \to \pi\omega) \sim BR(F \to \pi\phi) \tag{78}$$

would therefore amount to a very hefty blow to $1/N$ practitioners. Incidentally the same is true for the QCD sum rule approach.

(β) WA could be important in F decays due to the accidental presence of a nearby π-like resonance which then decays strongly, i.e.

$$F^+ \xrightarrow[\text{WA}]{} \text{“}\pi'\text{'' } \to f\ . \tag{79}$$

In that case G *parity* would forbid (or at least suppress) decays into an even number of pions while allowing decays into an odd number of pions:

$$F^+ \to \text{“}\pi' \not\to 4\pi \tag{80}$$

$$F^+ \to \text{“}\pi' \to 3\pi \tag{81}$$

Since these theoretical remarks might reflect just theoretical biases, it is important, if not even mandatory to search for $F \to \pi\omega$ as best as possible.

(iv) "What E691 taketh away with one hand, it giveth back with the other!" E691 has presented highly intriguing evidence for $F^+ \to (\pi^+\pi^-\pi^+)_{\text{non-res}}$ [3]

$$\frac{BR(F^+ \to (\pi^+\pi^-\pi^+)_{\text{non-res}})}{BR(F^+ \to \phi\pi^+)} = 0.29 \pm 0.07 \pm 0.05 \ . \tag{82}$$

As emphasized before, it is dangerous to draw firm conclusions from one or two decay modes. Nevertheless it is tempting to invoke (82) as evidence for WA playing a significant, though somewhat suppressed role (on the $\sim 20\%$ level) in F decays. For if (82) were produced via FSI from the Spectator process (or by resonance enhanced WA as discussed above Eq. (79)) why then is there no signal for $F^+ \to \pi^+\rho^0$? Keep in mind that WA is expected—as outlined in Section 2—to contribute more to three- than to two-body decays.

(v) No reliable conclusion on F decays can be drawn before a sizeable number of exclusive decay modes has been measured. More generally one would like to know whether two-body modes are equally important, i.e. dominant in non-leptonic F decays as they are in D^0, D^+ decays. Stech *et al.*[9] predict for $\tau(F) = \tau(D^0)$:

$$BR(D^0 \to PP,\ PV,\ VV) \sim 0.67 \pm 0.10 \tag{83a}$$

$$BR(D^+ \to PP,\ PV,\ VV) \sim 0.56 \pm 0.15 \tag{83b}$$

$$BR(F^+ \to PP,\ PV,\ VV) \sim 0.56 \pm 0.08 \tag{83c}$$

These numbers should be compared to the total non-leptonic decay widths. With $b_{SL}(D^0) \sim 0.075$,$b_{SL}(D^+) \sim 0.17$ and *assuming* $b_{SL}(D^0) = b_{SL}(F^+)$ one concludes from the data

$$BR(D^0 \to non - lept.) \simeq 0.85$$

$$BR(D^+ \to non - lept.) \simeq 0.65$$

$$BR(F^+ \to non - lept.) \simeq 0.85$$

Two-body final states are thus still expected to dominate non-leptonic F decays - yet by a somewhat smaller margin than in D^0/D^+ decays. Ignoring the VV final states one obtains in the Stech ansatz

$$BR(D^0 \to PP,\ PV) \sim 0.44 \tag{84}$$

$$BR(F^+ \to PP,\ PV) \sim 0.31 \tag{85}$$

whereas the Blok-Shifman treatment leads to [25]

$$BR(D^0 \to PP,\, PV) \sim 0.45 \tag{86}$$

$$BR(F^+ \to PP,\, PV) \sim 0.20\ . \tag{87}$$

Detailed data on F decays and more theoretical work is clearly needed to see whether the apparent difference between (84) and (85) or between (86) and (87) or between (85) and (87) is really significant or just reflects the theoretical uncertainties.

(vi) One can search semi-leptonic F decays for glueball candidates — in particular if (75) were found to hold— and one can look for $F^+ \to p\bar{n}$. These are admittedly "odd-ball" searches, but should be undertaken nevertheless.

B. Charmed Baryon Decays

Since the important points here have been nicely covered in Professor Huang's lecture [26] , I can be extremely brief and make mainly a comment on what *cannot* be learned in charmed baryon decays.

Recently the unfortunate claim has been made that a comparison between $\Gamma(\Lambda_c \to e\nu\Lambda)_{\mathrm{theor.}}$ and $\Gamma(\Lambda_c \to e\nu\Lambda X)_{\mathrm{exp.}}$ yields

$$|V(cs)| \sim 1/3 \tag{88}$$

in clear conflict with

$$|V(cs)|_{3\ \mathrm{fam.}} \simeq \cos\theta_c \simeq 0.97\ . \tag{89}$$

If true it would present strikingly clear, though indirect evidence for the existence of a fourth family. Such an analysis rests on three ingredients: (a) experimental data; (b) baryonic form factors; (c) the weak coupling $|V(cs)|$. It is obvious that Eq. (88) is completely inconsistent with all the data on D meson decays since it would lengthen decay times by a factor of 10! Therefore the baryonic form factors employed were probably grossly inadequate, the data used conceivably wrong and quite possibly both happened. The important point is that charm baryon decays allow interesting studies of hadronization effects—but not of weak couplings! Those are much better studied in the decays of charm mesons since they are much easier to produce and find and they also represent a simpler bound state.

Recently there has been a slight increase in the experimental value for $\tau(\Lambda_c)$, yet it is still considerably shorter than $\tau(D^0)$ [3] :

$$\left\langle \frac{\tau(\Lambda_c)}{\tau(D^0)} \right\rangle = 0.44 \begin{array}{c} +0.07 \\ -0.05 \end{array} . \tag{90}$$

This points to an important contribution of WA to Λ_c decays—which, as emphasized before—is not that surprising. The preliminary findings that

$$\Lambda_c \to p\pi K \tag{91}$$

is *not* dominated by K^* or Δ resonances is quite consistent with this picture.

C. Cabibbo Suppressed Decays

As already discussed in Section 2 (above Eq. (27)) color coherence leads to an enhancement of the Cabibbo suppressed modes $D^+ \to K\overline{K} + \pi$'s. One can employ the Stech *et al.* ansatz to translate this general expectation into a specific prediction and compare it to the MARK III data[9,3] :

$$\frac{BR(D^+ \to \overline{K}^0 K^+)}{BR(D^+ \to \overline{K}^0 \pi^+)} \simeq \begin{cases} 0.33 & \text{theoret.} \\ 0.317 \pm 0.086 \pm 0.048 & \text{MARK III} \end{cases} \tag{92}$$

$$\frac{BR(D^+ \to \pi^0 \pi^+)}{BR(D^+ \to \overline{K}^0 \pi^+)} \simeq \begin{cases} 0.04 & \text{theoret.} \\ < 0.15 (90\% \; C.L.) & \text{MARK III .} \end{cases} \tag{93}$$

Equation (92) represents a nice success, both qualitatively and quantitatively whereas more sensitive data are needed for $D^+ \to \pi^0 \pi^+$—a point to which we will return later.

A rather similar pattern seems to hold for $D^+ \to \overline{K}^{0*} K^+$ vs. $D^+ \to \overline{K}^{0*} \pi^+$—yet at present the data are far from conclusive.

The situation is less clear for Cabibbo suppressed D^0 decays. Theoretically there is no obvious, clear reason why such decays should be significantly enhanced like it is in the case of D^+ decays. If resonance enhanced WA were a major source of (Cabibbo allowed) D^0 decays one could entertain the idea that Cabibbo suppressed transition rates are even further decreased due to the absence of an appropriate resonance.

It is actually in these modes that charm decays revealed their first puzzling feature:

$$\frac{BR(D^0 \to \pi^+ \pi^-)}{BR(D^0 \to K^- \pi^+)} = 0.033 \pm 0.010 \pm 0.006 \lesssim tg^2 \theta_c \simeq 0.05 \tag{94}$$

$$\frac{BR(D^0 \to K^- K^+)}{BR(D^0 \to K^- \pi^+)} = 0.122 \pm 0.018 \pm 0.012 > tg^2 \theta_c \; . \tag{95}$$

Defining

$$R \equiv \frac{BR(D^0 \to K^-K^+)}{BR(D^0 \to \pi^-\pi^+)} \tag{96}$$

one might guess in a very naive spectator ansatz

$$R \sim \frac{|V(cs)V(us)|^2}{|V(cd)V(ud)|^2} \times \text{Phase space} \sim \text{Phase space} \lesssim 1$$

where we have used the tight constraints on the KM parameters that hold in a three family ansatz. Yet $SU(3)_{\mathrm{FL}}$ breaking, as exemplified by $f_K > f_\pi$, has to be included and one expects $R > 1$ on quite general grounds. More specifically one finds

$$R \sim \begin{cases} 1.44 & \text{"Stech"} \\ 1.4 & \text{"1/N"} \end{cases} . \tag{97}$$

(As already state, $SU(3)_{\mathrm{FL}}$ breaking has not been incorporated into the QCD sum rule approach yet.) Thus there is a very apparent conflict between theoretical expectations (97) and experimental findings (94,95). On the theoretical side this discrepancy could be fed from three different sources.

(*i*) an underestimate of $SU(3)_{\mathrm{FL}}$ breaking;

(*ii*) large FSI, rescattering, etc.

(*iii*) Penguin operators, which were ignored in "Stech" and in "$1/N$" since they are non-leading.

Source (iii) is clearly the most intriguing one. In addition to the usual Penguin operators there are the so-called "sideways" or "horizontal" Penguins. I will ignore those for now since in a $1/N$ expansion they are even more suppressed ($\sim 1/N\sqrt{N}$) than the ordinary Penguins ($\sim 1/\sqrt{N}$).

Penguin transitions are certainly not a major force in charm decays: they are always Cabibbo suppressed, and, as usual, the Penguin operator commands only a small coefficient in $\mathcal{L}_{\text{eff}}(\Delta C = 1)$. In addition there is the technical problem that Penguin operators are, strictly speaking, not local anymore since the mass of the virtual strange quark is considerably lighter than that of the external charm quark. In summary: no reliable prediction on the strength of Penguin transitions can be made; it is only clear that they are small. Yet it is important to keep two things in mind[27] :

(*a*) Penguin and Spectator transitions contribute coherently to $D \to K\overline{K}$, $\pi\pi$.

(*b*) The two processes contribute with the same sign to $D \to K\overline{K}$, but with the opposite sign to $D \to \pi\pi$ because of $|V(cd)| \simeq -\sin\theta_c$!

Including Penguin transitions thus further enhances R and, because of the coherence, their impact is magnified. For example a mere 20% Penguin amplitude raises R from its values in (97) to

$$R(0.2 \text{ Penguin}) \sim 3.2 \ . \tag{98}$$

If that were the resolution to the puzzle posed by (94,95) we would have learned a very important lesson with far-reaching consequences:

- it would be the first positive evidence for Penguins;
- it would have repercussions for estimates on ϵ'/ϵ_K;
- it would strengthen the belief in the relevance of Penguin operators in B decays, with important consequences for rare decay modes and CP asymmetries in them.

Yet before one jumps to such conclusions one has to address more mundane explanations for $R \sim 3-4$.

ad (i):

There are two modes that are not affected by FSI and/or WA: $D^+ \to \overline{K}^0\pi^+$, $\pi^0\pi^+$. Both final states are exotic in the isospin context, i.e. $I(\overline{K}^0\pi^+) = 3/2$, $I(\pi^0\pi^+) = 2$. Thus WA does not contribute at all and FSI are not expected to play a significant role. Therefore

$$\frac{\Gamma(D^+ \to \pi^0\pi^+)}{\Gamma(D^+ \to \overline{K}^0\pi^+)} = \frac{1}{2}\, tg^2\theta_c \times SB \tag{99}$$

where $SB \neq 1$ represents $SU(3)_{\rm FL}$ breaking.

ad (ii)

The decays

$$D^0 \to \overline{K}^0 K^0,\ \pi^0\pi^0 \tag{100}$$

can be produced from the Spectator process via rescattering. WA can generate them, too - in particular also, contrary to claims in the literature, $D^0 \to \overline{K}^0 K^0$ unless $SU(3)_{\rm FL}$ symmetry is obeyed in the hadronization process. Different authors [28] using a different ansatz obtain widely different predictions for $D^0 \to \overline{K}^0 K^0$, $\pi^0\pi^0$. Data are therefore required to settle this rather relevant issue.

D. Semi-leptonic Decays

In the Bauer-Stech-Wirbel description one finds [10]

$$\Gamma(D \to \ell\nu K) \sim (7.6-8.3)\times 10^{10}\,\mathrm{sec}^{-1} \tag{101}$$

$$\Gamma(D \to \ell\nu K^*) \sim (7.7-9.5)\times 10^{10}\,\mathrm{sec}^{-1} \tag{102}$$.

Adding them up results in a value that basically saturates the observed semi-leptonic width

$$\Gamma(D \to \ell\nu X) \sim (18.6 \pm 2.1)\ 10^{10}\,\mathrm{sec}^{-1} \ . \tag{103}$$

There is unfortunately one piece of data from MARK III[3] that casts a shadow over this picture: while their measurement of

$$\frac{BR(D \to K\pi e\nu)}{BR(D \to K\pi e\nu) + BR(D \to Ke\nu)} = 0.44 \begin{array}{l} +0.08 \\ -0.09 \end{array} \tag{104}$$

is quite consistent with (101,102),

$$\frac{BR(D \to K^* e\nu)}{BR(D \to K\pi e\nu)} = 0.55 \pm 0.13 \tag{105}$$

is not. A confirmation of (105), with smaller error bars, would raise some unpleasant theoretical concerns:

(i) From (104,105) one infers using central values only:

$$\frac{BR(D \to K^* e\nu)}{BR(D \to K\pi)} \sim 0.43 \tag{106}$$

i.e. quite different from (101,102). In the Grinstein *et al.* ansatz[29] it is even harder to accommodate. This does not mean that no model could be constructed that reproduces (106). But it implies that some of the overlap wavefunctions used by Stech *et al.* are inaccurate, in particular for the $D \to V$ matrix element. This would suggest that their predictions for $D \to VV$ modes are not quite reliable.

(ii) Finding a $\sim 20\%$ non-resonant contribution in semi-leptonic D decays is not really surprising. However, it suggests that the corresponding non-resonant contribution to semi-leptonic bottom decays will be even more important—which might or might not be a welcome result.

While $|V(cd)|$ is rather tightly constrained theoretically in a three family ansatz by $\sin\theta_c$, the experimental bounds on it as obtained from deep inelastic scattering are not overly impressive. Semileptonic D decays offer an independent handle on it when one has learned to deal with hadronization[9] :

$$\frac{BR(D \to \ell\nu\pi)}{BR(D \to \ell\nu K)} \sim (0.08 - 0.09)\,\frac{|V(cd)|^2}{0.05} \tag{107}$$

$$\frac{BR(D \to \ell\nu\rho)}{BR(D \to \ell\nu K^*)} \sim (0.07)\,\frac{|V(cd)|^2}{0.05}\ . \tag{108}$$

Additional dynamical information is gained from analyzing the shape of the energy spectrum of the charged leptons. This leads to the following challenge: is it possible to recover $|V(cd)|$ from the endpoint spectrum in $D \to \ell^+\nu X$ or does one have to rely on *exclusive* modes like $D \to \ell\nu\pi$, ρ. The answer to this challenge is of course of high relevance in B decays.

E. Doubly Cabibbo Suppressed D Decays (=DCSD)

Charm decays unlike K decays allow to observe a phenomenon that is quite typical for heavy flavor states—decays that are suppressed by more than one small KM angle:

$$\Gamma(D \to K^+\pi\text{'s}) \propto |V(cd)V^*(us)|^2 \simeq tg^4\theta_c\ . \tag{109}$$

As discussed in some detail in Section 6, the question of the real strength of DCSD is a crucial topic in searches for $D^0 - \overline{D}^0$ mixing. Here I want to raise another issue: looking for DCSD can be viewed as a search for New Physics. For Standard Physics, i.e. W exchange is highly suppressed in amplitude:

$$A(c \underset{\mathrm{W}}{\longrightarrow} d\bar{s}u) \propto tg^2\theta_c \ .$$

If there were New Physics mediated by charged Higgs fields for example one would guesstimate

$$A(c \underset{\mathrm{H}}{\longrightarrow} s\bar{d}u) \propto m_c m_d \qquad A(c \underset{\mathrm{H}}{\longrightarrow} d\bar{s}u) \propto m_c m_s \ .$$

Thus the "signal", i.e. New Physics, to "noise", i.e. Old Physics, ratio might quite possibly be greatly enhanced in DCSD, say by a factor $(m_s/m_d)^2/tg^4\theta_c \sim 4 \times 10^4$!

F. Applications in B Decays

(i) The value of $\tau(B^{\pm})/\tau(B^0)$ or $b_{\mathrm{SL}}(B^{\pm})/b_{\mathrm{SL}}(B^0)$ is both of intrinsic interest and serves as a crucial input parameter for many studies like $B^0 - \overline{B}^0$ mixing, etc. From our present understanding of $\tau(D^+)/\tau(D^0)$ I infer, quite conservatively I believe:

$$\frac{\tau(B^+)}{\tau(B^0)} \lesssim 1.3 \quad scaling \quad from \quad \tau(D) \tag{110a}$$

to be compared with the present CLEO findings[17]

$$\frac{\tau(B^+)}{\tau(B^0)} \lesssim 2.0 \tag{110b}$$

An improvement in the experimental bound is still highly desirable.

(ii) Considerable emphasis is sometimes (mis-) placed on comparing the experimental numbers for $b_{\mathrm{SL}}(B)$, averaged over $B^{\pm}, B^0$, which are around 11-12%, with quite naive theoretical guesstimates around 14%. Applying the lessons gained in D decays to B decays one finds that hadronization effects, questions on the proper value of ξ, etc. can quite naturally change $b_{\rm SL}(B)$ from 14% down to 12% or so.

(iii) Two-body modes do not dominate B decays as they did D decays. Nevertheless they are not insignificant and actually quite crucial when searching for rare decays and CP asymmetries[30] .

(iv) On general theoretical grounds one expects[24]

$$f_F > f_D \simeq \frac{3}{2} f_B \ . \tag{111}$$

Any bound on or number for f_F, f_D thus reflects on f_B which is a crucial parameter in describing $B^0 - \overline{B}^0$ mixing. It is amusing to note that the present MARK III upper limit

$$f_D \leq 340 \ MeV$$

leads to

$$f_B \lesssim 220 \ MeV$$

when invoking (111) which agrees with theoretical upper bounds obtained directly for f_B.

(v) Not too much is known on $|V(ub)|$ for certain beyond the statement that it is considerably smaller than $|V(cb)|$. As discussed in part **D.** detailed studies of semi-leptonic D decays will provide us with valuable insights into the relative merits of inclusive—$B \to \ell\nu X$—versus exclusive—$B \to \ell\nu\pi, \rho$—studies.

(vi) If Penguin transitions were identified in $D^0 \to KK$, $\pi\pi$ decays one would be on firmer ground to predict $BR(B \to K\pi, K\rho)$ and CP asymmetries in these modes.

6. Heresies: Exotic Decays, $D^0 - \overline{D}^0$ Mixing and CP Violation

A. Exotic D Decays

MARK III has searched for the decay $D^0 \to \mu e$ which is forbidden in the Standard Model and established an upper limit[3] :

$$BR(D^0 \to \mu e) < 1.5 \times 10^{-4} \qquad (90\%\ C.L.)\ . \tag{112}$$

Searches of this type might appear as a complete waste of time when one remembers that much smaller branching ratios (and even tinier decay widths) can be reached in K decays. Such a statement is overly pessimistic in general: firstly it ignores the possibility that exotic flavor changing currents could be produced by the exchanges of Higgs fields which couple much more strongly to a heavy flavor like charm than to strangeness; furthermore it is conceivable though not guaranteed that exotic flavor changing neutral currents possess much larger mixing angles to up-type quarks like $c\overline{u}$ than to down-type quarks like $s\overline{d}$.[31]

After this general pronouncement it has to be added however that $D^0 \to e\mu$ does not provide the most promising field to search for New Physics. This becomes obvious by comparing $D^0 \to e\mu$ to $D^+ \to \mu^+\nu_\mu$:

$$BR(D^0 \to e\mu) < 1.5 \times 10^{-4} \tag{113}$$

$$\Gamma(D^0 \to e\mu) \lesssim 3.4 \times 10^8\, \mathrm{sec}^{-1} \tag{114}$$

$$BR(D^+ \to \mu^+\nu) < 8.4 \times 10^{-4} \tag{115}$$

$$\Gamma(D^+ \to \mu^+\nu) \lesssim 8.4 \times 10^8 \, \mathrm{sec}^{-1} \ . \tag{116}$$

$D^+ \to \mu^+\nu$ is a Standard Model process due to real WA; it is tiny due to the combined effect of helicity suppression $(m_\mu/m_D)^2$ and wavefunction overlap suppression $(f_D/m_D)^2$. For $f_D = 200$ MeV one actually predicts $BR(D^+ \to \mu^+\nu) \simeq 2.9 \times 10^{-4}$.

Unfortunately quite analogous suppression factors enter in $D^0 \to e\mu$: the new interaction has to be local, therefore (f_D/m_D) again enters in the amplitude; if it were due to a spin-one exchange, the same helicity factor (m_μ/m_D) emerges; since a spin-zero coupling would violate chiral invariance one invokes general arguments to conclude that the coupling had then to be proportional to the fermion mass, i.e. m_μ!

Much more promising decay modes are therefore those that do not suffer from this double suppression, namely

$$D \to \pi\nu\overline{\nu},\ \rho\nu\overline{\nu},\ \pi\mu e \quad \text{etc.} \tag{117}$$

One might add that quite a few grand unified models that have gained popularity in the last years contain lepto-quarks, i.e. bosons that connect leptons and quarks; they can quite naturally generate the transitions in (117).

A. $D^0 - \overline{D}^0$ Mixing

(a) Phenomenology:

Mixing means that the flavor eigenstate D^0 is not a mass eigenstate; thus its time evolution is not described by a single exponential:

$$\left|D^0(t)\right\rangle = g_+(t)\left|D^0\right\rangle + \frac{q}{p}\, g_-(t)\left|\overline{D}^0\right\rangle$$

$$\frac{p}{q} = \frac{1+\epsilon}{1-\epsilon}; \quad g_\pm(t) = \tfrac{1}{2}\, e^{-\frac{1}{2}\Gamma_1 t} e^{i m_1 t}\left(1 \pm e^{-\frac{1}{2}\Delta\Gamma} e^{i\Delta m t}\right) \tag{118}$$

with $\Delta\Gamma = \Gamma_2 - \Gamma_1$, $\Delta m = m_2 - m_1$. Γ_i and m_i denote the width and the mass for the two mass eigenstates D_i. The rate for semi-leptonic D^0 decays leading to a "wrong-sign" lepton is then easily found to be

$$\Gamma(D^0(t) \to \ell^- X) \propto \left|\left\langle \ell^- X | D^0(t)\right\rangle\right|^2 \propto \left|\frac{q}{p}\right|^2 e^{-\Gamma t}(1 - \cos \Delta m t) \tag{119}$$

where in the last step we have assumed for simplicity $\Delta\Gamma = 0$—a point we will return to later.

It is this deviation from an exponential time dependence that represents the defining property for mixing.

A fixed target experiment like E691 can measure such a time evolution; in addition, as an extra bonus, D^* decays allow to flavor-tag the produced neutral D meson:

$$D^{+*} \to D^0\pi^+ \qquad D^{-*} \to \overline{D}^0\pi^- . \tag{120}$$

The most general expression describing $D^0 \to K^+\pi^-$ for small mixing reads [32]

$$\begin{aligned}\Gamma(D^0(t) \to K^+\pi^-) \propto e^{-\Gamma t}\Big\{ &(\Gamma t)^2(x^2+y^2) + 4tg^2\theta_c \left|\widehat{\rho}_f\right|^2 \\ &+ 4y(\Gamma t)\, tg^2\theta_c \,\mathrm{Re}\,\frac{p}{q}\widehat{\rho}_f - 4x(\Gamma t)\, tg^2\theta_c\, \mathrm{Im}\,\frac{p}{q}\widehat{\rho}_f \Big\}\end{aligned} \tag{121}$$

with the definitions

$$x = \frac{\Delta m}{\Gamma} \; ; \quad y = \frac{\Delta \Gamma}{2\Gamma} \; ; \quad \frac{A(D^0 \to K^+\pi^-)}{A(D^0 \to K^-\pi^+)} = tg^2\theta_c \widehat{\rho}_f \; . \tag{122}$$

The four terms in (122) are easily interpreted: the first one containing t^2 (in addition to the time exponential) is the pure mixing term. The second one with no additional time dependence represents DCSD. The third one linear in t is due to the interference between the amplitude for DCSD and $\Delta\Gamma$ mixing. The fourth one finally which is also linear in t describes interference between DCSD and Δm mixing; it represents a CP asymmetry which is evident from the appearance of $Im \frac{p}{q}\, \widehat{\rho}_f$. This fact can be understood immediately by recalling that $x\Gamma t = \Delta m t$ is just the first term in the expansion of $\sin \Delta m t$—a function that changes sign under $t \to -t$.

E691 has presented a preliminary analysis using *only the first two terms in* (122). They find[3]

$$r_D \simeq \tfrac{1}{2}\,(x^2 + y^2) \lesssim 0.5\% \tag{123}$$

$$|\widehat{\rho}_f|^2 \simeq \frac{1}{tg^4\theta_c}\,(0.025 \pm 0.014) \simeq 9 \pm 5 \tag{124}$$

A few comments are in order here:

(i) If CP invariance holds the fourth term in (122) has to vanish; however, the third term is still there in general. If it contributes with a positive sign, it will *lower* the upper bound in (123); if it interferes destructively it will *raise* it. I find the first possibility more likely though not guaranteed: if CP is not violated $\mathrm{Re}\,\frac{p}{q}\,\widehat{\rho}_f = -|\widehat{\rho}_f|$ holds due to $\epsilon = 0$, $V(cd) \simeq -\sin\theta_c$. Furthermore $\Delta\Gamma = \Gamma(D_-) - \Gamma(D_+)$ follows where $D_+[D_-]$ denotes the even [odd] CP eigenstate. Just counting the number of PP, PV channels available for D_+ versus D_- decays—the argument is not more sophisticated than that—suggests $\Gamma(D_+) > \Gamma(D_-)$ and therefore $y \,\mathrm{Re}\, \frac{p}{q}\, \hat{\rho}_f > 0$.

(ii) The value for the strength of DCSD in $D \to K\pi$, Eq. (124) appears much too high since a computation based on factorization yields[33]

$$|\hat{\rho}_f|^2 \simeq 2 \tag{125}$$

for $f = K\pi$.

Important cross checks can be obtained by studying $D^+(t) \to K^+\pi^-\pi^+$—where there can be no mixing—and $D^0(t) \to K^+\rho^-$; the strength of mixing—x and y—characterizes the decaying D meson and therefore should be equally present or absent in all appropriate decay modes. For DCSD the situation is quite different: for example one estimates[33]

$$|\hat{\rho}_f|^2 \sim 0.5 \tag{126}$$

for $f = K\rho$ to be contrasted with (125)!

When one lacks sufficient time resolution—as it happens in $e^+e^- \to \psi'' \to D\overline{D}$ or $e^+e^- \to \Upsilon(4s) \to B\overline{B}$—one can study only time-integrated quantities. Decays of D^0 mesons into wrong-sign leptons are then searched for as a sign for mixing:

$$r_D = \frac{BR(D^0 \to \ell^- X)}{BR(D^0 \to \ell^+ X)} \simeq \left|\frac{q}{p}\right|^2 \frac{x^2 + y^2}{2} \tag{127}$$

for $x^2, y^2 \ll 1$ as is appropriate for $D^0 - \overline{D}^0$ mixing.

Two complications arise here that should be kept in mind:

(α) Strictly speaking, $r_D \neq 0$ by itself does not establish mixing, it does so only within the framework on the Standard Model! Remember the old debate on the $\Delta S = \Delta Q$ rule in K physics. When we discuss D^0 decays into "wrong-sign" kaons we will recognize this remark as not purely academic.

(β) In e^+e^- annihilation or hadronic collisions there are always charm-anticharm pairs produced.

To deal with complication (β) one employs correlations between the $D\overline{D}$, etc., decays; as we will see this takes care also of complication (α), at least in principle. The simplest correlation is that between the charge of leptons originating in semi-leptonic charm decays:

$$y = \frac{N(\ell^\pm \ell^\pm X)}{N(\ell^+ \ell^- X)} = \begin{cases} r_D & p \text{ wave} \\ \frac{2r_D}{1+r_D^2} & s \,\&\, p \text{ wave} \end{cases} \tag{128}$$

where $N(\ell\ell X)$ denotes the number of direct leptons of a given charge in $D^0\overline{D}^0 \to \ell\ell X$ with the $D^0\overline{D}^0$ pair in a relative p wave—like in $\psi'' \to D\overline{D}$—or in a configuration where s and p waves contribute equally—like a GeV or more above threshold.

It is intuitively clear why the ratio is suppressed in a p wave: Bose statistics tells us that one cannot have a $D^0(t_1)D^0(t_2)$ state in a p wave for $t_1 = t_2$.

The whole complexity of this program comes into play in a theoretical analysis of the MARK III study on[3]

$$e^+e^- \to \psi'' \to D^0\overline{D}^0 \to K^\pm K^\pm + \pi\text{'s} \,. \tag{129}$$

Most importantly one has to note that even within the Standard Model there is no strict $\Delta C = \Delta S$ rule; DCSD produce $\Delta C = -\Delta S$

transitions. Comparing the rates for $D^0\overline{D}^0$ being in a p wave or in an s wave configuration will allow us to disentangle the two effects [33,34]

$$\frac{N(K^\pm\pi^\mp,K^\pm\pi^\mp)}{N(K^\pm\pi^\mp,K^\mp\pi^\pm)} \simeq \begin{cases} \frac{1}{2}(x^2+y^2) & (p), \\ \frac{3}{2}(x^2+y^2)+4tg^2\theta_c\,|\widehat{\rho}|^2+8ytg^2\theta_c\widehat{\rho}_f & (s). \end{cases} \tag{130}$$

It is important to note that without mixing the reaction

$$e^+e^- \rightarrow \psi'' \rightarrow D^0\overline{D}^0 \rightarrow (K^\pm\pi^\mp)(K^\pm\pi^\mp)$$

is forbidden by Bose statistics! The same can be shown to hold for $\psi'' \rightarrow D^0\overline{D}^0 \rightarrow (K^\pm\rho^\mp)(K^\pm\rho^\mp)$, $(K^{*\pm}\pi^\mp)(K^{*\pm}\pi^\mp)$, but not for $(K^*\rho)(K^*\rho)$ or $(K\pi\pi)_{non-res}(K\pi\pi)_{non-res}$. The large width of the ρ meson does not invalidate these results in principle, but of course dilutes their power considerably in practice.

Equation (130) also does not apply when $D^0\overline{D}^0$ decay into two different final states like $D\overline{D} \rightarrow (K\pi)(K\rho)$. There one finds when mixing is ignored, i.e. $x = y = 0$ [33] :

$$\frac{N(K^\pm\pi^\mp,\,K^\pm\rho^\mp)}{N(K^\pm\pi^\mp,\,K^\mp\rho^\pm)} = |\widehat{\rho}_{K\pi} - \widehat{\rho}_{K\rho}|^2\,tg^4\theta_c\ . \tag{131}$$

It does not vanish unless all DCSD had the same universal suppression factor $\widehat{\rho}tg^2\theta_c$. This is however very unlikely to hold. A computation involving factorizable contributions actually yields:

$$\widehat{\rho}_{K\pi} \simeq -1.4 \qquad \widehat{\rho}_{K\rho} \simeq -0.7\ . \tag{132}$$

This is another example of how dangerous it is to argue purely on the quark level and ignore modifications imposed by hadronization.

A last remark on future findings: comparing E691 and MARK III data and improving them might lead to the conclusion that there is no sign for $D^0 - \overline{D}^0$ mixing, say $r_D < 0.5\%$. At the same time we might find, say

$$\widehat{\rho}_{K\pi} \sim 3 \qquad \widehat{\rho}_{K\rho} \sim 2$$

i.e. considerably larger values than expected theoretically. As discussed in Section 5.E this might still signal the presence of New Physics.

(b) Standard Model Estimates:

On very general grounds one expects $D^0 - \overline{D}^0$ mixing to be small:

- $D^0 \to \overline{D}^0$ transitions are suppressed by $\sin^2\theta_c$; that is true also for $K^0 \to \overline{K}^0$ and $B^0 \to \overline{B}^0$ - however regular D decays, in contrast to K and B decays, are not suppressed by a small KM angle.
- Due to the GIM mechanism $D^0 \to \overline{D}^0$ transitions can proceed only due to $SU(3)_{\text{FL}}$ breaking. $K^0 \to \overline{K}^0$ $[B^0 \to \overline{B}^0]$ is driven by the much larger $SU(4)_{\text{FL}}$ $[(SU(6)_{\text{FL}}]$ breaking.

The remaining question is only "how small is small"? Like in the case of $K^0 - \overline{K}^0$ mixing there are two types of contributions to Δm_D, one generated by the simple quark-box diagram and one due to long-distance physics:

$$\Delta m_D = \Delta m_{\text{D, box}} + \Delta m_{\text{D, L.D.}} \,. \tag{133}$$

It has been recognized for a long time that $\Delta m_{\text{D, box}}$ is very tiny, a major reason being that $SU(3)_{\text{FL}}$ breaking enters via the ratio $(m_s^2 - m_d^2)/M_W^2 \ll 1$.

The situation is much less clear-cut for $\Delta m_{D,L.D.}$. One finds that there are at least 12 diagrams contributing with different signs, that soft gluons enter, etc. It is clearly hopeless to attempt a purely theoretical calculation.

Instead of giving up one can employ more phenomenological prescriptions to arrive at an estimate at least [33,35]. Let us consider

$$D^0 \to \text{“}PP \to \overline{D}^0$$

i.e. transitions mediated by a pair of two in general virtual pseudoscalar mesons $P = K, \pi$. This amplitude contains four parts representing the different intermediate states:

$$A(D^0 \to \overline{D}^0) = \sin^2\theta_c \left\{[K^+K^-] + [\pi^+\pi^-] - [\pi^+K^-] - [K^+\pi^-]\right\} \tag{134}$$

the K^+K^- $[\pi^+\pi^-]$ pair couples to both D^0 and $\overline{D}^0$ with strength $\sin\theta_c[-\sin\theta_c]$; $K^+\pi^-[\pi^+K^-]$ on the other hand couples to D^0 with $-\sin^2\theta_c[\cos^2\theta_c]$ and to $\overline{D}^0$ with $\cos^2\theta_c[-\sin^2\theta_c]$.

In the limit of $SU(3)_{\rm FL}$ symmetry the amplitude in (134) vanishes, yet remember

$$\frac{BR(D^0 \to K^+K^-)}{BR(D^0 \to \pi^+\pi^-)} \sim 3 - 4$$

as evidence for large symmetry breaking in exactly these modes!

To obtain a rough guesstimate one proceeds as follows: starting from the experimental numbers $BR(D^0 \to K^+K^-) \sim 0.6\%$, $BR(D^0 \to \pi^+\pi^-) \sim 0.2\%$ to which one adds 50% to include the

$K^0\overline{K}^0$ and $\pi^0\pi^0$ modes one arrives at

$$BR(D_+ \to K\overline{K} + \pi\pi^-) \simeq 2BR(D^0 \to K\overline{K} + \pi\overline{\pi}) \sim 0.025$$

(with $|D_+\rangle = CP\,|D_+\rangle$) where we have used the coherent nature of mixing. Thus very roughly

$$\Delta m_D \sim 0.03\,\Gamma_D \qquad \Delta\Gamma_D \sim 0.03\,\Gamma_D \qquad r_D \sim \mathcal{O}(10^{-3})\ . \quad (135)$$

Needless to say the real value of r_D due to long distance dynamics could be considerably smaller. The point of this exercise was to show that values like those given in (135) are not clearly ruled out in the Standard Model.

There is a well-known strategy for improving the quality of our estimate:

- derive a dispersion relation;
- evaluate it using measured $\pi\pi$, $K\overline{K}$, $K\pi$ phase shifts.

However the actual execution of this program is rather non-trivial and is therefore unlikely to be undertaken unless $D^0 - \overline{D}^0$ mixing is found on the $r_D \sim 10^{-3}$ level

To sum up: the Standard Model might allow for

$$x_D \sim 0.03\,, \quad y_D \sim 0.03\,, \quad r_D \sim 10^{-3}$$

but *not* for $r_D \sim 10^{-2}$.

C. New Physics Scenarios

It is very hard to see how new quarks, etc., that are then included in the box contribution can push r_D appreciably beyond the 10^{-3}

level. The situation changes if there are genuine flavor changing neutral currents on the tree level as one can encounter them in non-minimal Higgs models. It is quite possible to construct models where these exotic currents are mediated by the exchange of Higgs scalars with a mass of around a few TeV in such a way that they are not clearly identifiable in Δm_K, but produce

$$x_D \sim \mathcal{O}(0.1)\,, \quad y_D \lesssim 0.03\,. \tag{136}$$

Such models contain quite naturally CP violation in their Higgs sector that appears superweak in K decays, but is much stronger in heavy flavor decays.

D. CP Violation in Charm Decays

Observing CP violation in charm decays would be fascinating per se and would at the same time establish the presence of New Physics. The good news is that there are theoretical scenarios where D^0 decays could exhibit CP asymmetries of up to 10%, i.e. much larger than in K^0 decays. The bad news is that even under such favorable circumstances the experimental searches will be extremely challenging.

(a) D^0 decays:

the most promising decay modes, as we will see, are

$$D^0 \to K_S K^+ K^-,\ K^+ K^- \tag{137}$$

each commanding a branching ratio of roughly 0.5%. They are special in this context since these final states represent CP eigenstates. The time evolution of their transition rates takes a very interesting form

$$\Gamma(D^0(t) \to f) \simeq e^{-\Gamma t}\left(1 - \sin \Delta m t \, \mathrm{Im} \frac{p}{q} \rho_f\right) \qquad (138)$$

$$\Gamma(\overline{D}^0(t) \to f) \simeq e^{-\Gamma t}\left(1 + \sin \Delta m t \, \mathrm{Im} \frac{p}{q} \rho_f\right) \qquad (139)$$

with

$$\rho_f = \frac{A(\overline{D}^0 \to f)}{A(D^0 \to f)}, \quad f = K^+K^-, \quad K_sK^+K^- .$$

We have assumed $\Delta\Gamma = 0$ for simplicity. It is easy to show that either relation (138) or (139) establishes CP violation. Considering just (138) for small mixing, i.e. $\Delta m \ll \Gamma$ one gets

$$\Gamma(D^0(t) \to f) \simeq e^{-\Gamma t}\left(1 - (\Gamma t) \; x \mathrm{Im} \frac{p}{q} \rho_f\right) . \qquad (140)$$

$D^0 - \overline{D}^0$ mixing somewhat below the 1% level, say $r_D \simeq 5 \times 10^{-3}$ implies $x \simeq 0.1$. Such a mixing strength requires the presence of New Physics as discussed above which quite naturally leads to CP violation as well; thus $Im \frac{p}{q} \rho_f \sim 0.5$ could hold and

$$\Gamma(D^0(t) \to f) \simeq e^{-\Gamma t}(1 - 0.05(\Gamma t)) \qquad (141)$$

is not a ludicrous scenario. For $t \sim 2\tau_{D^0}$ this represents a 10% CP asymmetry. Furthermore, the time dependence is rather peculiar and not easily faked by backgrounds. Being able to resolve the time evolution is therefore highly desirable, though not essential in general.

An E691 and ARGUS type experiment is well suited for such an analysis:

(α) It has sufficient time resolution.

(β) It can employ the D^* trick to distinguish between D^0 and $\overline{D}^0$ decays. The final states obviously do not allow this distinction and summing over D^0 and $\overline{D}^0$ decays, see Eqs. (138) and (139), washes the effect out.

The situation is less favorable in this respect when one measures D decays just above production threshold.

(α) Since one cannot resolve the time evolution one has to rely on correlations like

$$D^0\overline{D}^0 \to (\ell^+/K^+ + X^-) + f \quad \text{vs.} \quad D^0\overline{D}^0 \to (\ell^-/K^- + X^+) + f \tag{142}$$

(β) Unfortunately these correlations cannot show a CP asymmetry in $\psi'' \to D^0\overline{D}^0$, i.e. when $D^0\overline{D}^0$ are produced in a p wave[30]. A CP asymmetry can become observable if $D^0\overline{D}^0$ form an s wave state as in, e.g.

$$e^+e^- \to D\overline{D}^* + \text{h.c.} \to D^0\overline{D}^0\gamma \, . \tag{143}$$

(b) $\psi'' \to D^0\overline{D}^0$:

If both neutral D mesons were seen to decay into a CP eigenstate of the same CP parity, e.g.

$$\psi'' \to D^0\overline{D}^0 \to (K^+K^-)_D(K^+K^-)_D \tag{144}$$

then CP violation would have been established. For the initial state is CP even whereas the final state (p wave!) is CP odd. Note that we are talking about a rate here and *not* a difference between two CP conjugate rates.

There is of course, as always, a drawback to this method as apparent from a different representation of the reaction chain in (144):

$$\psi'' \to D_+ D_- \underset{\text{Mix.}}{\longrightarrow} D_+ D_+ \to (K^+K^-)_D (K^+K^-)_D \ .$$

The rate (in almost all cases) requires $D^0 - \overline{D}^0$ mixing to have occurred and therefore very roughly

$$\text{relative rate} \sim x^2 BR(D \to f_1)\ BR(D \to f_2) \qquad (145)$$

where f_1, f_2 are CP eigenstates of the same CP parity. If one can employ only one such decay mode with a branching ratio of 1% one estimates a relative rate of at best 10^{-6} for $r_D \simeq 0.5\%$. Yet if one could sum over many appropriate channels like $D^0 \to K_S\omega$, $K_S\eta$, etc., one might obtain—very optimistically—up to 10% for each of the two branching ratios in (145) leading to a relative rate of $\sim 10^{-4}$!

(c) D^+, F^+, Λ_c^+ decays:

For CP asymmetries to become observable one needs the intervention of non-trivial FSI to generate different phase shifts for different isospin amplitudes. That (remember the discussion in Section 3, etc.) does not pose any problem. Secondly, one needs sizeable violations of weak universality; I consider that rather unlikely, but not impossible to happen. Therefore one should make an effort to compare as many CP conjugate rates as possible. For more detailled discussions, see the papers by L.-L. Chau and co-workers.

7. Summary and Outlook

Weak decays of kaons have been studied experimentally as well as theoretically for more than 30 years now. A pessimist might point out that despite all the efforts spent we still have not come up with an explanation for a basic and striking phenomenon, the $\Delta I = \frac{1}{2}$ rule which produced huge decay rate differences: $\tau(K^+)/\tau(K_s) \sim 135$!

Charm decays have been studied for a considerably shorter time span; namely, 10 to at most 15 years. It would be unjustified to claim that we understand charm decays fully; nevertheless I clearly believe that we have developed a decent overall understanding of charm decays which is actually considerably better than that in strange decays (including hyperon decays). The reason for that does not lie in a sudden jump in the intellectual power of theorists working in this field—after all, the $\Delta I = \frac{1}{2}$ rule has not been explained yet —but in the concurrence of three factors:

(*i*) Nature was apparently kind enough to provide us with a relatively simple dynamical system. See for example the two-body dominance in D decays.

(*ii*) Experimentalists worked hard enough to provide us with good, *comprehensive* data.

(*iii*) There was and still is a lively feed-back between theorists and experimentalists.

Nevertheless there cannot be any space for complacency. The basis for the statement that we have obtained a very decent understanding of charm decays is still unstable: a two sigma change in the data here and a two sigma change there and soon we would be in real trouble.

As far as experimental capabilities are concerned, charm physics is a mature field. As far as its theory is concerned, we are still in late adolescence. It is therefore both possible and very important

(a) to secure and widen the basis on which we place our claim of success

- by obtaining more precise data on D decays like $D \to VV$ modes or $D^0 \to \overline{K}^0 \omega, \eta, \eta'$
- by obtaining more data on F and charm baryon decays (semi-leptonic and non-leptonic branching ratios).

(b) to probe deeper by more detailed data on

- Cabibbo suppressed modes like $D^0 \to \overline{K}^0 K^0, \pi^0\pi^0, D^+ \to \pi^0\pi^+, D \to \ell\nu\pi, \rho$
- doubly Cabibbo suppressed decay like $D^+ \to K^+\pi^-\pi^+$.

The important lesson to remember from the MARK III experience is that just a few even well measured branching ratios are not sufficient to subject theoretical treatments to sensitive tests, that a comprehensive analysis involving many different channels is crucial.

Finally we should not let orthodoxy blind us for the potential for still making major new discoveries in charm decays like $D^0 - \overline{D}^0$ mixing, exotic rare decays or even, most ambitiously, CP violation. There are two historical lessons from K decays of importance here. Firstly

(*i*) Parity violation first exhibited itself in K decays;

(*ii*) $K^0 - \overline{K}^0$ mixing was discovered leading to postulating charm;

(*iii*) CP violation which showed, among many other things, the need for top and bottom quarks.

All of this represented New Physics at that time.

Secondly, the time scale over which real intellectual progress develops in this kind of physics is not one or two years—it is one or two decades!

The future of charm physics is therefore still full of promise - yet it can be realized only after very hard and patient work.

Acknowledgements:
It is a pleasure to thank Profs. Ye and Huang for inviting me to participate in such a stimulating and well organized meeting. I have very much enjoyed the many discussions and contacts I had with my Chinese colleagues. One note of apology might be in order: due to the very tight deadline for submitting the manuscript I was unable to include as extensive a list of references as I would have liked or regarded as appropriate.

References

1. See the lectures by F.Gilman, these Proceedings, for details.
2. For a very nice review with references to earlier work, see: R.Rueckl, Habilitation Thesis, 1983.
3. See the lectures by D.Hitlin and M.Witherell, these Proceedings.
4. S.P.Rosen, Phys.Rev.Lett. 44(1980)4.
5. H.Fritzsch and P.Minkowski, Phys.Lett.90B(1980)455; W.Bernreuther et al., Z.Physik C4(1980)257; I.Bigi, Z.Physik C5(1980)313.
6. M.Bander et al., Phys.Rev.Lett. 44(1980)7,962(E).
7. Having been born and raised in Bavaria, I am entitled to being conservative and stubborn. I invoke this privilege to stick to the notation $F = (c\bar{s})$.
8. I.Bigi, Phys.Lett.90B(1980)177.
9. D.Fakirov and B.Stech, Nucl.Phys.B133(1978)315; M.Bauer, B.Stech, Phys.Lett.152B(1985)380; M.Bauer, B.Stech and M.Wirbel, preprint HD-THEP-86-19, to appear in Z.Physik C.
10. M.Wirbel, B.Stech and M.Bauer, Z.Physik C29(1985)637.
11. L.-L.Chau and H.-Y.Cheng, Phys.Rev.Lett.56(1986)1655; preprint UCD-86-32-R.
12. For a related, though not identical analysis, see: A.N.Kamal, Phys.Rev.D33(1986)1344.
13. X.-y.Li, X.-q.Li and Ping Wang, preprint AS-IIP-87-007; Ping Wang, these Proceedings.

14. I.Bigi and M.Fukugita, Phys.Lett.91B(1980)121.

15. J.H.Kuehn, R.Rueckl, Phys.Lett.135B(1984)477.

16. I.I.Bigi, Phys.Lett.169B(1986)101.

17. For a recent review on B decays, see: B.Gittelman and S.Stone, Cornell preprint CLNS 87/81.

18. A.J.Buras et al., Nucl.Phys.B268(1986)16; A.J.Buras, in: Proceedings of the International Symposium on Production and Decay of Heavy Hadrons (=ISPDHH), Heidelberg, 1986, eds. K.R.Schubert and R.Waldi, pg.179.

19. V.A.Novikov et al., Phys.Rep.41(1978)1.

20. Reinders et al., Phys.Rep.127(1985)1.

21. A.R.Zhitnitsky et al., Yad.Fiz.38(1983)1277;
T.Aliev, V.Eletsky, Yad.Fiz.38(1983)1537.

22. H.Krasemann, Phys.Lett.96B(1980)397;
S.Godfrey, Phys.Rev.D33(1986)1391.

23. E.Golowich, Phys.Lett. 91B(1980)271.

24. M.A.Shifman, in: Proceedings of ISPDHH, Heidelberg, 1986, pg. 199.

25. B.Blok, M.A.Shifman, preprints ITEP-9,17,37 (1986).

26. T.Huang, these Proceedings.

27. A.I.Sanda, Phys.Rev.D22(1980)2814; X.Y.Pham, Mod.Phys. Lett. A(1986)619.

28. X.Y.Pham, preprint PAR LPTHE 87-16; A.N.Kamal, R.Sinha, preprint Alberta Thy-1-87.

29. B.Grinstein et al., CALT-68-1311(1986).

30. I.I.Bigi, A.I.Sanda, Nucl.Phys.B281(1987)41.

31. For a more detailled discussion, see:
W.Buchmueller and D.Wyler, Nucl.Phys.B268(1986)621.

32. I.I.Bigi, in: Proceedings of the XXIII International Conference on High Energy Physics, Berkeley, 1986, ed.S.C.Loken, World Scientific, pg.857.

33. I.I.Bigi and A.I.Sanda, Phys.Lett.171B(1986)320.

34. D.-S. Du, D.-D.Wu, Chin.Phys.Lett.3(1986)389.

35. L.Wolfenstein, Phys.Lett.B164(1985)170; J.Donoghue et al., Phys.Rev.D33(1986)179.

36. L.-L.Chau, in: Proceedings ISPDHH, Heidelberg, 1986.

CHARM PHYSICS IN FIXED TARGET EXPERIMENTS

M. S. WITHERELL
Department of Physics
University of California, Santa Barbara, CA 93117

Abstract I give a review of charm physics results from fixed target experiments. There is an introduction to the difficulties and strengths of such experiments. Most of the paper consists of experimental results on charm lifetimes and decays from Fermilab E-691. I also discuss charm production, charmonium spectroscopy and prospects for open charm spectroscopy. Finally, I review the experiments which will be running in the near future.

Since I am discussing the development of charm physics at fixed target experiments, I should start with the reasons why the progress in this area has been so difficult. The most important reason is that the cross section for charm production is only about 0.5% of the total cross section for photons, and about 0.1% for hadrons. In addition, the multiplicity is higher than at SPEAR. As a result of these conditions, the D^+ peak in the $K^-\pi^+\pi^+$ mass spectrum sits on a background which is about 30 times as large as the signal in photoproduction, and about 200 times as large as the signal in hadroproduction. The only property of charm which is distinctive enough to allow such a small signal to be seen clearly is the separated charm decay vertex.

To resolve the charm vertex as separate from the primary vertex the charm decay length L_D must be much larger than its error, δL_D. The error δL_D is approximately equal to σ_t/θ, where σ_t is the transverse measurement error on each track, and θ is the typical angle between the decay tracks and the direction of the parent D-meson. The decay distance is, on the average

$$L_D = c\tau_D \; P/M \simeq c\tau_D/\theta.$$

So our requirement for good separation becomes

$$\sigma_t << c\tau_D.$$

For the D^o, $c\tau = 130\mu m$, and for the D^+ it is $320\mu m$. The measurement errors for wire drift chambers are typically $200\mu m$, which is too large to resolve the charm vertex. The lexan bubble chamber, LEBC, which was developed at CERN for the purpose of studying charm, has a measuring error of around $20\mu m$. More recently, silicon microstrip detectors (SMD) which obtain resolutions of $10 - 15\mu m$ have been used in charm experiments.

The advantage of LEBC is that the charm identification is clean and efficient.[1] The disadvantage is that the rate it can take is limited. A good run of about 10^6 pp interactions contains about 1000 charm events of which a few hundred can be identified. This technique is useful for measuring cross sections and the D lifetimes, but it is not useful for measuring individual decay modes.

The silicon microstrip detector is now widely used in charm experiments.[2] A detector is usually a $300\mu m$ thick silicon wafer with strip width of $25 - 50\mu m$. The ionization from the charged particle produces a signal of 25,000 electrons at the anode. Solid state detectors had been in use in nuclear physics for some time, but it took recent progress in fast, low-noise amplifiers to make them suitable for high energy physics experiments. A pioneer in applying the use of SMD hodoscopes to physics experiments was the ACCMOR group, in experiment NA11 and NA32.[3] In a good run of these experiments, about $3 \cdot 10^7$ interactions with $3 \cdot 10^4$ charm events could be recorded. This was enough to see a significant number of events in individual decay modes. In the best modes, $D^o \rightarrow K^-\pi^+$ and $D^+ \rightarrow K^-\pi^+\pi^+$, there were about 1000 events produced. After losses for spectrometer acceptance and vertex separation efficiency, there were about 30 detected events in these modes. The mass peaks showed very little background, with signal-to-background ratios of about 10, which were used for measuring the charm meson lifetimes.

E-691

At the time that E-691 was proposed in December of 1983, fixed-target charm experiments had been useful primarily for cross sections and lifetimes. The goal of the proposed experiment was to combine very high statistics and clean charm separation, in order to produce physics results in the areas of charm decay, spectroscopy, lifetime, and production.

There were a number of factors which contributed to the large charm sample observed in E-691. In photoproduction the fraction of events with charm is about 0.5%, higher than in hadroproduction. The data acquisition system was designed to handle 100 events per second with 30% deadtime, so that a full 6250 BPI tape could be filled with 40,000 events in about 8 minutes of beam time. The trigger, which required a scalar sum of transverse energy (E_T) greater than 2.2 GeV, was very efficient for charm. The Tevatron represented a large improvement over earlier accelerators in two respects. The duty factor was about 30%, rather than about 7% in the 400 GeV operation at Fermilab. For an experiment of this type, limited by instantaneous data rate, this means a factor of 4 improvement in average data rate. In addition the 800 GeV protons were necessary to produce a high-rate 260 GeV electron beam. Finally, the Advanced Computer Program (ACP) multiprocessor system was required to be able to handle the large computing load produced by E-691.

There were 300 interactions per second during the data-taking for E-691, of which 100 satisfied the E_T requirement in the trigger. During 10^6 seconds of live time a total of 10^8 events were written to tape, of which somewhat over 10^6 contained charm. Thus there were

$5 \cdot 10^4$ events per good decay mode ($D^o \to K^-\pi^+$ or $D^+ \to K^-\pi^+\pi^+$). The spectrometer acceptance is around 15% and the vertex efficiency is about 40%, resulting in about 3000 events for each of these modes.

A 800-GeV/c proton beam with $2.5 \cdot 10^{12}$ protons per 22-second spill struck the primary target. Photons from π^o-decay were used to produce electron-positron pairs in a converter. The resulting electron beam had an average energy of 260 GeV and an intensity of $6.5 \cdot 10^7$ particles per spill. This beam was focused on a tungsten target with thickness 20% of a radiation length, followed by a magnet. The photon energy was measured by observing the degraded electron energy in shower counters. The photon beam followed a bremsstrahlung spectrum (1/k) from 80 GeV to 260 GeV, with an average energy of 145 GeV.

The Tagged Photon Spectrometer which was used for E-691 is shown in Figure 1.[4] The TPS is a large-aperture spectrometer with two magnets, 35 drift chamber planes, and two Cerenkov counters for particle identification. The electromagnetic and hadronic calorimeters were used for electron selection, photon detection, and the transverse energy measurement needed for the trigger. The region near the target is shown in Figure 2. There were 9 planes of silicon detectors, three in each of three views. The strip width was $50\mu m$, and each of the 6840 strips were read out with discriminator-shift register electronics.

Lifetimes

Before describing the lifetime measurements, I will review briefly the general picture of charm decays. There is more detail in the lectures of I. Bigi[5] and D. Hitlin[6] in these proceedings. Fig. 3 shows the four quark diagrams that describe the Cabibbo-allowed decays of charmed mesons. The first two are both spectator diagrams in which the light antiquark is not one of the four fermions at the weak vertex. Dia-

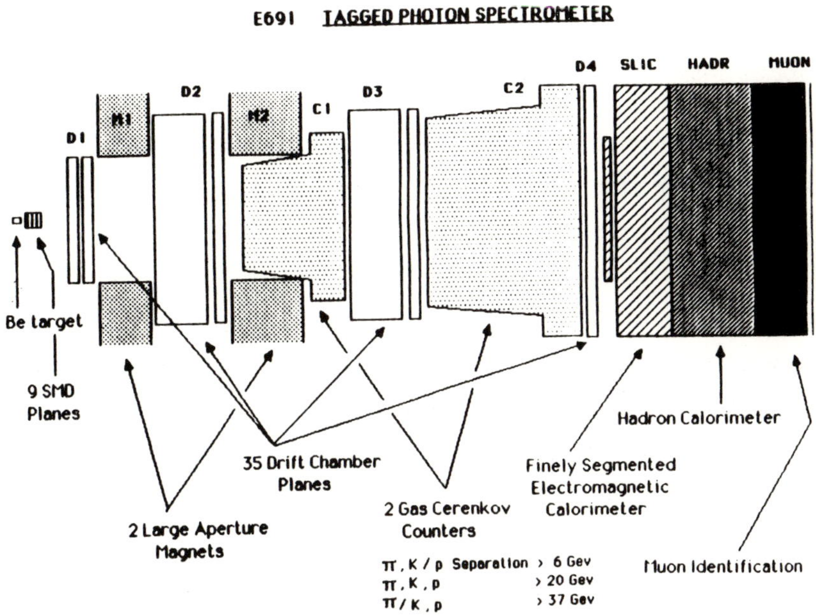

Fig. 1. The experimental apparatus for E-691.

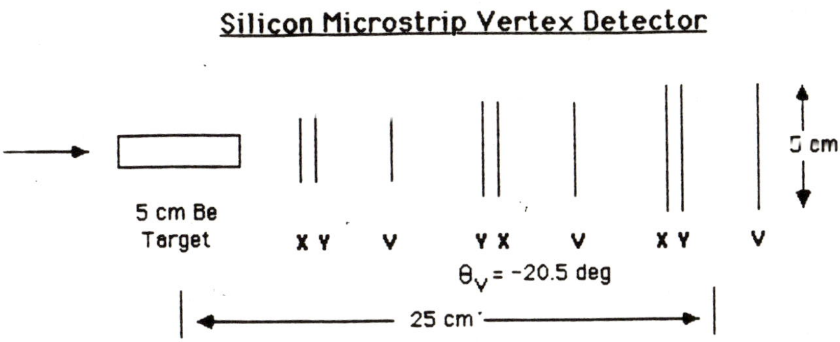

Fig. 2. The target area.

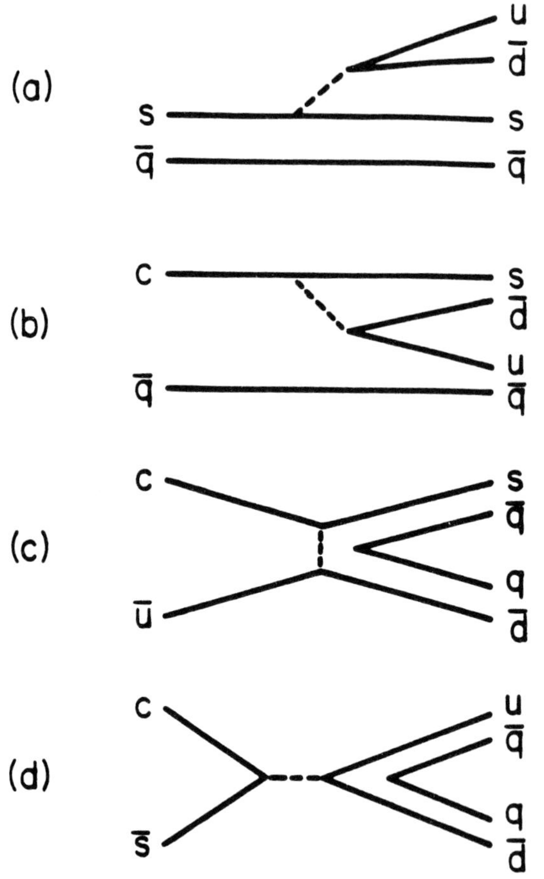

Fig. 3. The quark diagrams for Cabibbo-allowed decays of charmed mesons (a) spectator diagram with automatic color alignment; (b) spectator diagram without color alignment; (c) W-exchange diagram in D^o decay; and (d) W-annihilation diagram in D_s^+ decay.

gram (a) can be called "color-aligned" because the two $q\bar{q}$ pairs are necessarily color singlets. In diagram (b) the $q\bar{q}$ pairs are not always color singlets, and would be very strongly suppressed without QCD corrections. The last two diagrams are non-spectator or annihilation diagrams. Diagram (c) is the W-exchange diagram and contributes to D^o decay only. Diagram (d) is W-annihilation and produces a Cabibbo-allowed decay only in the case of the D_s^+.

For some time we have known that the lifetime of the D^+ is much longer than the D^o. According to the simplest model, the charm quark would decay with little effect from the light antiquark, and the ratio of lifetimes would be near unity. How can the ratio be as large as 2.5? This is the subject of much of Bigi's lectures in this proceedings. To simplify greatly, there are two major candidates to explain the effect: annihilation decays and interference.

The annihilation diagrams contribute to D^o, D_s^+, and Λ_c^+ decays, but not to the D^+ decays except at the Cabibbo-suppressed level. The annihilation diagrams were expected to be small for the mesons because of helicity suppression.[5] The Mark III limit on the mode $D^+ \rightarrow \mu^+\nu_\mu$ confirms this picture.[6] Unless the decay constant ($\tilde{f}_D$ in Bigi's notation) measured in nonleptonic decays is much larger than f_D measured in leptonic decays, the annihilation diagrams are only responsible for a small fraction of the D^o and D_s^+ decay rates. One decay which was expected to be due to annihilation is $D^o \rightarrow \overline{K}^o\phi$, which is discussed extensively in other talks here. The other is the $D_s \rightarrow \rho\pi$ decay, which I will cover in some detail.

The second possible cause of large differences in D lifetimes is suppression of the D^+ decay rate due to interference[5] The two spectator diagrams in Figure 3 can contribute to identical final states for the D^+. There is a negative interference between the two amplitudes, raising the D^+ lifetime. Although in the simple quark calculation, this suppression is not as large as needed for the lifetime difference, adding up the particular two-body final states leads to a lifetime ratio of 2 or more. Such a large coherent effect is not necessarily maintained for nonresonant multibody decays, but since the total decay rates seem to be dominated by two-body decays, calculating the lifetime ratio this way should be reasonably reliable.

The basic parameters of the models are fixed by the ratio $\tau(D^+)/\tau(D^o)$, which is now known to about 5%. Predictions can then be made for the other charm lifetimes. In models in which the interference causes the lifetime difference, the ratio $\tau(D_s^+)/\tau(D^o)$ is very close to 1. On the other hand, if annihilation decays dominate, this ratio is usually expected to be smaller than one although a wide range of values is possible.[5] The ratio $\tau(\Lambda_c^+)/\tau(D^o)$ is usually expected to be in the range 0.5-0.8, and to have a large contribution from the W-exchange diagram.

The method of lifetime analysis used in E-691 was chosen to minimize the corrections to the true decay time spectrum, $\exp(-t/\tau)$. Three vertex cuts defined the charm sample. The charm decay tracks had to originate from a common vertex with a good χ^2. The momentum vector of the charm candidate had to agree with the direction from the primary vertex to the secondary. Finally, for each event a minimum decay length L_M was defined as $5-10\sigma$, depending on decay mode. (Here σ is the error in calculating the decay length L_D.) We required that $L_D > L_M$ for the event to be included in the charm lifetime sample. The proper time was calculated using the point along the decay path a distance L_M from the primary vertex as t=0:

$$t = (L_D - L_M)/(cP/M).$$

The minimum decay length L_M should be chosen independently of the actual decay length, L_D. It should also be large enough that the point defined as t=0 is always downstream of the true production point. If these conditions are both met, the only correction to the $exp(-t/\tau)$ behavior is due to any time dependence of the efficiency of detecting the charm vertex. Choosing to study the time evolution of the charmed particle only after a minimum decay length isolates the measurement from uncertainties in the primary vertex position, at the cost of some efficiency.

The function used for the maximum likelihood fit to the time spectrum was

$$Nf(t)exp(-t/\tau)/\tau + B(t).$$

Here B(t) is the normalized time distribution for the background, as determined from the region of the mass plot outside the signal region. The two parameters allowed to vary in the fit were N, the number of events in the charm signal, and τ, the charm lifetime. The function f(t), which was obtained from the Monte Carlo, corrects for the effects of absorption, resolution, acceptance, and efficiency. The entire correction represented by f(t) shifted the lifetime by about 10% or less.

For the D^o lifetime study we used three samples (throughout this paper, the charge conjugate states are implicitly included):

$$D^{*+} \to \pi^+ D^o,\ D^o \to K^-\pi^+ \qquad (A)$$

$$D^{*+} \to \pi^+ D^o,\ D^o \to K^-\pi^+\pi^-\pi^+ \qquad (B)$$

$$D^o \to K^-\pi^+,\ (\text{no } D^{*+}) \qquad (C)$$

Events which satisfied the requirements for set (A) were excluded from set (C) so the samples were independent. For sets (A) and (B) the mass difference $m(D^{*+}) - m(D^o)$ was required to be between .1435 and .1475 GeV/c^2. Table I shows the number of signal and background events used for each mode, and the lifetime results obtained from each fit. The mass distributions for the events which satisfied the vertex cuts are shown in Figure 4. The proper time distribution

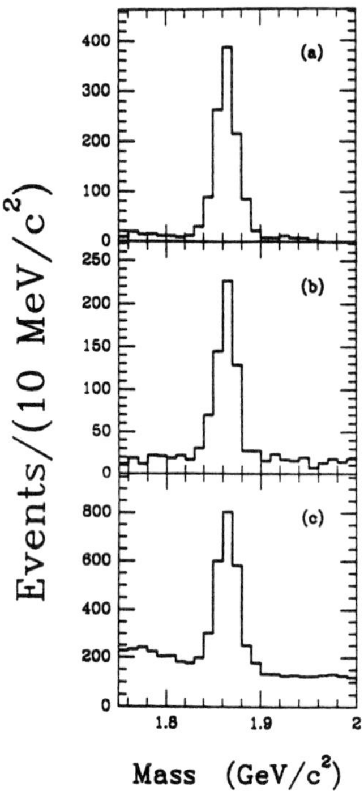

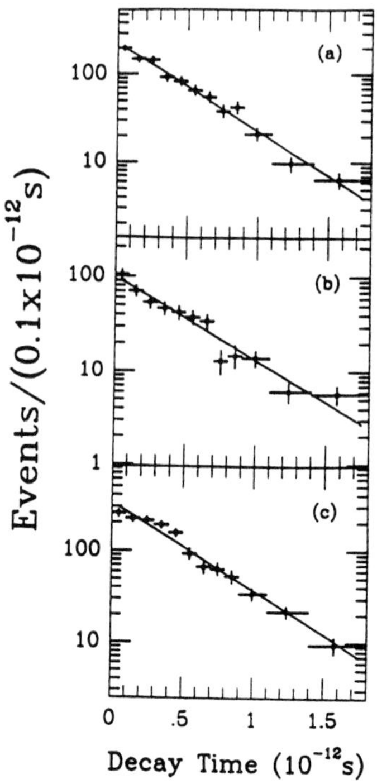

Fig. 4. Invariant mass spectra for three D^o channels with vertex cuts as described in the text. (a) $D^{*+} \rightarrow \pi^+ D^o$, $D^o \rightarrow K^-\pi^+$, (b) $D^{*+} \rightarrow \pi^+ D^o$, $D^o \rightarrow K^-\pi^+\pi^-\pi^+$, and (c) $D^o \rightarrow K^-\pi^+$, no D^{*+}.

Fig. 5. Proper spectra for the three D^o channels, in the same order as Fig. 1. The data points are shown with background subtracted. The error bars represent the statistical error, including that on the background. The smooth curve represents the best fit as described in the text.

for the three D^o channels, shown in Figure 5, follow an exponential form.

Table I *Characteristics of the D^0 samples; the errors on the signal and background events are statistical, and those for the lifetimes are statistical and systematic, respectively.*

Mode	L/σ	No. Signal	No. Background	Lifetime (ps)
(A)	5	964 ± 32	30 ± 3	0.424 ± 0.016 ± 0.006
(B)	7	488 ± 22	78 ± 5	0.441 ± 0.024 ± 0.009
(C)	8	1489 ± 39	1011 ± 17	0.418 ± 0.015 ± 0.008

The lifetimes for the three modes agree within statistical errors. The best value comes from a global fit to all three samples:

$$\tau(D^o) = 0.425 \pm .010 \pm .010 \ ps.$$

This number is based on an analysis of 80% of the full data sample. We have published a result[7] based on 30% of the data sample, and it was consistent with the present result. The world average from the Particle Data group was $0.43^{+.05}_{-.04}$ ps.[8]

The decay mode $D^+ \rightarrow K^-\pi^+\pi^+$ was used for the D^o lifetime study. The mass spectrum for the accepted events, using a minimum decay length $L_M = 10\sigma$, is shown in Figure 6. There are 2050 D^+ events in the selected D mass region between 1.848 and 1.890 GeV/c^2. The number of expected background events determined from the number of events outside the D^+ region, is 931 ± 16. The D^+ time distribution is shown in Figure 7, and the measured lifetime is $\tau(D^+) = 1.07^{+0.04}_{-0.03} \pm 0.025$ ps. The 1986 world average was $0.92^{+.13}_{-.10}$ ps[8] The ratio of the lifetime is

$$\tau(D^+)/\tau(D^o) = 2.52 \pm .11 \pm .08,$$

which agrees well with the ratio of semileptonic branching ratios from Mark III.[9]

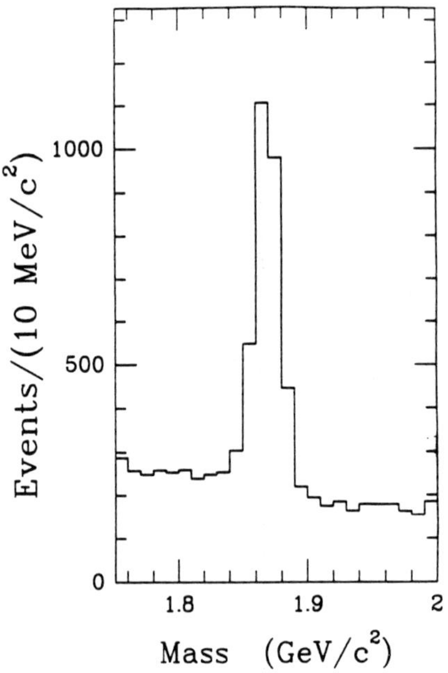

Fig. 6. Invariant mass spectrum for $K^-\pi^+\pi^+$.

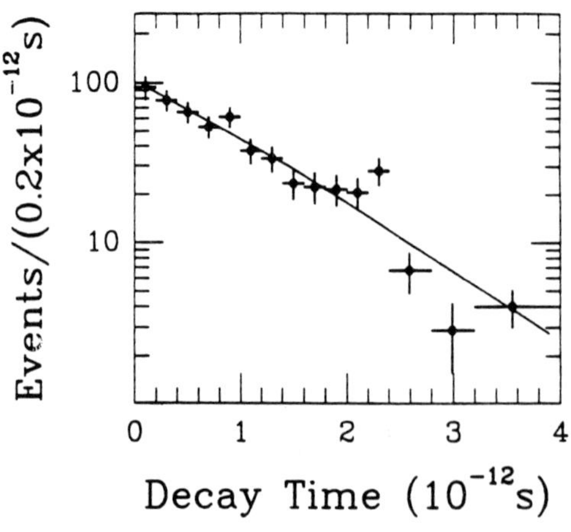

Fig. 7. Proper time spectrum for D^+ events. The data points are shown with background subtracted, as in Fig. 5. The smooth curve represents the best fit as described in the text.

For the D_s lifetime analysis, we used the channels $D_s^+ \to \phi\pi^+$ and $D_s^+ \to \overline{K}^{*o}K^+$. The angular distribution for the decay $\phi \to K^-K^+$ is $dN/d(cos\theta) = A\ cos^2\theta$, where θ is the angle between the K^- and π^+ in the ϕ rest frame. We required $|\ cos\theta\ | > 0.3$ for the $\phi\pi$ sample, which kept 97% of the signal and 70% of the background. The K^+K^- mass range chosen was 1.012-1.027 GeV/c^2. We also required $|\ cos\theta\ | > 0.3$ for the $\overline{K}^{*o}K^+$ sample, where θ is the decay angle between the K^- and K^+ in the K^* rest frame. We required the $K^-\pi$ mass to be in the range 0.845-0.945 GeV/c^2. Table II shows the characteristics of the two D_s^+ samples.

Table II *Characteristics of the D_s^+ samples; the errors on the signal and background events are statistical, and those on the lifetimes are statistical and systematic, respectively.*

Mode	L/σ	No. Signal	No. Background	Lifetime (ps)
(D)	6	118 ± 14	41 ± 4	$0.52^{+0.07}_{-0.06} \pm 0.017$
(E)	10	56 ± 9	20 ± 3	$0.43^{+0.08}_{-0.06} \pm 0.013$

The mass spectra are shown in Figure 8. In each plot there are clean peaks for the D_s^+ and for the Cabibbo-suppressed D^+ decay. The decay time distributions for events in the mass range 1.953-1.985 GeV/c^2 are shown in Figure 9. A joint fit to both distributions is used for the best measurement.

$$\tau(D_s^+) = 0.48^{+.05}_{-.04} \pm .02ps.$$

The ratio $\tau(D_s^+)/\tau(D^o) = (1.13 \pm .12 \pm .04)$ is consistent with 1. This result is also based on 80% of the full data sample, somewhat larger than the published result.[9] Our lifetime result is larger than the 1986 world average of $0.28^{+.16}_{-.07}$, but is consistent with it. Another recent measurement of the D_s^+ lifetime with good errors by NA 11/32 was

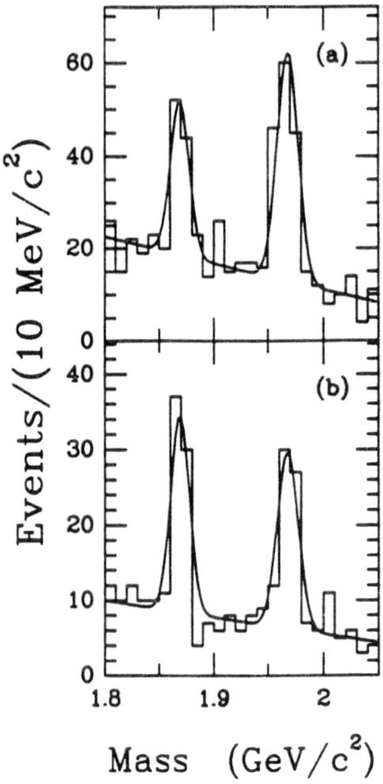

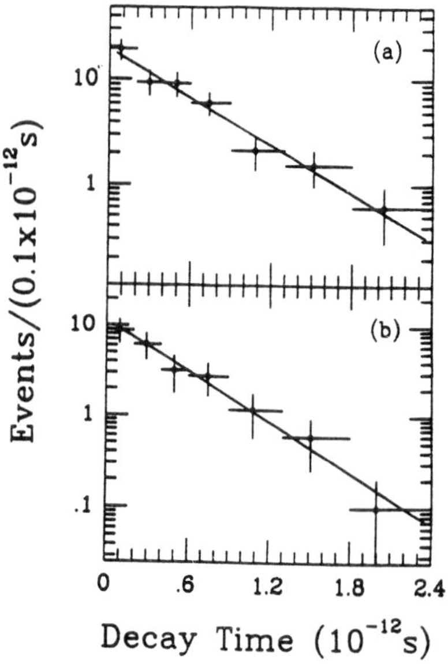

Fig. 8. Invariant mass spectra for two D_s^+ channels with vertex cuts as described in the text. (a) $D_s^+ \to \phi\pi^+$, $\phi \to K^+K^-$ and (b) $D_s^+ \to \overline{K}^{*o}K^+$, $\overline{K}^{*o} \to K^-\pi^+$.

Fig. 9. Proper time spectra for the two D_s^+ samples, in the same order as for Fig. 8. The data points are shown with background subtracted. The error bars represent the statistical error, including that on the background. The smooth curve follows the best fit as described in the text.

$\tau(D_s^+) = 0.33^{+.10}_{-.06}$ ps.[10] Table III shows a summary of the D lifetimes from E-691.

Table III *Summary of the D Lifetimes*

Particle	Lifetime (ps)	stat. error (ps)	sys. error (ps)
D^0	0.425	±0.010	±0.010
D^+	1.07	$^{+0.04}_{-0.03}$	±0.025
D_s^+	0.48	$^{+0.05}_{-0.04}$	±0.02

To summarize, the D_s^+ lifetime is consistent with the D^o, and the early indication that $\tau(D_s^+)$ might be significantly less than $\tau(D^o)$ has not been confirmed. This result is consistent with models in which the lifetime difference is not dominated by annihilation diagrams.

Our knowledge of the Λ_c^+ decays is not nearly as good as for the D-mesons. There is a W-exchange diagram $c + d \rightarrow s + u$ the contribution of which is probably not as small as in the meson case. There is no helicity suppression, and the wave function is probably not as small at zero distance between the two quarks. The signatures of large exchange contributions in baryon decay are

$$\tau(\Lambda_c^+) < \tau(D^o)$$
$$\tau(\Xi_c^+) > \tau(\Lambda_c^+)$$
$$\tau(\Xi_c^o) \simeq \tau(\Lambda_c^+).$$

We studied the decay $\Lambda_c^+ \rightarrow pK^-\pi^+$. The mass spectrum is shown in Figure 10, with a clear signal at the Λ_c^+, containing 68 ± 12 events over a background of 56 events. The decay time spectrum with background subtracted, is shown in Figure 11. The lifetime result

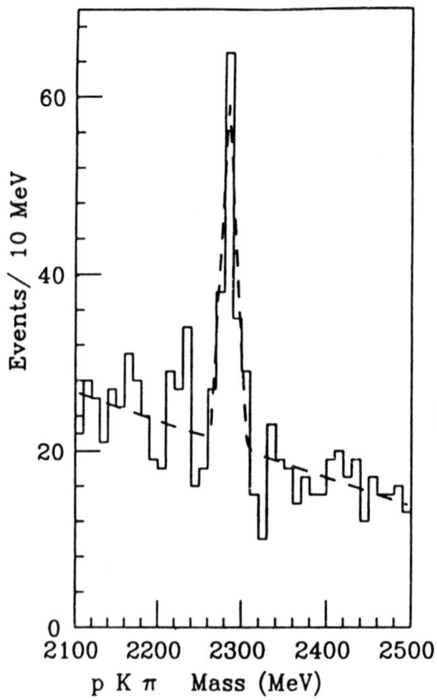

Fig. 10. Invariant mass spectrum for $pK^-\pi^+$ events.

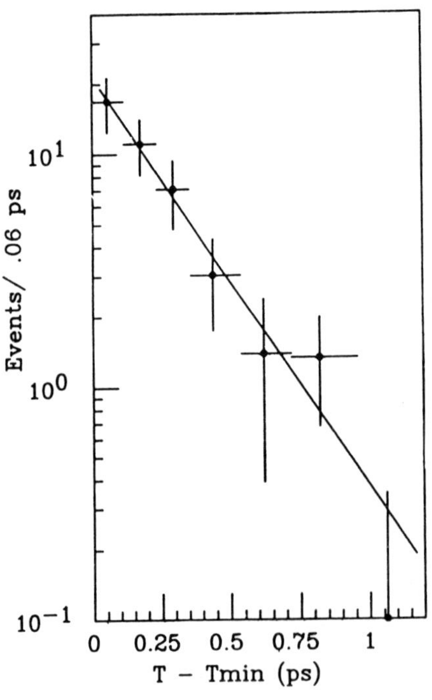

Fig. 11. Decay time spectrum for the $\Lambda_c \to pK^-\pi^+$ events, with background subtracted. The error bars represent the statistical error, with background subtracted. The smooth curve follows its best fit to the form described in the text.

from the maximum likelihood fit is

$$\tau(\Lambda_c^+) = 0.24 \pm 0.04 \pm 0.02.$$

A series of measurements of the Λ_c^+ lifetimes are shown in Table IV.

Table IV Λ_c^+ Lifetime Summary

Group	Reference	Λ_c^+ Lifetime (ps)
NA16	11	$.19^{+.14}_{-.07}$
WA58	12	$.23^{+.09}_{-.06} \pm .05$
E-531	13	$.20^{+.07}_{-.05}$
NA32	14	$.14^{+.05}_{-.03} \pm .03$
NA27	15	$.12^{+.05}_{-.03}$
E-691		$.24 \pm .04 \pm .02$

The weighted average of these results is $\tau(\Lambda_c^+) = .19^{+.03}_{-.025}$ ps, corresponding to $\tau(\Lambda_c^+)/\tau(D^o) = .44^{+.07}_{-.05}$. Thus the Λ_c^+ lifetime is significantly smaller than that of the D^o, probably due to the fact that W exchange is large. There is still variation of about a factor of 2 in the measurements, so the error on the weighted average may be an underestimate. Data recently collected by NA32 in a recent run should improve their statistical error significantly. Because Λ_c^+ decays may be dominated by the W-exchange diagram, it is particularly important to study the pattern of branching ratios. Already it has been noticed that the $pK^-\pi^+$ decay is not dominated by the $p\overline{K}^{*o}$ mode.[16]

In summary, the lifetimes are consistent with a picture in which spectator diagrams are dominant for meson decay. The relatively long D^+ lifetime can be explained by interference effects. In this picture, it is natural that the Λ_c^+ lifetime is short, due to the unhindered annihilation diagrams. We should be cautious about how uncritically we believe in this picture, however. Detailed measurements on more decays of the D_s^+ must be made before we can say that the annihilation contribution is completely understood.

Charm Decay Studies

The only fixed target experiment to date which has contributed significantly to the measurement of charm branching ratios has been E-691. There will be a number of new experiments from fixed target experiments in the next couple of years, however. It is therefore worth looking at which decay modes are easier for fixed-target experiments. In general, decays which have two or more prongs from the decay vertex are the ones for which the vertex cuts are most effective. For example, modes like $D^o \to K^-\pi^+$, $D^+ \to K^-\pi^+\pi^+$ and $D_s^+ \to K^+K^-\pi^+$ are very clean. Modes with a K^o or Λ^o and one charged track, such as $D^+ \to \overline{K}^o\pi^+$ and $\Lambda_c^+ \to p\overline{K}^o$ only have one prong from the charm vertex. There is some background reduction gained by requiring the single prong to be separated from the primary vertex, but it is not nearly as favorable as the 2- or 3- prong modes. Modes such as $D^o \to \overline{K}^o\pi^+\pi^-$ and $D^+ \to \overline{K}^o\pi^+\pi^-\pi^+$ do benefit from the full power of background suppression, on the other hand.

There are a few additional lessons learned from E-691 about charm decay studies in fixed-target experiments. The sensitivity is very high, which is important for modes with very small branching ratios but little background. The backgrounds, after tight vertex cuts, are due primarily to charm events. In fact, much of the combinatorial background from charm events can be excluded. Thus the signal-to-background is better than e^+e^- at high energies or even at SPEAR, unless double-tag events are used. The effect of the vertex cuts on background is greater at high multiplicity. For example, the background suppression for $D^+ \to K^-\pi^+\pi$ is a factor of about 10 greater than for $D^o \to K^-\pi^+$. For all of these reasons, photoproduction experiments will continue to be a good place to study difficult charm decays with two or more charged tracks from the charm vertex.

The vertex cuts used for lifetime studies were chosen to suppress background without distorting the exponential time dependence. In that analysis, we required that (1) the secondary vertex gave a good fit, (2) the charmed particle momentum vertex pointed from the primary vertex to the secondary, and (3) the decay length was greater than a minimum decay length. For decay studies, because we need better signal-to-background ratios, we apply additional cuts which reduce background, but are strongly biased against short lifetime decays. We require that no track other than those from the charm decay mode should pass within $80\mu m$ of the secondary vertex. We also require that no track from the charmed particle decay should pass closer to the primary than to the secondary. Typically these cuts reduce the background by an extra factor of about six.

Mark III has done a great deal of work studying D^o and D^+ decays, running at the $\psi(3770)$. The areas in which E-691 can most easily contribute to our understanding of charm decays are those which are complementary to what Mark III has already done. D_s^+ and Λ_c^+ decays are good candidates, because they are more difficult to study with high rates at SPEAR. Among the D^o and D^+ decays, $D^o - \overline{D}^o$ mixing, semileptonic decays, rare decays, and high multiplicity decays all have characteristics which make them good topics for study on E-691. In the discussion here, I will give results from some decay analyses that demonstrate various strengths of an experiment of this type: D_s decays, $D^o - \overline{D}^o$ mixing, and semileptonic decays.

D_s Decays

The simplest D_s decay mode to study is the $K^+K^-\pi^+$ final state. The ratio of $\Gamma(D_s^+ \to \overline{K}^{*o}K^+)/\Gamma(D_s^+ \to \phi\pi^+)$ is a measure of the

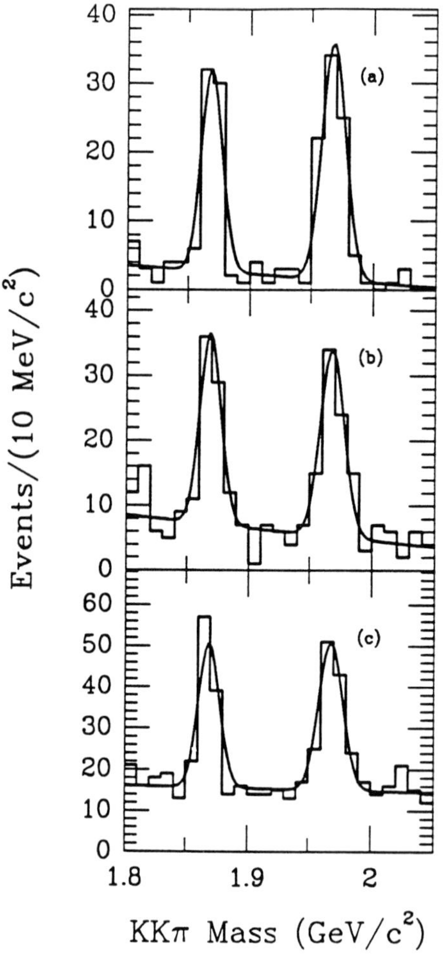

Fig. 12. The $K^-K^+\pi^+$ mass spectra for the (a) $\phi\pi^+$, (b) $\overline{K}^{*o}K^+$, and (c) nonresonant $K^+K^-\pi^+$ final states. All three samples have vertex cuts as described in the text. There is an additional cut on the decay time of $t > 0.20$ ps for sample (c). The curves represent fits with Gaussian peaks for the D^+ and D_s^+ and a linear background.

ratio of the two spectator diagram (see Figure 1). It is also the first opportunity to measure the size of nonresonant decays relative to the two-body decays. We know that for the D^o and D^+ the two-body modes dominate, which is important for predictions of the lifetime ratio, and we would like to check that pattern for the D_s^+. To see the nonresonant decay requires a combination of statistics and cleanliness that is difficult to achieve in e^+e^- experiments.

For the $K^+K^-\pi^+$ event sample, we used three tracks with a joint Cerenkov probability greater than 0.30. We applied the series of vertex cuts described above, including the vertex isolation cuts not used for lifetime studies. Figure 12(a) and (b) show the $\phi\pi^+$ and $\overline{K}^{*o}K^+$ mass spectra obtained. As expected, the backgrounds are even lower than in the sample used for lifetime studies, because of the extra isolation cuts. Fig. 12(c) shows the $K^+K^-\pi^+$ events not included in either of the resonance categories, with an additional cut on lifetimes of 0.2 ps to reduce the background further. There are peaks at both the D^+ and D_s^+ masses in the nonresonant sample. Although the peaks included some contribution from $\overline{K}^{*o}K^+$ decays in which the $K\pi$ mass is outside the K^* mass cut, that source can only account for about 25% of the D_s^+ peak. The nonresonant $D_s^+ \to K^+K^-\pi^+$ decay is therefore observed for the first time.

To calculate the rates for the three components of D_s^+ (or D^+) decays we do a Dalitz-plot analysis using the time dependence of the signal and background. Each event is characterized by $M^2(KK\pi)$, $M^2(K\pi)$, $M^2(KK)$ and the decay time t. We use a likelihood function for the fit which includes the known shapes for the $\phi\pi$ and K^*K, and a nonresonant term which is flat in the Dalitz plot. Figure 13 shows the Dalitz plots with a cut in proper time $t > 0.15$ ps, in addition to the minimum vertex separation cut, $L_D > 6\sigma$. In both the D^+ and D_s^+ plots, the bands due to the $\phi\pi$ and $\overline{K}^*K$ decays are evident. The

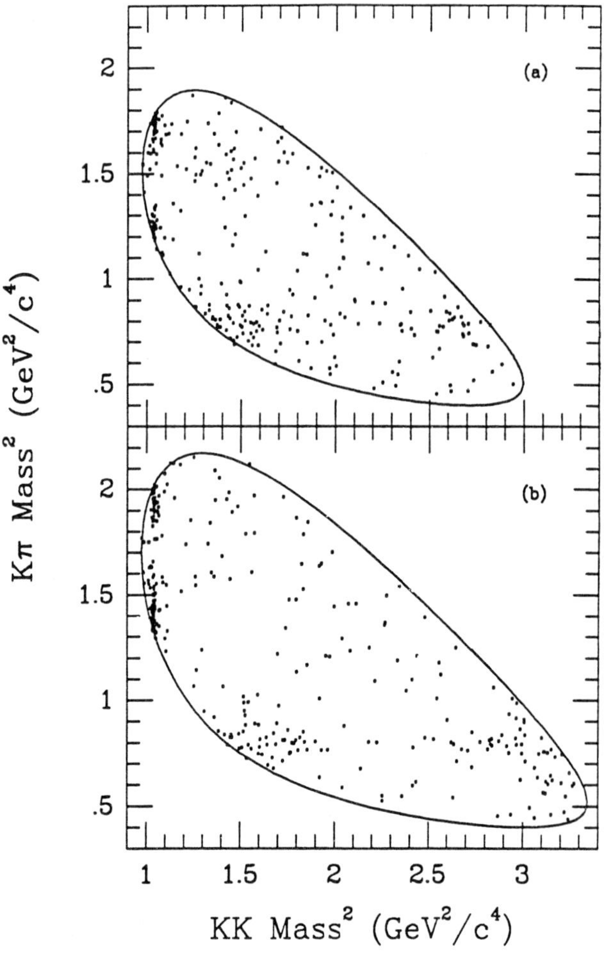

Fig. 13. The Dalitz plots for the (a) D^+ and (b) D_s^+ regions.

$\phi\pi$ events are collected near the ends of the band with $M^2_{KK} = M^2_\phi$, because of the $cos^2\theta$ distribution. Similarly, the $\overline{K}^* K$ events are at the ends of the broad K^* band. There are also events distributed throughout the plot, which are primarily due to nonresonant charm decays, rather than background. The results of the fit are shown in Tables V and VI.

Table V $D^+ \to K^+K^-\pi^+$ Results

Mode	# of Events	$\frac{\textit{Branching ratio}}{B(D^+ \to K^-\pi^+\pi^+)}$
$D^+ \to \phi\pi^+$	72 ± 9	$.103 \pm .014 \pm .009$
$D^+ \to \overline{K}^{*o}K^+$	68 ± 10	$.074 \pm .012 \pm .007$
$D^+ \to K^-K^+\pi^+$(nr)	67 ± 13	$.050 \pm .009 \pm .006$

Table VI $D_s^+ \to K^+K^-\pi^+$ Results

Mode	# of Events	$\frac{\textit{Branching ratio}}{B(D_s^+ \to \phi\pi^+)}$
$D_s^+ \to \phi\pi^+$	104 ± 11	1
$D_s^+ \to \overline{K}^{*o}K^+$	84 ± 11	$.74 \pm .12 \pm .06$
$D_s^+ \to K^-K^+\pi^+$(nr)	41 ± 12	$.23 \pm .07 \pm .07$

The relative branching ratios are corrected for efficiency and for the fraction of ϕ and K^* decaying into the observed state.

The $\overline{K}^{*o}K^+$ decay mode of the D_s^+ is almost as common as the $\phi\pi$, showing that the suppression of the diagram in Figure 1(b) relative to that in Fig. 1(a) is small. The three-body decay is only about 20% of the $K^-K^+\pi^+$ total, which is consistent with the picture that two-body decays dominate. It is encouraging that we see this small nonresonant mode clearly, even though the branching ratio is small and there are no resonant cuts to reduce the background. The D^+ branching ratios all agree with the Mark III measurements,[17] with smaller errors.

The $D_s^+ \to K^+K^-\pi^+$ sample also allows a good measurement of the mass of the D_s. When the mass is varied in the fit, the best value is 1967.7 ± 0.8 MeV/c^2. Our absolute mass scale is only known to about ± 1.5 MeV, so we can improve our measurements by tying it to the well-known D^+ mass of 1869.3 ± 0.6 MeV/c^2.[18] Our measurement of the D^+ mass from the $K^-\pi^+\pi^+$ and $K^+K^-\pi^+$ modes is 1868.9 ± 0.3 MeV/c^2 (statistical error only). Using this to correct the D_s^+ mass, our best value is

$$M(D_s^+) = 1968.1 \pm 0.8 \pm 0.6 \ MeV/c^2$$

The systematic error is dominated by the uncertainty in the D^+ mass. The 1986 world average from the Particle Data Group[7] was 1970.5 ± 2.5 MeV/c^2.

A major question in charm decay has been the extent to which the non-spectator or annihilation diagrams are responsible for the lifetime differences. The decay mode which is often proposed as the best one to measure the contribution is $D_s \to \rho\pi$.[19] If the c and $\bar{\text{s}}$ quarks annihilate at the weak vertex, the final state only has strange particles a small fraction of the time. Thus if the annihilation decay rate were equal to the spectator decay rate, the ratio $B(D_s^+ \to \rho\pi^+)/B(D_s^+ \to \phi\pi^+)$ would be near 1.

The analysis was very similar to that used for the $K^+K^-\pi^+$ analysis, but with somewhat tighter cuts. The minimum decay length, L_M, was chosen to be 15σ to reduce the background as much as possible, and identical cuts were applied to the $K^+K^-\pi^+$ sample for normalization. Figure 14 shows the $\pi^+\pi^-\pi^+$ mass spectrum for all events passing the vertex requirements. There are clear peaks at the

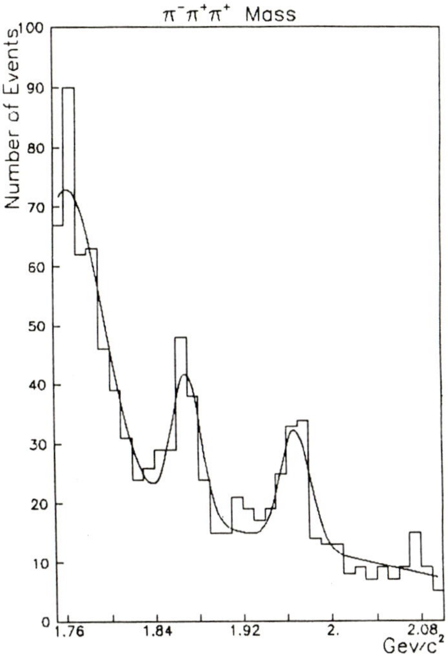

Fig. 14. The $\pi^+\pi^-\pi^+$ mass spectrum for all events passing vertex cuts. The curve represents a fit with a linear background, Gaussian peaks for the D^+ and D_s^+, and a peak for the misidentified $D^+ \to K^-\pi^+\pi^+$ events.

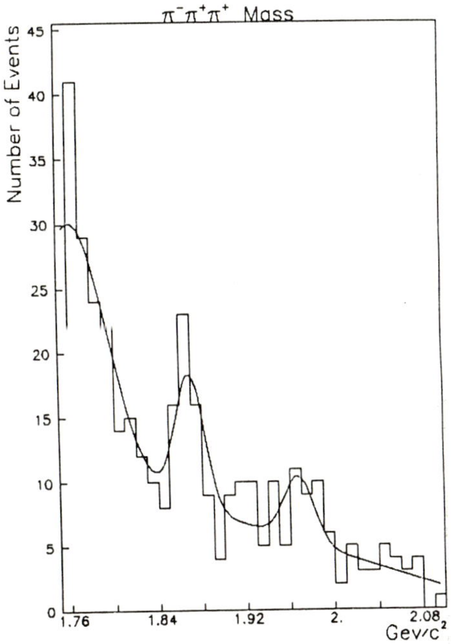

Fig. 15. The $\pi^+\pi^-\pi^+$ mass spectrum for events consistent with $\rho^o\pi^+$ decay. The curve is similar to that in Figure 14.

D^+ and D_s^+ mass. In addition, there is a broad enhancement peaking near 1.75 GeV/c^2 due to misidentified $D^+ \to K^-\pi^+\pi^+$ decays. The smooth curve in the figure is the result of a maximum likelihood fit to the data. The shape of the misidentified peak is known from studying the $D^+ \to K^-\pi^+\pi^+$ events with the Monte Carlo and with the large sample of such decays in the data. The masses and widths of the D^+ and D_s^+ peaks are known from the Monte Carlo and from the peaks in the $K^+K^-\pi^+$ sample. The background other than that from the false peak is parameterized by a linear term, which fits all of our three-body modes. The results are a total of 80 ± 15 D^+ events and 65 ± 12 D_s^+ events. The significance of the D_s^+ peak is about 6 standard deviations.

To determine the fraction of decays which come from the $\rho^o\pi^+$ state, we made a selection of events consistent with $\rho\pi$ decay. For this sample we accepted events with one of the two $\pi^-\pi^+$ combinations within ± 1 full width of the ρ mass. We also required $| \cos\theta | > 0.3$, where θ is the angle analogous to that defined in the $K^+K^-\pi^+$ analysis. This sample included about 89% of the $\rho\pi$ events and about 31% of the events from a flat Dalitz plot distribution. The mass spectrum for the $\rho\pi$ sample is shown in Figure 15. It is apparent that a significant fraction of the D^+ mode remains. There is a small peak at the D_s^+, but it is consistent with the number expected if all the decays were nonresonant. The maximum likelihood fit gives 49 ± 9 D^+ and 20 ± 7 D_s^+. After subtracting the number of nonresonant events, the number of events from the $\rho\pi$ mode is 34 ± 12 D^+ and -2 ± 10 D_s^+. After correction for efficiencies, the preliminary results are (the $\pi^+\pi^-\pi^+$ results include the $\rho\pi$ contribution)

$$\frac{B(D_s^+ \to \pi^+\pi^-\pi^+)}{B(D_s^+ \to \phi\pi^+)} = .29 \pm .07 \pm .05$$

$$\frac{B(D_s^+ \to \rho\pi^+)}{B(D_s^+ \to \phi\pi^+)} < .08 (90\% C.L.)$$

$$\frac{B(D^+ \to \rho\pi^+)}{B(D^+ \to K^-\pi^+\pi^+)} = .041 \pm .008 \pm .004$$

$$\frac{B(D^+ \to \rho\pi^+)}{B(D^+ \to \pi^+\pi^-\pi^+)} = .022 \pm .008 \pm .003.$$

The systematic errors are dominated by the uncertainty in the relative Cerenkov efficiencies.

The D_s^+ branching ratios show that states with no strange particles are significantly suppressed. Therefore annihilation does not appear to be the major source of the $D_s^+ - D^+$ lifetime difference. It is a surprise that less than about 35% of the $D_s \to \pi^+\pi^-\pi^+$ decays comes from the only resonant mode, $\rho^o\pi^+$. Perhaps the pattern for annihilation decays is not similar to that for spectator decays, and the 2-body modes are not dominant. In light of this result, it is especially important to look for $D_s \to 4\pi$ and 5π states. As in the cases of the D^o and D^+, there are surprises in the pattern of D_s^+ decays.

$D^o - \overline{D}^o$ Mixing

In analogy with the $K^o - \overline{K}^o$ systems, the $D^o - \overline{D}^o$ system can be characterized by two CP eigenstates, D_1 and D_2, with mass difference $\Delta m = m_2 - m_1$, and width difference $\Delta\Gamma = \Gamma_2 - \Gamma_1$. Then the time-dependent component with wrong sign strangeness, such as $K^+\pi^-$, in a beam which was originally D^o, is given by

$$I(D^o \to K^+\pi^-) = exp(-\Gamma t)\ \{\frac{t^2}{4}[(\Delta M)^2 + (\frac{1}{2}\Delta\Gamma)^2] + |\rho|^2 + t\rho[(\frac{1}{2}\Delta\Gamma)Re\ \ell \mp (\Delta m)Im\ \ell]\}$$

where $\ell = (1-\epsilon)/(1+\epsilon)$ and ϵ is the parameter familiar from K^o decay. The $\mp$ sign is taken for wrong sign $D^o(\overline{D}^o)$ decays, and is unimportant unless CP violation is large. The first term is due to mixing, and the second is due to doubly Cabibbo-suppressed decays (DCSD) described by the ratio of amplitudes

$$\rho = A(D^o \to K^+\pi^-)/A(D^o \to K^-\pi^+)$$

The ratio ρ is expected to be of order tan $^2\theta$. The third term takes into account possible interference. The parameter $r = B(D^o \rightarrow K^+\pi^-)/B(D^o \rightarrow K^-\pi^+)$ is equal to $[(\Delta M)^2 + (\frac{1}{2}\Gamma)^2]/2\Gamma^2$ if the wrong-sign decays are due to mixing alone, and $r = |\rho|^2$ if they are due to DCSD. The mixing events have a time distribution proportional to $t^2 exp(-\Gamma t)$, while the DCSD distribution is the usual $exp(-\Gamma t)$. Thus if we cut at $t > 2\tau_{D^o}$, 68% of the events due to mixing are kept, but only 14% of the DCSD events are retained. Such a cut also reduces the non-charm background to a negligible level.

The standard calculations of D^o mixing based on the box diagram[5] predict Δm to be small and $\Delta\Gamma$ to be very small, both below the current experimental sensitivity. Long range effects[20] could give rise to observable mixing, however. The calculation of such effects depends on the intermediate meson states, and is not very precise.

One limit on mixing is $r < .006$ from an experiment studying like-sign dimuon production by pions.[21] This limit depends somewhat on the production cross section and kinematic distribution assumed. Another limit of $r < .023$ comes from $D^{*+} \rightarrow \pi^+ D^o$ events produced in e^+e^- interactions.[22] The result which has focused much attention on the mixing problem is an observation of 3 events by Mark III[23] in which 2 charm decays are seen with the same strangeness in the final state. If these events are due to mixing, it corresponds to $r \simeq 1\%$.

The only way to increase the sensitivity for D^o mixing is to understand the DCSD component. We have been able to do that by using the decay time information in the process $D^{*+} \rightarrow \pi^+ D^o$, $D^o \rightarrow K^-\pi^+$. The charge of the π tags the initial state as D^o or $\overline{D}^o$, and the combination of the D^* requirement and vertex cuts produce a very clean D^o signal. In addition, the mixing events have the very different time dependence noted above.

Figure 16(a) shows a scatter plot of Q versus $M(K\pi)$ for the right

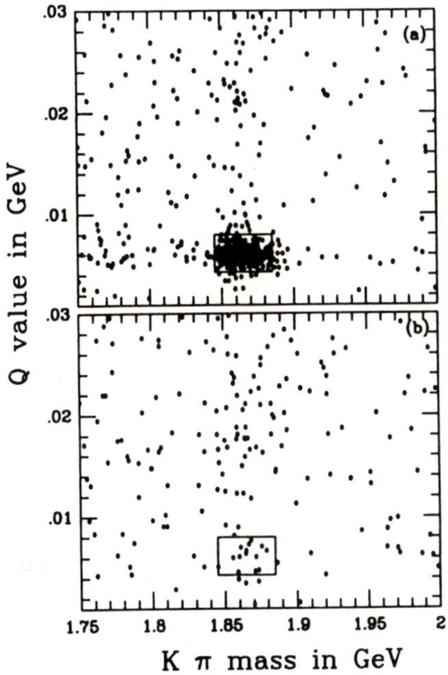

Fig. 16. (a) The scatter plot of Q = $M(K\pi\pi) - M(\pi) - M(K\pi)$ vs $M(K\pi)$ for the $K^-\pi^+\pi^+$ (and charge conjugate) sample. There is a requirement that $t > 0.22$ ps, where t is the proper decay time. (b) The same plot for $K^-\pi^+\pi^-$ events.

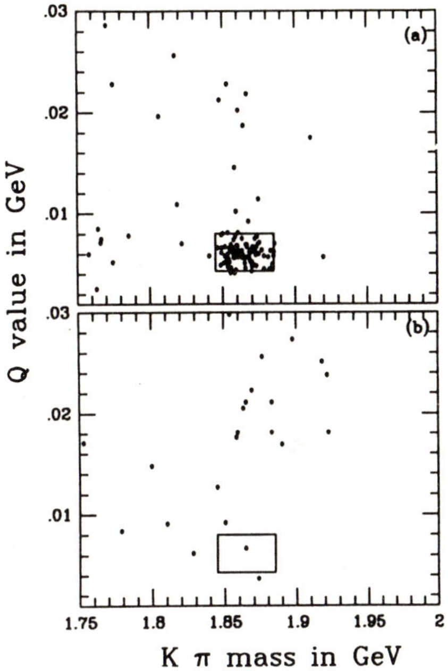

Fig. 17. (a) The scatter plot of Q vs $M(K\pi)$ for $K^-\pi^+\pi^+$ events, with the requirement $t > 0.88$ ps. (b) The plot for $K^-\pi^+\pi^-$ events with $t > 0.88$ ps.

sign ($D^o \to K^-\pi^+$) sample with decay time $t > 0.22$ psec, or about $\frac{1}{2}$ D^o lifetime. $Q = M(K^-\pi^+\pi^+) - M(K^-\pi^+) - M(\pi^+)$ is the available rest mass energy in the D^{*+} decay. Inside the boxed region are 326 events, including about 3 events from D^o's combining with random pions and 2 other background events. In the vertical direction above the signal region, a band of D^o events combined with random pions is visible. A band of D^o events in which the D^o does not decay into the $K\pi$ state (i.e., one of the decay products is misidentified or not detected) is evident in the horizontal direction to the left of the boxed area. Figure 16(b) shows the same scatter plot for wrong sign events ($K^+\pi^-\pi^+$). There is a small excess of events in the D^* region, at about the 1.8 standard deviation level. This could easily be a statistical fluctuation, or it could be due to mixing or DCSD.

Figure 17 shows the analogous plots with $t > 0.88$ ps $\simeq 2\tau_{D^o}$. In Figure 17(a) there is still a strong D^* signal from the long-lived tail of the right-sign decays. In Figure 17(b) there should be 1-2 events in the boxed region if there is no mixing, due to D^o's with random pions and the small fraction of DCSD events with large decay times. There is only one event, consistent with no mixing. If the contribution to r from mixing, r_m, were 1%, there would be 6 events at the D^*. Using a maximum likelihood fit to the right sign events using Q, $M(K\pi)$ and t for each events, we find 597 ± 38 D^* events, corrected for events with $t < 0.22$ ps. The corresponding members for the wrong-sign sample are 9.6 ± 4.4 DCSD events and -0.7 ± 2.0 events from mixing. Thus we find $r_m = (-0.7 \pm 2.0)/(597 \pm 38) = -.0012 \pm .0034$. This leads to the limit

$$r_m < .005 \quad (90\%\ C.L.).$$

There is also the possibility of interference between the DCSD amplitude and mixing. This is a large effect if $\mid \Delta\Gamma \mid$ is large but $\mid \Delta M \mid << \mid \Delta\Gamma \mid$, and the DCSD amplitude is larger than expected.

If $\Delta\Gamma > 0$ our limit improves, but if $\Delta\Gamma < 0$ it becomes worse because the two amplitudes tend to cancel each other near $t \simeq 2\tau_{D^o}$. This case can be studied by looking at both $K\pi$ and $K\pi\pi\pi$ decay modes, since the DCSD amplitudes, if large, are not necessarily equal, and by looking at shorter decay times.

Exclusive Semileptonic Decays

The study of exclusive semileptonic decays is particularly interesting because they are easier to interpret than nonleptonic decays. The Cabibbo-favored decays can proceed only through the spectator diagrams for all three D-mesons. In addition, there is no complication due to either final state interactions or interferences. To explain exclusive semileptonic decays, one must take into account the available hadronic states. The s quark produced in the charm decay is left in a fairly low mass $s\bar{q}$ state with the spectator antiquark. From knowledge of the wave functions one should be able to predict the rate for the decays $Ke\nu$ and $K^*e\nu$.[5] For example, $D \rightarrow Ke\nu$ is a pure vector interaction, with a rate $\Gamma \propto G^2 \mid V_{cs} \mid^2 \mid f_+(t) \mid^2$. Here $t = M^2(e\nu)$ is the four-momentum transfer squared and $f_+(t)$ is the vector form factor. Thus the exclusive rate checks the normalization of $f_+(t)$, assuming $V_{cs} \simeq cos\theta_c$.

We originally made an early analysis of the $D \rightarrow Ke\nu$ mode because it also checked an important step in the measurement of absolute branching ratios. The Mark III result, $\Gamma(D^o \rightarrow K^-e^+\nu)/\Gamma(D^o \rightarrow K^-\pi^+) = 0.71 \pm .16$,[24] depends on the $D^o \rightarrow K^-\pi^+$ rate from the fit to all of the hadronic branching ratios.[6]

Experimentally, exclusive semileptonic decays can be difficult to study because there is no narrow mass peak to suppress the background. In our case, we must use the electron identification and the charm vertex separation to produce a clean signal. For the electron identification, we required that the energy deposit in the electromagnetic calorimetry agreed with the reconstructed charged track in energy and position. We also demanded that the electromagnetic shower had the proper shape and that there was no energy in the hadronic calorimeter. These cuts were enough to produce a clean electron signal. In addition we wanted to remove electrons that came from pair production. To do this we used only electrons with momentum greater than 12 GeV/c^2 and vetoed those electrons in which the opposite-sign electron from the pair was observed in either the tracking or the calorimetry.

After selecting clean electrons, we looked for K^-e^+ candidates that came from a good vertex well separated from the primary vertex. We used these to look for the process $D^{*+} \to \pi^+ D^o$, $D^o \to K^-e^+\nu$. The position of the primary and secondary vertices were used to make a rough measurement of the D^o direction. Assuming the $D^o \to Ke\nu$ decay process, we calculated the neutrino momentum vector, which has two solutions. We then combined both of these with each π^+ in the event to calculate $M(\pi K e\nu)$, and chose the smallest solution. The resulting mass plot, in Figure 18, shows a peak just above threshold at the D^* mass. The similar plot for wrong sign events, in Figure 19, shows no peak, and can be used to estimate the background. In a fairly broad region, 2.000-2.025 GeV/c^2, we find 110 events, with 38 background. Comparing this with our $D^o \to K^-\pi^+$ signal and correcting for efficiency, we find $\Gamma(D^o \to K^-e^+\nu)/\Gamma(D^o \to K^-\pi^+)$ $= 0.77 \pm .13 \pm .12$ (preliminary). This agrees with the old Mark III measurement, but it also agrees well with the revised number,[6] of 0.93 ± 0.21.

There will be a number of improvements in the final analysis of the $D^o \to K^-e^+\nu$ decay, including three times as much data. There has been a revision of the electron identification algorithm to provide higher efficiency and smaller error on this efficiency. Since this was the

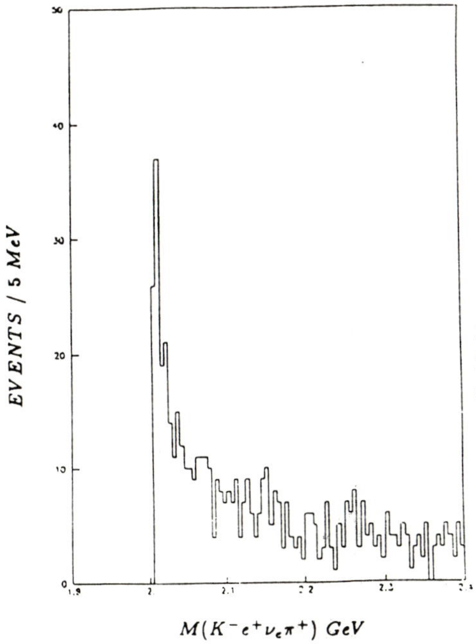

Fig. 18. The $K^-\pi^+e^+\nu_e$ mass spectrum, assuming $D^o \to K^-e^+\nu_e$ decay.

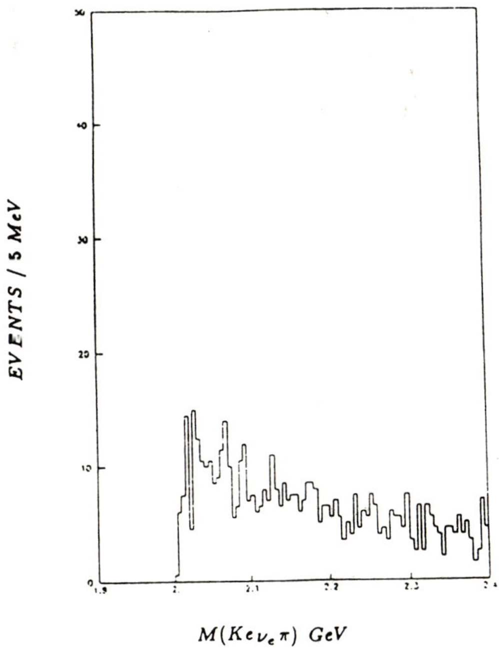

Fig. 19. The $K\pi e\nu$ mass spectrum for wrong sign combinations

dominant source of systematic error, that error will be significantly reduced.

There are many other interesting semileptonic decays that are being studied. The easiest to see is the $D^+ \to K^-\pi^+e^+\nu$ decay. Although there is no D^* decay seen to suppress background, the long D^+ lifetime and the extra power of the vertex cuts for 3-prong decays result in a very clean signal. A total of at least 200 $D^+ \to K^-\pi^+e^+\nu_e$ decays will be available, with little background. The most interesting question is the fraction of K^* in the $K\pi$ final state. On the basis of about 45 $D \to K\pi e\nu$ decays, Mark III saw $55 \pm 11\%$ were from the K^*.[6] This is somewhat surprising, in that it says the ratio of $K^*e\nu/Ke\nu$ is less than calculations predict.[5] We will make a more precise check of this ratio. Other interesting decays are $D_s^+ \to \phi e^+\nu$, $D^o \to \pi^-e^+\nu$, and $D^+ \to \rho^o e^+\nu$. Whether we have enough sensitivity to see them is not yet clear. They may take a higher-rate experiment, either in the fixed-target or e^+e^- environment.

Hadroproduction

The hadronic production of charm is interesting for two reasons. First, the production of at least some heavy quarks should be calculable from perturbative QCD. In particular, the bottom quark mass should certainly be large enough to make such calculation reliable. For charm production, the measurements are easier but the theoretical interpretation is more difficult, because the charm quark mass is light enough that higher-order terms are large. There may also be an unexpected source of charm (e.g., due to charm content of the proton) which is not included in the existing models. The second reason for studying hadroproduction of charm is that it is a possible source of heavy quarks for decay studies. Hadroproduction has not yet been able to compete with photoproduction of charm, although that may be different for baryons. In any case, hadroproduction will be necessary to see large samples of b-quarks at either fixed-target accelerators or hadron colliders. It is therefore very important to understand the production process.

The first order QCD diagrams are $q\bar{q} \to c\bar{c}$ and $gg \to c\bar{c}$. The calculations of charm production from these diagrams, using a standard fragmentation scheme to provide the charmed particles, is called the fusion model. The model predicts most of the kinematic distributions for charm production well, but the calculated cross section is lower than the data by a factor of about 5. Second-order calculations help reduce the discrepancy, but are still a factor of two or three low.

The fusion model is useful in looking for distributions in which the data does not seem to fit the model. There is some evidence of a leading particle effect, in which charmed particles that share quarks with the beam particles are produced more forward than others. There has also been evidence, mostly from ISR experiments, that the charm cross section rises even faster with energy than predicted.

The LEBC experiment NA27[25] gives a fairly complete picture of charm production at $\sqrt{s} = 27$ GeV, although with low statistics. They measure $\frac{1}{2}[\sigma(D)+\sigma(\overline{D})] = 17 \pm 2\mu b$, using a hydrogen target for which no A-dependent correction is necessary. The p_T spectrum follows the form $d\sigma/dp_T^2 \propto exp(-ap_T^2)$, where $a \simeq 1(GeV/c)^{-2}$, so that $< p_T^2 > \simeq 1(GeV/c)^2$. If the fusion model is used with a fragmentation model in which the partons have $< k_T^2 > = 0.6(GeV/c)^2$, a p_T dependence similar to the data is obtained.

The x_F dependence is not as straightforward as the p_T dependence. (Here $x_F = p_z/(p_z)_{max}$.) Charm production is usually fit to the form $dN/dx_F \approx (1 - x_F)^n$. A value of n around 4 is predicted by the fusion model, and corresponds to central production. A leading particle effect would lead to much smaller values of n for Λ_c, $\overline{D}^o$, and D^- production in pN interactions, or D^o and D^- production in

π^-N interactions. The experimental evidence is not consistent. The ISR experiments[26] see forward Λ_c production by protons, and NA27 sees forward D^- and D^o production by π^-. The ISR results are contradicted by NA27, however, which sees no forward production using proton beams. In π^-N interactions, NA32[27] sees no forward production of D^- and D^o, which is in conflict with NA27. Experiments with better statistics are needed to resolve these problems.

The energy dependence of the charm cross section appears to make more sense than it once did. Figure 20 shows the D cross section in pp interactions, normalized to its value at $\sqrt{s} = 27$ GeV, as a function of $\sqrt{s}$. The shaded region shows the reasonable limits obtained from the fusion model. The recent measurement of E-743[28] gives the ratio of $1.7^{+0.6}_{-0.5}$ for $\sigma_{D\overline{D}}(\sqrt{s} = 39\ GeV)/\sigma_{D\overline{D}}(\sqrt{s} = 27\ GeV)$, which is well within the predicted range. The ISR results correspond to a value of about 20 on that plot, which is a factor of about 15 times the value at $\sqrt{s} = 39$ GeV. With the exception of the ISR results, all experiments lie within this range.

The total picture in hadroproduction is somewhat confusing. There is good agreement of the average p_T^2 with the fusion model. There is no clear evidence of large forward cross sections, since every positive result is contradicted by another experiment. The evidence against a very rapid rise in the charm cross section with energy is quite strong. Finally, the $D/\overline{D}$ cross section at $\sqrt{s} = 27$ GeV is in the range 15-20 μb, which is larger than predicted by the model. We need a high-statistics, clean charm sample from a hadroproduction experiment, in which the signals are significant even after being finely divided. This would make possible a decisive study of the leading particle effect, the A-dependence, and correlation of charm pairs within one event. Unless some surprises emerge, it appears that charm decay studies at fixed target machines will continue to be dominated by photoproduction.

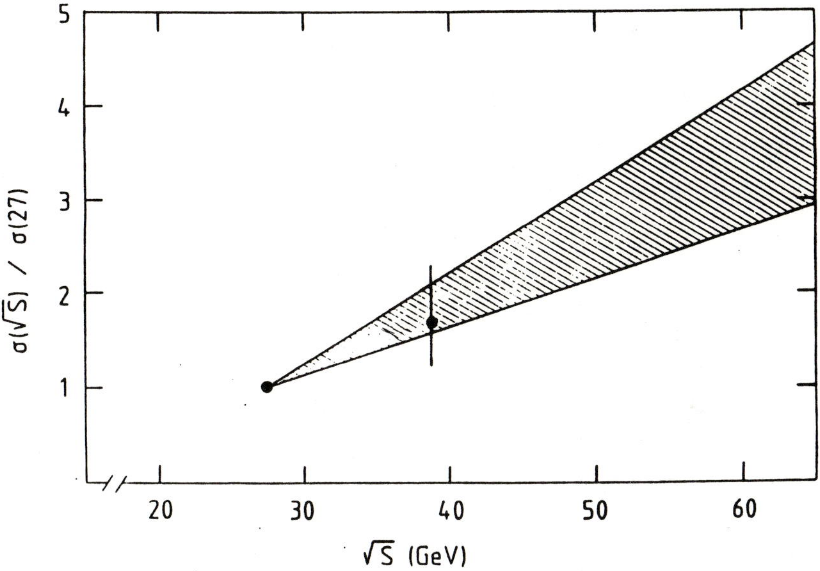

Fig. 20. Total $D/\overline{D}$ cross section, normalized to the value at $\sqrt{s} = 27.4$ GeV, versus $\sqrt{s}$ (center-of-mass energy). The measurements are from LEBC. The band covers the range expected from fusion model calculations.

Photoproduction

The dominant diagram in photoproduction of charm is photon-gluon fusion (PGF). For many reasons, the perturbative calculation in this case should be more reliable than for hadroproduction. There is only one gluon distribution, so that the predictions are less sensitive to its choice. The higher order corrections are smaller by a large factor, because of the photon in the initial state. The photon has the full beam energy, which means that the collision is somewhat harder than for hadron beams of the same energy. Finally, there is no leading particle effect possible, since there are no spectator quarks in the beam particle. Unlike hadroproduction, there is a high-statistics, clean charm sample from photoproduction. E-691 has about 10^4 events in the modes $D^o \to K^-\pi^+$, $D^o \to K^-\pi^+\pi^-\pi^+$, and $D^+ \to K^-\pi^+\pi^+$. The background is low, and it is possible to see significant signal even after the data are subdivided.

The ratio of $\overline{D}/D$ is an interesting quantity to measure, since it depends on how the hadronization occurs. Because the quark forms a color singlet with the diquark in the nucleus, the charmed particle may be produced at lower x_F than the antiparticle. A particular case of this is the final state $\overline{D}\Lambda_c X$, which gives an excess of $\overline{D}$ over D. The SLAC hybrid bubble chamber experiment measured $\sigma(\overline{D}\Lambda_c^+ X)/\sigma(\overline{D}X) = (71 \pm 11 \pm 6)\%$,[29] while WA58 saw $(28 \pm 23)\%$ for the same ratio.[30] In E-691, we measured $\overline{D}/D = 1.02 \pm .05$ in the forward region, which says that $\overline{D}\Lambda_c X$ is small for $E_\gamma \simeq 170$ GeV.

The p_T^2 distribution from E-691 is shown in Figure 21. The $< p_T^2 >$ of D mesons is seen to be $1.34 \pm .05 \pm .08$ $(GeV/c)^2$, which agrees with a Monte Carlo using the PGF model and Lund fragmentation. The x_F distribution is given by $(1 - x_F)^n$ where n $= 2.7 \pm 0.1 \pm 0.6$, which also agrees with the PGF model. The charm is somewhat more forward produced than in hadroproduction. Figure

22 shows the ratio of cross sections $\sigma_{D\overline{D}}(\sqrt{s} = 200\ GeV)/\sigma_{D\overline{D}}(\sqrt{s} = 100\ GeV) = 1.83 \pm 0.29 \pm 0.20$, which is consistent with previous experiments,[31] and the PGF model.

The cross section $\sigma(\gamma Be \rightarrow c\bar{c})$ is measured by E-691 to be $(3.17 \pm 0.22) \pm 25\%$ $\mu b/Be$ nucleus. The total cross section for D^o and D^+ was measured in the range $x_F > 0.2$. The Monte Carlo was used to correct for acceptance, and estimates of the D_s^+ and Λ_c^+ production were used to estimate the full charm cross section. Using as a reasonable estimate that the cross section is proportional to $A^{0.92}$, we get $\sigma = (0.42 \pm .03) \pm 25\%$ μb/nucleon. This agrees with the PGF model using $m_c \simeq 1.5\ GeV/c^2$. Since the total charm cross section is sensitive to m_c and the energy rise and x_F dependence to the gluon structure function, it should be possible to separately measure these quantities. In addition there will be a large sample of events in which both charm events are measured, providing further comparisons with the model.

Charmonium Spectroscopy

The area of charm physics that has been most completely dominated by experiments at e^+e^- colliders is charmonium spectroscopy. It is surprising, then, that a method of studying some of the more elusive charmonium states in a fixed target environment has been developed. The first experiment to use the method was ISR experiment R704 which studied antiproton-proton formation of charmonium. The "fixed" target was actually a gas jet target which can be used in the storage ring without degrading the vacuum. The cooled antiproton beam developed for $p\bar{p}$ collisions at higher energies gives very small $\Delta p/p$, and therefore good mass resolution. Such formation is interesting because $c\bar{c}$ states of all possible spin-parity combinations are accessible through $\bar{p}p$. I will briefly describe R704 and the proposed follow-up experiment E-760 at Fermilab.

The physics of charmonium has been discussed in the talks by F. Gilman[32] and M. Chanowitz[33] at this conference. The states that are interesting for $\overline{p}p$ formation are the η_c, the χ states, and the $^1P_1$ state. These have masses in the range 2.9-3.6 GeV/c^2, which corresponds to a momentum range 3.5-6.0 GeV/c for the antiproton beam. 90° in the center-of-mass frame corresponds to 40° in the laboratory. At the ISR, the momentum spread of the antiproton beam was $\Delta p/p \simeq \pm 0.001$, corresponding to a resolution in $\sqrt{s}$ of $\pm$ 1 MeV.

The experimental design is something quite novel.[34] The processes being looked for are very small fractions ($10^{-9} - 10^{-7}$) of the total cross section, so the channels must be very selective. The channels chosen were

$$\overline{p} + p \to \chi \to \gamma + \psi, \psi \to e^+ + e^-$$

and

$$\overline{p} + p \to \eta_c, \chi_j \to \gamma + \gamma.$$

These all involve high-mass, two body decays into e^+e^- or $\gamma\gamma$. Such electromagnetic states are extremely rare in strong interactions, and working at 90° in the center-of-mass further reduces backgrounds. Figure 22 shows the R704 experimental set-up. In the χ search they saw about 30 χ_1, and about 50 χ_2 decays, with little background. The masses and widths are given in Table VII:

Table VII χ State Parameters from R704

State	Mass (MeV/c^2)	Width (MeV/c^2)
χ_1	$3511.3 \pm 0.4 \pm 0.4$	< 1.3
χ_2	$3556.5 \pm 0.4 \pm 0.5$	$2.6^{+1.4}_{-1.0}$

The new experiment, Fermilab E-760, will begin to take test data at the accumulator ring this year. It will have 5 times the acceptance of

R704, partly because it will have complete azimuthal coverage. The luminosity will eventually be 10^{31} $cm^{-2}sec^{-1}$, 7 times that at the ISR. This gives a total of about 1000 $\eta_c \rightarrow \gamma\gamma$ events in a 300 hour run. Only 15 such events were seen at the ISR. With such a large sample, it would be possible to measure $\Gamma(\eta_c \rightarrow \gamma\gamma) \cdot \Gamma(\eta_c \rightarrow \overline{p}p)$, $\Gamma_{tot}(\eta_c)$ and $M(\eta_c)$ quite accurately. Good predictions exist for the Γ_{tot} and $\Gamma(\eta_c \rightarrow \gamma\gamma)$. E-760 hopes to collect samples of about 10^3 events for each of the χ states, as well as a few hundred events of the type $\overline{p} + p \rightarrow {}^1P_1 \rightarrow \psi + \chi$. This would mark a great advance in our knowledge of the charmonium system.

Open Charm Spectroscopy

The status of baryon spectroscopy is not very advanced. Baryons are not easy to study at SPEAR because of their high mass. They are also difficult at fixed target experiments because of small cross sections and short lifetimes.

The best-measured baryon is the Λ_c^+, which has received much attention recently. Although the early experiment saw lower masses, the present world average[7] is 2281.2 ± 3.0 MeV$/c^2$. There are three more recent measurements, which agree quite well, shown in Table VIII.

Table VIII Λ_c^+ Mass Measurements

Group	# Events	# Back.	Mass
$ARGUS$[35]	479	2500	$2283.5 \pm 2.6 \pm 2.1$
$NA32$[14]	14	0	2286 ± 2
E-691	68	56	$2284 \pm 2 \pm 1$

Now that the Λ_c mass is well established it will be less likely for fluctuations to be accidentally interpreted as Λ_c^+ peaks.

The lifetime we discussed earlier after discussing the charmed meson lifetimes. The lifetime average, around 0.19 ps, is much shorter than any of the mesons. Very little is known about specific decay modes, except that the $pK^-\pi^+$ mode is always the easiest to observe. Mark II[16] reported that the $pK^-\pi^+$ final state is not dominated by $p\overline{K}^{*o}$ and $\Delta^{++}K^-$. Other modes observed are $\Lambda\pi^+\pi^-\pi^+$, $p\overline{K}^o$, and $p\overline{K}^o\pi^+\pi^-$, but there is still not a single ratio measured very well.

There is an interesting question raised by the Λ_c^+ branching ratios. The reported Mark II absolute branching ratio is $B(\Lambda_c^+ \rightarrow pK^-\pi^+) = 2.2\%$.[36] This compares with $B(D^+ \rightarrow K^-\pi^+\pi^+) = 9\%$, and $B(D^o \rightarrow K^-\pi^+\pi^+) = 13\%$. Even taking into account decays with Λ^o's, which we know are not too large, the absolute Λ_c branching ratios look too small. We need another measurement of the absolute branching ratio for the Λ_c^+.

ARGUS[35] uses their Λ_c sample to look for the $\Sigma_c^{++} \rightarrow \pi^+\Lambda_c^+$ and $\Sigma_c^o \rightarrow \pi^-\Lambda_c^+$ decays. They see 82 ± 27 $\Sigma_c^o \rightarrow \pi^-\Lambda_c^+$ on 500 backgrounds, with a mass difference $\Delta M = 165.8 \pm 0.7$ MeV/c^2. For the Σ_c^{++} they see 92 ± 26 events with $\Delta M = 168.3 \pm 0.8$ MeV. Fermilab E400 has reported preliminary results showing quite different ΔM values, however.[37] There must be cleaner Σ_c signals before their properties can be said to be well measured.

Finally, the Ξ_c^+ was observed in 135 GeV/c Σ^- + Be interactions.[38] A peak was seen in the $\Lambda K^-\pi^+\pi^+$ mass spectrum at 2460 ± 25 MeV. They have used this signal to measure the lifetime, and get $\tau = 0.48^{+.21+.20}_{-.15-.10}$ ps. The cross section times branching ratio is large, especially for 135 GeV/c. The production is also very forward, described by $(1 - x_F)^{1.7}$, which does not look like $p + N \rightarrow \Lambda_c + X$. There are some troubling aspects, which keep the Ξ_c from being considered well established.

There is some new data from Fermilab E-400, which looked at

$n + N \rightarrow \Xi_c^+ \rightarrow \Lambda K^- \pi^+ \pi^+$.[39] Although this is very preliminary, it gives some support to the earlier sighting. There is still a large inconsistency, however. The E-400 mass spectrum includes a second peak, at lower mass, which they ascribe to the decays $\Xi_c^+ \rightarrow \Sigma^o K^- \pi^+ \pi^+$, with $\Sigma^o \rightarrow \Lambda^o \gamma$. The problem is that the ratio of these two peaks must be almost constant, and not depend on experiment.

In general then, the charmed baryons need good measurements. The Λ_c^+ is well established, and its mass and lifetime are well measured. Its branching ratios must be studied, however, especially because Λ_c decays may be dominated by W-exchange. The Σ_c states are probably there, but the masses are not well established. The Ξ_c needs some more work before it can be considered solid. Finally, the Σ_c^+, Ξ_c^o, and Ω_c^+ still need to be filled in. There is a lot of interesting physics in the charmed baryons, if we can figure out how to get at it.

The charmed meson spectroscopy is in a state similar to that of the charmed baryons. There is only one observed peak, which cannot be identified with a particular meson. The $\ell = 1$ $c\overline{q}$ states (D^{**}) were discussed in F. Gilman's talks. There are 4 mesons, 1 with $J^P = 0^+$, 2 $J^P = 1^+$, and 1 $J^P = 2^+$. Because the 2 1^+ states are not eigenstates of charge conjugation, the $^1P_1$ and $^3P_1$ states can mix to give states in which the total angular momentum of the light quark, j_q, is the important quantum number[40], rather than S. In either case, the masses of the 4 states will tell us a great deal about the $c\overline{q}$ potential.

ARGUS[41] has seen the only candidate for a D^{**} state. They observe a peak in the $D^{**}\pi^-$ mass spectrum, with $M \simeq 2426\ MeV/c^2$ and $\Gamma \simeq 70\ MeV/c^2$. Either of the two 1^+ states or the 2^+ state can decay into $D^*\pi$. In most models the states are only separated by about 50 MeV/c^2 and the widths are 20-70 MeV/c^2.[40] Thus the broad state seen by ARGUS may be any mixture of the three.

The way to sort out the issue is to look at the $D\pi$ system into which only the 0^+ and 2^+ can decay. Because most D^o's come from D^* decay, the best place to look is in the $D^+\pi^-$ spectrum. E-691 has a sample of about 3000 clean D^+'s with no constraint on the mass of the parent. This will be a difficult business in present experiments, however. An interesting question is whether BEPC is a good place to study D^* and D^{**} spectroscopy, using $e^+e^- \to \overline{D}D^*$ or $\overline{D}D^{**}$.

Future Fixed Target Experiment

Finally, I will give a quick review of the fixed target experiment which will be contributing to charm physics in the next couple of years. All of them have silicon vertex detectors. E-687 is a photoproduction experiment with a spectrometer similar to that of E-691 but with a higher energy beam. The higher energy means more charm in the data, and a higher interaction rate is possible.

The spectrometer used for E-691 will be taking data for E-769, a hadroproduction experiment. It will use a π/K beam of momentum 300 GeV/c, and will collect up to $3 \cdot 10^8$ events. This should serve to solve many of the outstanding hadroproduction questions, including the existence of leading particle effects.

E-653 uses a compact spectrometer with one magnet and no Cerenkov counter. The present experiment uses a 800 GeV proton beam on an emulsion target, upstream of the silicon detectors. P-780 is a proposal to run at a much higher rate without the emulsion. They would like to use a trigger which requires a jump in charged multiplicity just downstream of a thin target. For the longer-range future, a new spectrometer to study charmed baryons produced in a hyperon beam at Fermilab has been discussed. Both of these proposals hope to make use of a possible enhancement of charmed baryons in production by baryons. This enhancement has not yet been established but could be very important for baryon studies if it is there.

Conclusion

In a series of talks as broad as these, one should have very short conclusions. There will be much new charm physics coming from fixed-target experiments, in the areas of production, lifetimes, decays, charmonium spectroscopy, and open charm spectroscopy. It will be important for the physicists working at BEPC to monitor the progress of these experiments, so that they can decide what is the best physics for them to pursue.

Another clear conclusion is that the field of charm physics is changing very fast. There is every sign that it will continue to be one of the most active and productive areas of particle physics.

REFERENCES

1. M. Aguilar-Benitez *et al.*, Phys. Lett. 123B, 98 (1983).
2. For an introduction to modern strip detectors, see Erik Heijne, Muon Flux Measurement with Silicon Detectors in the CERN Neutrino Beams, CERN Yellow Report 83-06 (CERN, Geneva, 1983).
3. R. Bailey *et al.*, Zeitschrift *für* Physik C30, 51 (1986).
4. R. Morrison *et al.*, Proceedings of the VII Vanderbilt Conference, edited by R. Panvini and T. Weiler (World Scientific, Singapore, 1987).
5. I. Bigi, in these Proceedings.
6. D. Hitlin, in these Proceedings.
7. J.C. Anjos *et al.*, Phys. Rev. Lett. 58, 311 (1987).
8. Particle Data Group, Phys. Lett. 170B, 1 (1986).
9. J.C. Anjos *et al.*, Phys. Rev. Lett. 58, 1818 (1987).
10. H. Becker *et al.*, Phys. Lett. 184B, 277 (1987).
11. M. Aguilar-Benitez *et al.*, submitted to Phys. Lett.
12. M. Adamovich *et al.*, CERN preprint CERN/EP 86-77, submitted to Phys. Lett.
13. N. Ushida *et al.*, Phys. Rev. Lett. 56, 1767 (1986).
14. S. Barlag *et al.*, Phys. Lett. 184B, 283 (1987).
15. S.R. Amendolia *et al.*, submitted to the 23rd International Conference on High Energy Physics, Berkeley, 1986.
16. J.M. Weiss, Proceedings of Baryon 80, Toronto (1980).
17. R.M. Baltrusaitis *et al.*, Phys. Rev. Lett. 55, 150 (1985).
18. G.H. Trilling, Physics Reports 75, 57 (1981).
19. See the discussions and references in reference 5.
20. J.F. Donoghue, E. Golowich, and B.R. Holstein, Phys. Rev. D33, 179 (1986).
21. C. Biino *et al.*, Phys. Rev. Lett. 56, 1027 (1986).
22. N. Kwak, in Proceedings of the 23rd International Conference on High Energy Physics, edited by Stewart C. Loken (World Scientific, Singapore, 1987), p. 752.
23. G. Gladding, presented at Physics in Collision V, Autun, 1985. Also see reference 6.

24. D.M. Coffman, Ph.D. thesis, California Institute of Technology (1986). Also see reference 6.

25. M. Aguilar-Benitez *et al.*, Phys. Lett. 161B, 400 (1985); and M. Aguilar-Benitez *et al.*, Z. Phys. C31, 1491 (1986). For further discussion see S. Reucroft in Proceedings of the 6th International Conference on Physics in Collision, (World Scientific, 1987).

26. K.L. Giboni *et al.*, Phys. Lett. 85B, 437 (1979); W. Lockman *et al.*, Phys. Lett. 85B, 443 (1979); and D. Drijard *et al.*, Phys. Lett. 85B, 452 (1979).

27. R. Bailey *et al.*, Z. Phys. C30, 51 (1986); and R. Bailey *et al.*, Phys. Lett. 132B, 230 and 237 (1983).

28. R. Ammar *et al.*, Phys. Lett. 183B, 110 (1987).

29. K. Abe *et al.*, Phys. Rev. D33, 1 (1986).

30. M. Adamovich *et al.*, CERN preprint CERN/EP 86-89 (1986).

31. J.J. Aubert *et al.*, Nucl. Phys. B213, 31 (1983); and A.R. Clark *et al.*, Phys. Rev. Lett. 45, 682 (1980).

32. F. Gilman, in these Proceedings.

33. M. Chanowitz, in these Proceedings.

34. C. Baglin *et al.*, Phys. Lett. 171B, 135 (1986); C. Baglin *et al.*, Phys. Lett. 172B, 455 (1986); and C. Baglin *et al.*, Phys. Lett. 187B, 191 (1987).

35. D.B. McFarlane in Proceedings of the 23rd International Conference on High Energy Physics, edited by Stewart C. Loken (World Scientific, Singapore, 1987), p. 664.

36. G.S. Abrams *et al.*, Phys. Rev. Lett. 44, 10 (1980).

37. P. Garbincius, private communication.

38. S.F. Biagi *et al.*, Phys. Lett. 122B, 455 (1983); and S.F. Biagi *et al.*, Phys. Lett. 150B, 230 (1985).

39. P. Coteus, presented at Meeting of the Division of Particles and Fields of the American Physical Society, Salt Lake City, Utah (1987).

40. J.L. Rosner, Proceedings of the 1985 International Symposium on Lepton and Photon Interactions at High Energies, edited by M. Konuma and K. Takahashi (Nissha Printing, Kyoto, 1986), p. 448.

41. H. Albrecht *et al.*, Phys. Rev. Lett. 56, 549 (1986). Also see reference 35 for more recent data.

Present Status of Beijing Spectrometer

Ye Minghan and Zheng Zhipeng

Centre of High Energy Physics and Synchrotron Radiation,
CCAST (World Laboratory)
and
Institute of High Energy Physics, Academia Sinica

1. Introduction

BES, Beijing Spectrometer, is being constructed as a general purpose magnetic spectrometer to be situated at one interaction point of Beijing Electron-Positron Collider (BEPC).

The schematic layout of BEPC is shown in Figure 1. The injector is an electron linear accelerator, 200 meters in length, which provides electrons and positrons from 1.4 to 1.55 GeV for injection into the storage ring. The injector consists of four main parts : 1) an electron gun and a 30 MeV pre-injector, 2) a 120 MeV

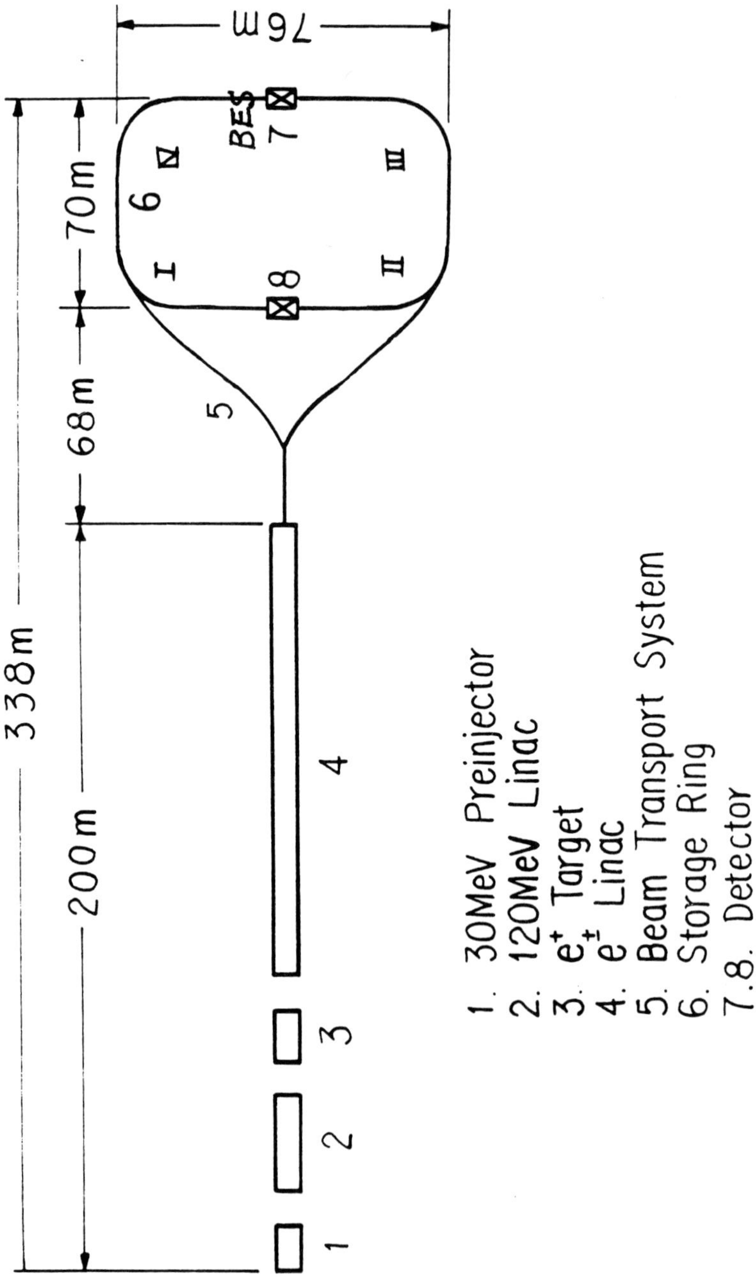

FIGURE 1 Schematic diagram of BEPC

accelerating section, 3) a positron production target, and 4) a 1.4 to 1.55 GeV electron-positron linear accelerator for final acceleration.

The circumference of storage ring is 240.4 meters. And, the luminosity is optimized at 2.8 GeV. The designed peak luminosity is $1.7\times10^{31} cm^{-2} s^{-1}$ assuming a maximum 0.04 beam-beam tune shift, 65 mA circulating current per beam and a vertical beta function of 0.1 m at the interaction point. The expected luminosity is proportional to E^2 between 1.5 and 2.8 GeV.

Because of the limited budget, only one detector, BES, is being built at the first stage.

2. Physics goals

The energy range around 3 - 5.6 GeV has been explored by MARK III, DM2, CLEO, and MARK II etc. They have yielded fruitful harvest. But they also got a few interesting results at odds with each other, and left many other important things unsettled. Thus this is still a rich golden mine for BES to work on. Some of the most prominent physics problems in this energy range are the following:

1). The searching for glueball is a fundamental test of QCD, as discussed in Dr.Chanowitz's lecture at this workshop. Although the theorists have not been able to calculate the exact masses of these glueball states, but for the experimental physicist, the radiative decays of J/4 and 4' are the best place to work for. This will be done by BES. As a matter of fact, there have already been a number of states observed in these decays which are thought to be glueball candidates, such as ℓ (1460), θ (1730), and ξ (2230).

The ℓ (1460) is one of the most exciting results found in J/4 decays. It is the oldest gluonic meson candidate, although its interpretation as a radially excited pseudoscalar also seems possible, and its present experimental situation is confusing. With the confirmation of the η (1275), it is hard to understand the mass pattern and production rates in J/4$\longrightarrow\nu$ (ℓ,η(1275)) ; if both particles are the isoscalar states of the radially excited pseudoscalar $q\bar{q}$ nonet. It seems that the ℓ(1460) would be a gluonic meson or at least has a substantial gluon content. The θ(1730) can hardly be explained as a $q\bar{q}$ resonance. Its likely interpretation is a gluonium resonance. There is strong indication that θ(1730) is really something new. The very existence of (2230) is controversial. If it does exist, such a narrow, high mass state may be a normal $q\bar{q}$ meson, a gluonic or hybrid meson. With high statistics, the quantum numbers and decay properties of these three states will be studied by BES, so that their very nature can be determined.

2). The decay of charmed mesons (D^o, $D^{\pm}$, $D_s^{\pm}$) is one of the major subject to be studied by BES. Here some of the Kobayashi-Maskawa matrix will be determined. The leptonic decay constants of $D^{\pm}$ and $D_s^{\pm}$ will be measured. The non-leptonic decay channels are important sources of information, for which a consistent theoretical frame which explains all the data, is to be established. Many, but not all, of Cabbibo favored channels have been measured by MARK III, but the errors of many channels are still unsatisfied. BES will be able to get these data to high accuracy, make it possible for precise theoretical analysis. While most of the Cabbibo favored channels seem to agree with the recent theoretical calculation, some of the Cabbibo suppressed channels seem to be hard to explain, like the ratio $D^o \rightarrow K^+K^-$/ $D^o \rightarrow \pi^+\pi^-$. The accurate measurement of all other Cabbibo suppressed channels by BES will be essential for advancing our understanding of these decay mechanism. Looking for signals of Cabbibo double suppressed decay at BES will serve the same goal. And also, not the least important, is the observing D^o - $\bar{D}^o$ mixing and CP-violation in D^o - $\bar{D}^o$ system. This is the crucial check of our current theory of weak interaction involving quarks and strong force. If the mixing of D^o - $\bar{D}^o$ does happen at a rate indicated by the events of MARK III, it will be a great challenge to particle theory.

3). BES will be able to do a good deal of work on the τ physics. This includes getting a more stringent limit on the mass, precise measurement of the non-leptonic branch ratios, and searching for unusual couplings of decays.

4). With high statistics, BES will be able to study precisely the charmonium spectrum. This will make further understanding of the nature of the potential. In particular, there is still a predicted charmonium state below threshold, the $_1P^1$ (J^{pc}=1^{+-}) undiscovered. It is BES's job to find it.

5). There is a whole set of charmed baryons (Λ_c,

Σ_c, etc.) which now only have some primitive imfor- mation. Their lifetimes and decay channels will be measured by BES. This will also be crucial for the progress of the theory for weak interactions involving quarks.

6). BES will study the behavior of R= (e+e- $\rightarrow$ hadrons)/6(e+e- $\rightarrow \mu^+\mu^-$), especially in the region of 4 - 4.5 GeV, where clarification of the cc spectrum will be useful.

With so many interesting and important physics, the work of BES may come out as a contribution in particle physics.

3. The description of BES

The design goals of BES are as follows:

a) large solid-angle coverage.

b) good charged-particle momentum resolution.

c) good charged-particle identification.

d) high detection efficiency for photons with lower energy cutoff, with good position resolution and moderate energy resolution.

Figure 2 is the schematic drawing of BES. The main parts are as follows:

1) Beam pipe: Beryllium is the ideal material for the beam pipe, jut the technology to manufacture such pipes does not yet exist in China. A beam pipe of thin aluminium with a wall thickness of 0.3 mm reinforced by 2 mm thick carbon fibre will be used; the radiation length of the wall is about $10.4 \times 10^{-3} X_o$. The outer diameter of the pipe is 15.4 cm. And a beam pipe of thin beryllium is expected to be used in second phase.

2) Central drift chamber: The central drift chamber has an inner diameter of 18.8 cm, an outer diameter of 29.6 cm, and is 110 cm long. The solid-angle coverage is 96% Of 4π. There are four concentric layers of sense wires, each layer having 48 wires; they are offset by ealf a cell in alternate layers. Each sense wire is surrounded by six field wires positioned in hexagonal form. The designed space resolution 6_x is about 150μm. The z-direction measurement is obtained by charge division, 6_z is about 2 cm. The chamber uses an 89% argon, 10% CO_2, and 1% methane gas mixture.

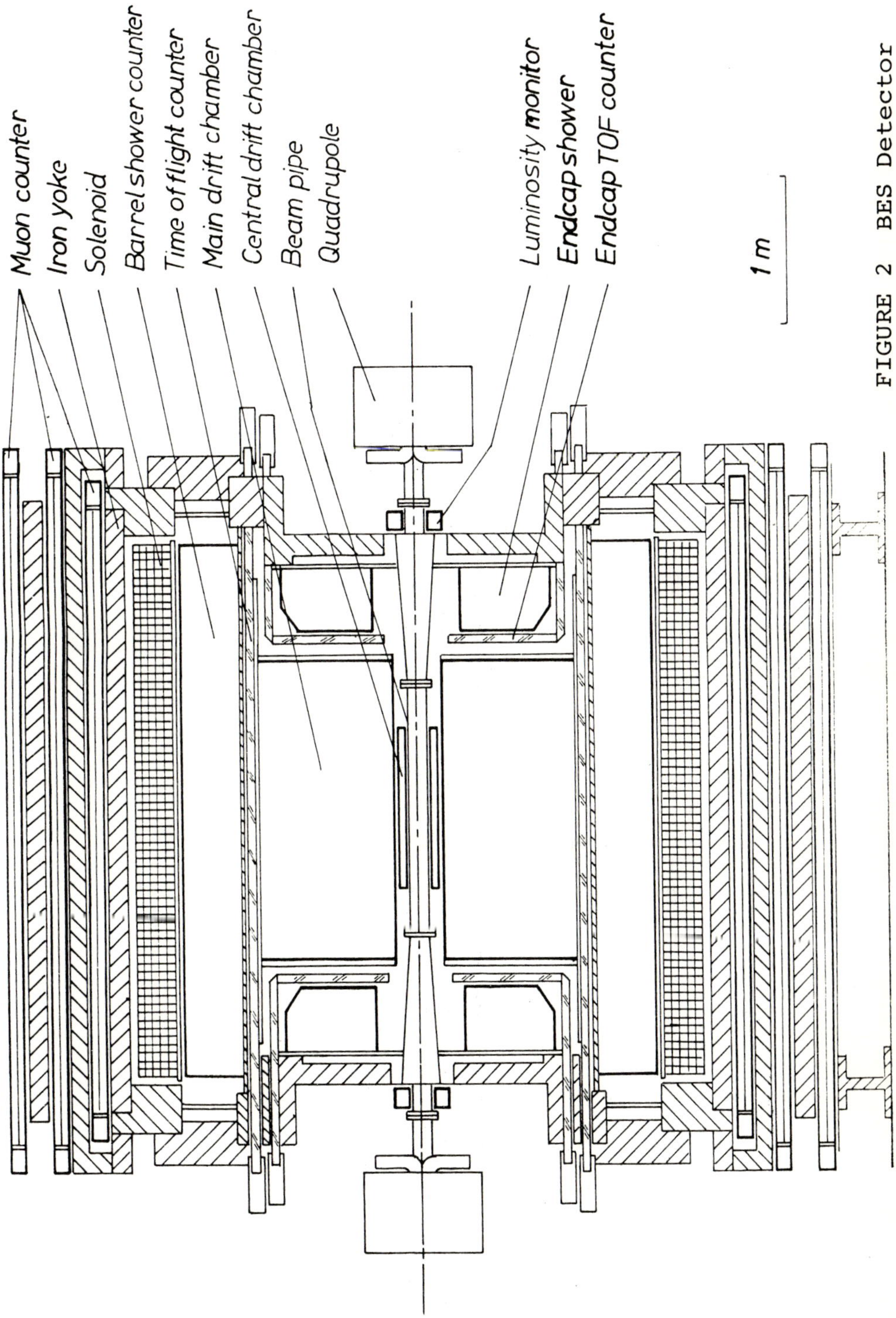

FIGURE 2 BES Detector

3) Main drift chamber has an inner diameter of 31 cm, an outer diameter of 230 cm and is 220 cm long. Multi-sense wire cell design is adopted. Each cell has four sense wires, and each wire is staggered altrnately 350μm from the mid-plane of the cell. This provides local left-right ambigity resolution. Four hits or three hits within a cell can give a small vector of track, which makes track reconstruction easier. The sense wires are 1 cm apart, and potential wires are placed between them in order to reduce cross-talk. Altogether there are ten layers. The solid-angle coverage of the second layer is 96% of 4π. The number of cells per layer counted from the innermost layer is 48, 48, 60, 48, 58, 68, 78, 88, 98, and 108, for a total amount of 702 cells with 2,808 sense wires. The total number of wires is 19,380. The z measurement is obtained by stereo wire layer. There are five axial layers and five stereo layers. Time-of-flight information and $\Delta E/\Delta x$ are measured simultaneously from each sense wire. The designed space resolution is about 200μm; 6_z is about 3 - 4 mm. The total number of samples of $\Delta E/\Delta x$ per track is 40, expected $6_E/E$ is about 10%. (6_E is rms of $\Delta E/\Delta x$; E is mean value of $\Delta E/\Delta x$); for 2 $6\pi/k$ separation, $p \leq 0.7$ GeV/c; for 2 $6k/p$ separation, $p \leq 1.2$ GeV/c. Figure 3 shows the schematic

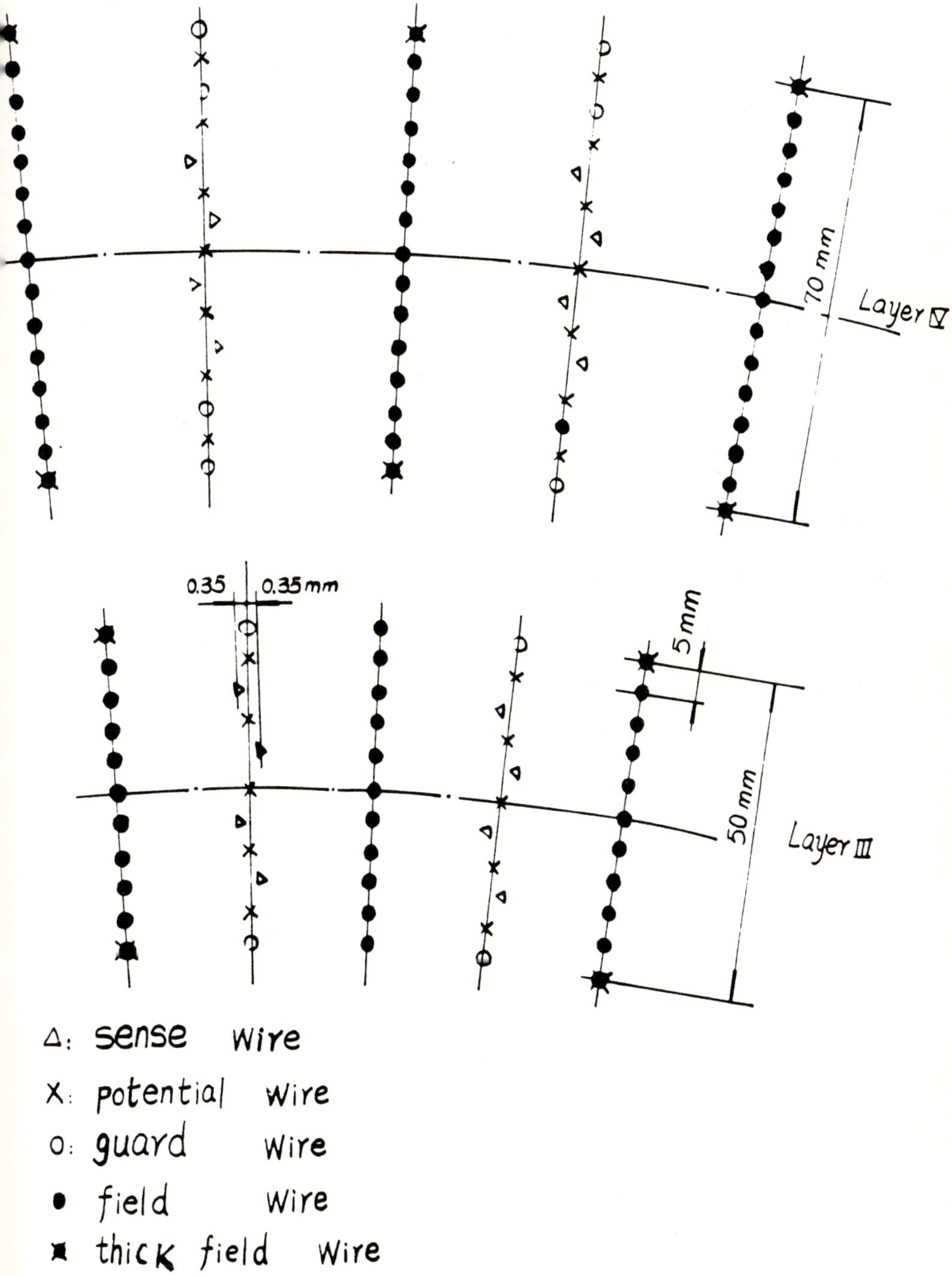

FIGURE 3 Cell Structure of main drift chamber

drawing of the cell structure. The gas mixture is the same as the central drift chamber. The momentum resolution is about:

$$\left(\frac{6_p}{P}\right)_{\text{space}} \sim 0.7\%\ P$$

$$\left(\frac{6_p}{P}\right)_{\text{multiple scattering}}^{0} \sim 1.2\%$$

4) Time-of-flight counter (TOF): In order to get good time resolution, we have investigated the relationship between the transmission efficiency of the scintillation light and the thickness of the scintillator. For the barrel TOF counter(284 cm long), 5 cm is the optimum choice of thickness (see Figure 4).

The TOF counter system consists of a barrel TOF counter and two end-cap TOF counters, with solid-angle coverage of 80% and 15%, respectively. For the barrel TOF counter, we have used 48 units of 5 cm thick NE110 scintillator, each 15cm x 284cm. The two end-cap TOF counters made of 2.5 cm thick NE102A scintillator, comprise 2 x 24 sectors forming two disks, each with an inner diameter of 70 cm and outer diameter of 211 cm. The photomultipliers used are XP2020.

The designed time resolution of the barrel TOF

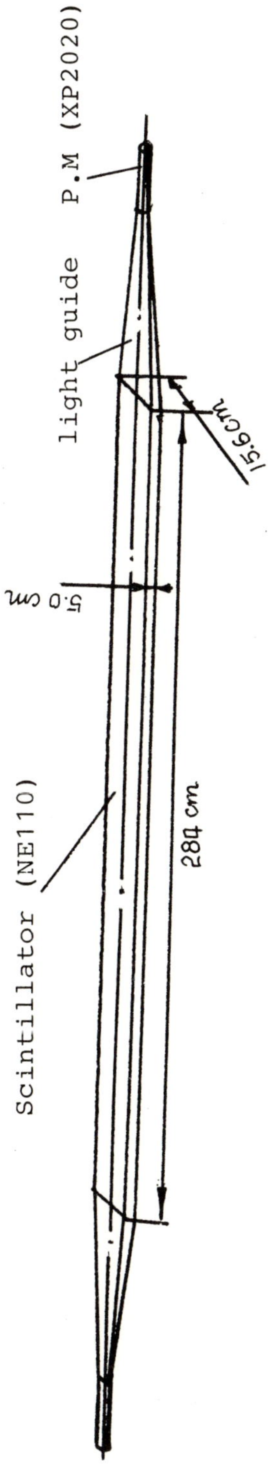

FIGURE 4 Barrel TOF Counter

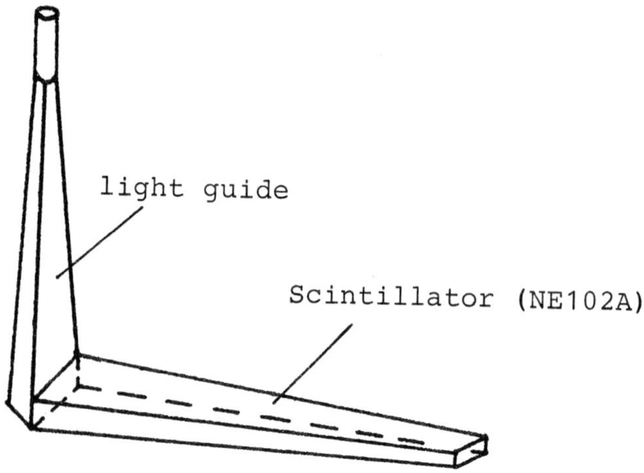

FIGURE 5 End-Cap TOF Counter

counter is 200 ps. For this time resolution and a flight distance of 1.17 m, $p \leq 1.07$ GeV/c for a 2 6 separation of π/k and $p \leq 1.78$ GeV/c for k/p separation. The expected time resolution of end-cap TOF counter is about 250 ps. Figure 5 shows the shape of end-cap TOF counter.

5) Shower counter: This counter consists of 24 sandwich layers of aluminium-clad lead plate and gas-discharge sampling counters. The total radiation length is 12 X_o. The self-quenching streamer discharge (SQS) mode is used. For the gas-discharge counter, the gas used is a mixture of argon(25%)+CO_2(50%)+n-pentane(25%). The output on 50Ω impedance is about 50 mV; the signal is much bigger than that from the proportional discharge mode, so the readout electronics are much simplified. One disadvantage is that the gas mixture is inflammable and hazardous. Figure 6 shows the charge for the various discharge modes. Figure 7 shows the structure of barrel shower counter.

The shower counter system consists of a barrel shower counter and two end-cap shower counters. The solid angle coverage is 80% and 13%, respectively, and the expected energy resolution $6_E/E$ is about 15% E (E in GeV). The z measurement is obtained by charge

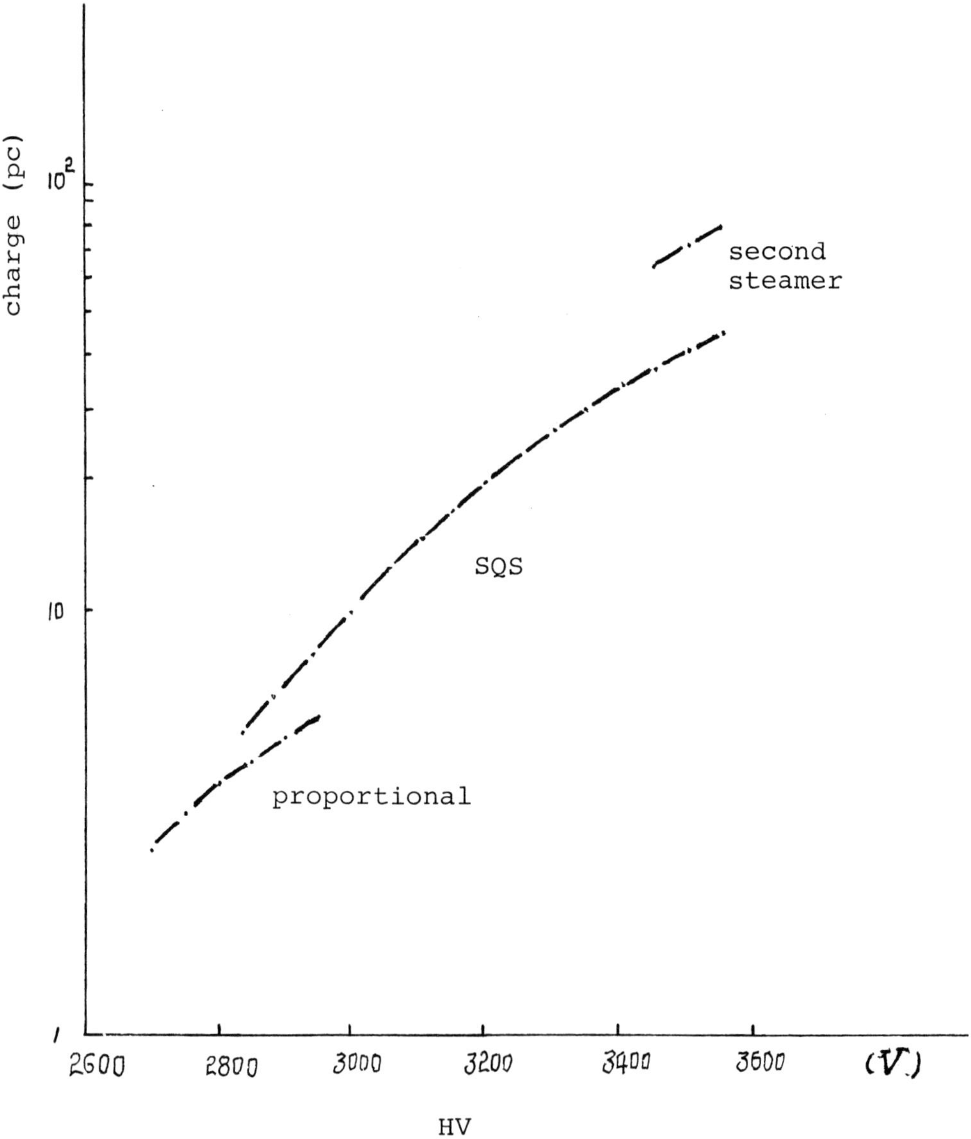

FIGURE 6 Characteristics of gas amplification

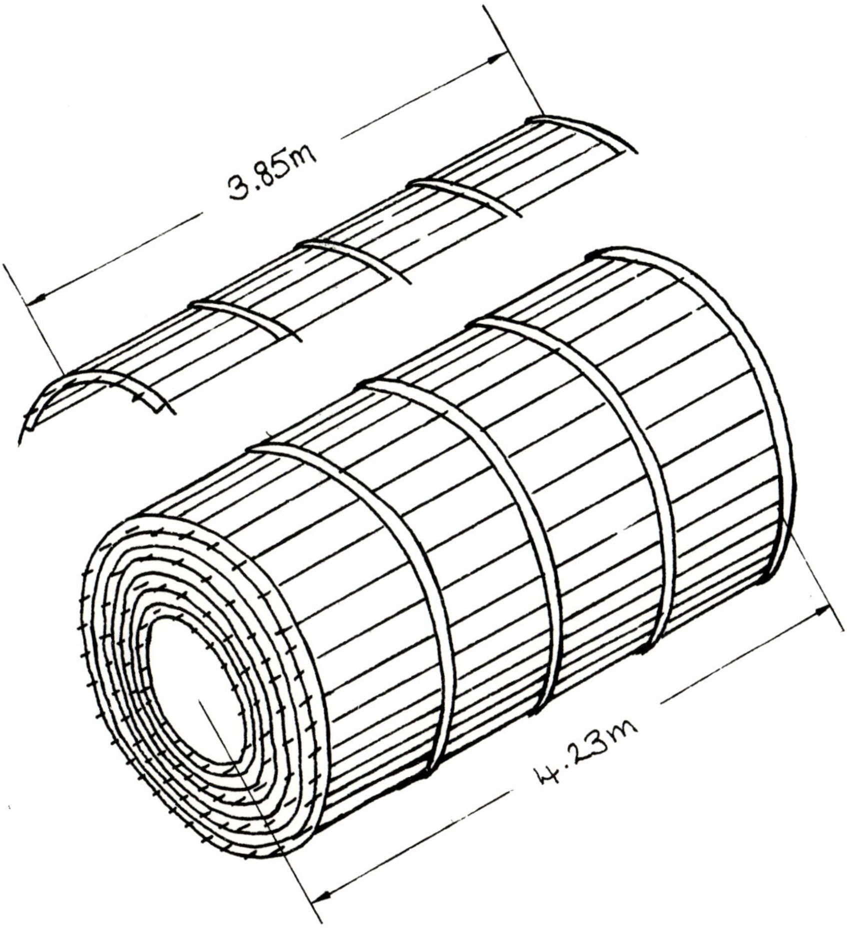

FIGURE 7 Barrel Shower Counter

division; $6_Z < 5$ mm is expected. Each layer of the barrel shower counter consists of 560 cells, to a total of 13,440. The end-cap shower counter have 190 cells in each layer, totaling 9,120 cells. In order to decrease the number of readout channels, some cells are grouped together. Thus the actual readout channel for the barrel is 6,720 and 5,520 for the end-cap. The lower energy cutoff is about 100 MeV with 100% efficiency. The angle resolution in ϕ is about 5 mrad.

6) Solenoid magnet: The solenoid magnet is a conventional one with a maximum field strength of 4.5 kG. It has an inner diameter of 348cm, an outer diameter of 414 cm, and the length is 360 cm. Aluminium conductor of 52 mm x 52 mm cross-section with a hole of 27 mm diameter is used. The iron yoke is in three layers, the thickness of each layer is 12 cm, 14 cm and 14 cm, respectively, counted from the inner one.

7) Muon identifier: Three layers of muon counter will be placed on the three layers of iron yoke. Each muon counter consists of aluminum extrude with length of 4.4 m (inner layer),or 4.68 m (outer layer). There are two rows with two cells for each extrude. The

cross-section is 60.9 mm x 50.8 mm for each cell. The total number of cells is 1,534 for muon identifier. The proportional discharge mode is used. The gas consists of 90% argon and 10% methane.

8) Luminosity monitor: The luminosity monitor consists of four pieces, two at each end of the beam pipe. Each piece consists of $24x76x3mm^3$ of scintillator (called P counter), $55x90x5mm^3$ of scintillator (called C counter), and $87x120x130mm^3$ sandwiched lead sheets and plastic scintillator shower counter (total 12 X_o, called S shower counter). The P counters are placed nearest to the interaction point and used to detect the Bhabha scattering particle. The C counters are placed behind the P counters and used to detect the other Bhabha scattering particle (recoil from the particle which enters the P counters at the opposite end of the beam pipe). the S shower counters are used to measure the energy of the Bhabha events. The upper and lower limits of acceptance of the luminosity monitor are 73 mrad and 29 mrad, respectively.

9) Readout electronics and on-line acquisition: The readout system adopts the following scheme: the signal from the detector, after it has been amplified or discrim-

inated, or converted (i.e. time signal converted into analog signal), is to be kept in a sample-and-hold circuit, and is to be read out later by a multiplexed intelligent ADC (i.e. the BADC of SLAC). The total number of readout channels of BES is 18,644. A VAX 11/785 computer will be used for on-line data acquisition. We follow the SLAC MARK III on-line system, with a VAX-CAMAC Channel (VCC).

10) Trigger: In order to reduce the dead-time, the BES trigger logic utilizes two levels. The first-level decision for charged-particle events will use the TOF counters with a 30 ns time window for rejecting more than 96% of the cosmic-ray background. The second level, which has a decision time of less than two cycles is a fast programmable track-finding processor using RAM chips. Cell hits from the innermost four layers of the main drift chamber are used for the track finding logic.

For the neutral events, we sum the signals from 24 radially aligned cells of the shower counter and use the corresponding radial energy deposit cut as a first-level decision. The second level for the neutral events will use the total energy cut of the shower counter and muon counter veto. For each level, a Boolean Logic Unit is used in the master trigger to allow up to 8 trigger modes and 16 input trigger conditions. Every input can be used as "must", "veto", and "don't care" in any mode. The trigger condition table can be loaded via CAMAC. The "OR" output will send out a master trigger signal.

Summary: BES is on basis of MARK III detector, but also there are some differences between BES and MARK III, such as:

1. configuration of main drift chamber,

2. setting up end-cap TOF counter to increase 15% of 4π solid angle coverage,

3. increase of 40 cm length for barrel shower counter and making end-cap shower counter close to interaction point by 15 cm to suppress the dead space (or "hole") between barrel shower counter and end-cap shower counter,

4. increasing the number of cells in each layer to improve position resolution in barrel shower counter,

5. using SQS mode instead of proportional mode for shower counter for saving of preamplifiers,

6. three layers of magnet iron yoke and muon counter instead of two layers in MARK III,

7. trigger logic using TOF counter with 30 ns window and RAM chips.

4. Progress and status of BES

In 1986, a beam test of the drift chamber prototype was made at KEK. The space resolution 6_x obtained is about 160μm. The assembly of the central drift chamber has been completed, all the wires have been strung. The beam test of the main drift chamber prototype was also made at KEK in 1986. The test results are as follows. The space resolution obtained is 156 to 208μm depending on the drift distance. For and e at 0.5 GeV/c, the energy resolution $6_E/E$ (6_E is rms of ΔE/Δx; E is mean value of ΔE/Δx) is 16% and 15%, respectively. 16,000 wires of the main drift chamber have been strung. Sealing and testing of both chamber will be finished before October this year.

For the prototype of barrel TOF counter, time resolutions obtained are 175 ps (at the end) and 196 ps (at the center) with real size by cosmic rays after time correction And for the prototype of end cap TOF countd ber, time resolution obtained is 217 ps.

Assembly of the TOF counters started in the end of 1986. 26 units of the barrel TOF counter have been completed (48 units altogether).

A beam test of the prototype of both barrel and end

cap shower counters was also conducted at KEK in 1986. The energy resolution $6_E/E$ tested is about $15\%E^{-\frac{1}{2}}$. The mechanical assembly and the stringing of wires of the barrel shower counter is progressing. The stringing of wires for the end cap shower counter has begun.

The beam pipe is ready. A long term testing of vacuum will be started soon. The construction of the muon counter has just begun. The luminosity monitor is ready and testing at KEK presently.

Most of the modules for the trigger system are ready. The TOF trigger unit and the track-finding processor have been tested. The trigger system is expected to be completed this year.

The iron yoke and coil have been assembled in the interaction hall. The magnetic field measurement machine has been assembled in laboratory and is now in trial operation. Assembly of magnet power supply is in progress.

For the on-line data acquisition system, the VCC and CAMAC modules have been connected with VAX 11/785. The program frame of BES data acquisition software is same as MARK III's. The MARK III software environment has been built on our VAX 11/785. Its on-line calibration software has been studied and tested successfully on VAX 11/785.

Testing of calibration system, including prototype of

shower counter and 240 channels of electronics has been operated.

1/5 of the readout for the drift chamber and 1/25 for the shower counter have been finished. The prototype readout systems for the TOF counters and the muon identifier have been completed. All the readout systems, except those for the muon counter and $\Delta E/\Delta x$ of the drift chamber, will be finished in early 1988; the rest will be ready by the end of 1988.

As far as off-line data analysis system is concerned, the supporting utility library and the frame of BES off-line analysis program system including Monte-Carlo programs have been established. The program package of reconstruction in the drift chamber has been put into operation and tested with Monte-Carlo method.

The EGS user code for shower counter has been implemented. The other parts of code for the shower counters, TOF and muon counters are in developing. It is expected to develop more physics tools for physicists, particle identification, more physics generator and kinematic fitting etc. We expect to finish all remain parts and improve some parts of system by June in the next year. We will put more manpower to increase the speed of development of on-line and off-line system from now.

It is expected that the drift chamber, the TOF counters and the shower counters will have been assembled and tested separately in the laboratory, and that the magnetic field of the solenoid coil will have been measured by October this year. This will be followed by the general assembly of the spectrometer and the cosmic-ray test. If everything proceeds as expected, BES will move into the interaction point in late 1988.

5. Plan for BES overall checkout and physics runs from 1989 to 1991

From Jan. to June in 1989, we expect to test and calibrate BES.

From July to Dec. in 1989, we expect to take overall checkout for BES. If the luminosity can be reached to $5\text{x}10^{29}\text{cm}^{-2}\text{sec}^{-1}$ in 1989, expected $7\text{x}10^{6}$ J/4 events will be obtained.

From Jan. to Dec. in 1990, we expect to take first physics run to accumulate $8\text{x}10^{7}$ J/4 events. If the luminosity can be reached to $5\text{x}10^{29}\text{cm}^{-2}\text{sec}^{-1}$, it will permit us to make further study on the characteristics of $\xi(2230)$, especially the spin and parity of $\xi(2230)$.

From Jan. to Dec. in 1991, we expect to take second physics run to study D physics at 4" energy.

6. Acknowledgements

We appreciate very much all the help from the world high energy physics community. The PRC/US Joint Committee for High-Energy Physics has made a big contribution to the progress of our project. We would like to thank, on the US side: Argonne, BNL, FNAL, LBL, and SLAC. Special thanks must go to SLAC, since BEPC is similar to their SPEAR, and BES is similar to MARK III. Many visitors from SLAC have given us lots of help on both physics and technique in the recent years.

We would also like to express our gratitude to CERN, KEK, DESY, RAL, TRIUMF, and other institutes. KEK is very kindly providing the beam for the test of our prototype drift chamber, shower counters, and luminosity monitor.

Professor T.D.Lee of Columbia University has made very valuable contributions to this project --without his painstaking efforts this project could hardly have moved. Special thanks also go to Professor W.K.H.Panofsky, who has given us much important help and advice.

Table 1 shows the main component parameters of BES.

Table 1. The main component parameters of BES

Components	Structure characteristics	Performance	Size
Beam pipe	0.3 mm wall thickness Al tube reinforced with 2.0 mm thick carbon-fibre	mass thickness: 10.4x10 R.L.	outer diameter: 15.4 cm Length: 150 cm
Central drift chamber	4 layers, double-field-wire cell, cylindrical chamber, z position determination by charge division	solid angle: 96% position resolution: 6_x: 150 μm 6_z: 2 cm	inner diameter: 18.8 cm outer diameter: 29.6 cm length: 110 cm
Main drift chamber	10 layers, multi-sense-wire cell, cylindrical chamber, 40 samples provide tracking and dE/dx information along radial direction. z position determination by stereo layers.	solid angle (second layer): 96% position resolution: 6_x:200 μm 6_z:3 - 4 mm momentum resolution: (6p/p)space ≃ 0.7%p (6p/p)m.s. ≃ 1.2% energy resolution: (for 40 samples of E/ x) /E 10%	inner diameter: 31.0 cm outer diameter: 230 cm length: 220 cm
Time-of-flight counters (barrel)	48 counters with 5 cm thick NE110 scintillator	solid angle: 80% time resolution 200ps	id. 231.4 cm od. 241.4 cm length of scintillator 284 cm
Time-of-flight counters (end-cap)	24 counters with 2.5 cm thick NE102 scintillator (each end-cap)	solid angle: 15% time resolution: 250ps	circular annulus form id. 70 cm od. 211 cm

Barrel shower counter	24 layers of pb/Al plate sandwich absorber, SQS mode gas sampling detector	solid angle: 80% , energy resolution: 6E/E=15%/ $E^{\frac{1}{2}}$	id. 247 cm od. 338.1 cm length 385 cm
End-cap shower counter	same as barrel	solid angle: 13% energy reso-tion: 6E/E=16%/ $E^{\frac{1}{2}}$	circular annulus form id. 65 cm od. 190 cm thickness 42 cm
Magnet	normal aluminum solenoid, providing maximum 4.5 kG axial field.	field uniform to 2% over the volume of tracking chamber	id. 348 cm, od. 414 cm, length 360 cm, weight of coil 31 ton. weight of yoke iron 400 ton.
Muon identi-fier	3 layer chambers, mounted outside 3 yokes individually. proportional tubes	solid angle 58%	distance from interaction region 462,510, 568(cm);length 435,470,470(cm), respectively.
Luminosity monitor	4 position counters. each consists of 24x76x3 mm^3 scintillator, 55x90x5 mm^3 scintillator and 87x120x130 mm^3 scintilla-tor calorimeter	minimum acceptance 29 mrad, maximum acceptance 73 mrad	distance from interaction region 192 cm

CHARM PHYSICS AT BEPC

TAO HUANG*
Center of Theoretical Physics, CCAST(World Laboratory) and Institute of High Energy Physics, Academia Sinica

Abstract We have discussed charm physics at BEPC (Beijing Electron-Positron Collider). In the first few years the main physics will be J/ψ, D and D_s mesons. τ lepton and charmed baryon will be studied later.

I. INTRODUCTION

We have been building an e^+e^- colliding machine (BEPC) to cover the region of center of mass energy from 3-5.6 GeV. This region has been studied by CEA, SPEAR, DORIS and ORSAY, and is still being explored by SPEAR. In the past years, many important advances in particle physics have been made in this region and many interesting unsolved questions are now under investigation with the Mark III detector at SPEAR. Here we begin by reviewing the results from the Mark III.

The Mark III is a general purpose magnetic solenoidal detector that was installed at SPEAR in July 1981. The initial checkout was completed in the fall of 1981 and the first data run was done at the J/ψ in the spring of 1982. This was followed in the fall of 1982 by running at the ψ''. In the spring of 1983 the first half of the running was

*Written with collaboration from Tongze Ruan, Ping Wang, Taijie Wang, Chi-min Wu, Dan-di Wu, Wuguang Yan, Hong Yu and Zhang-zhun Zhang

again devoted to running at the J/ψ, accumulating a total (1982+1983) of 2.7×10^6 produced J/ψ decays. The results separate into studies on the η_c, radiative decays and hadeonic decays. During the second half of the 1983 run at the ψ'' and by the winter of 1984, the Mark III had accumulated 9.4 p_b^{-1} at the $\psi''(3770)$ resonance to study the production and decay of charmed D mesons. A total of 5.8×10^6 J/ψ had been collected up to 1985.

In 1986 the Mark III completed a run at $\sqrt{S} = 4.14$ GeV. This center mass energy was chosen in order to search for the D_s and D_s^* at SPEAR. They obtained an integrated luminosity of 6.3 ± 0.46 p_b^{-1}. The data is now being reconstructed. They yield 26.7 ± 5.2 signal events of $D_s \to \phi\pi$ above 5.6 background events.

In addition, it is useful to recall the usage of the IBM 3081-K by the Mark III group. Approximately 1/4-1/3 of the CPU time of the IBM 3081-K (2000 hours/year) was allocated to this group for events reconstruction data analysis.

The Mark III group has obtained many very important results, such as (1) the first spin-parity determination of the η_c and observation of nine hadronic decay modes;[1] (2) The observation of glueball condidates (ι and θ) from the radiative decay of J/ψ;[2] (3) The observation of a narrow $K\bar{K}$ state called the $\xi(2.2)$;[2] (4) Detailed studies of other radiative decays of J/ψ, for example $J/\psi \to \gamma VV$ and $J/\psi \to \gamma\gamma V$ ($V=\rho,\omega,\phi$);[3] (5) The study of the J/ψ hadronic decays, especially the $J/\psi \to \{\omega,\phi\} X$;[4] (6) The determination of the polarization structure of the $f_2(1270)$, $f_2'(1525)$ and $\Theta(1720)$;[5] (7) The determination of the D meson lifetime and the study of D semileptonic decay: (8) The study of the pattern of the weak nonleptonic decays of D mesons (Cabbibbo-favored processes and some Cabibbo-suppressed processes);[6] (9) The measurements of the τ leptonic branching fractions;[7] (10) The study of D_s meson decay modes.[8]

In summary, the results from Mark III are evidences that many important physics problems remain to be studied in this region. Thus, we have chosen as our starting point an energy region that is still productive. We list some of the unsolved e^+e^- physics questions at medium energies that can be studied at BEPC.

(1) Radiative J/ψ decay gives us the best window on glueball (gg) and on hybrid states ($q\bar{q}g$), which is an unique prediction of QCD. Detailed studies are needed of previously discovered states, such as $\iota(1450)$, $\theta(1700)$, and $\xi(2230)$, and other channels, especially $J/\psi\to\gamma\gamma V$ and $J/\psi\to\gamma VV$ ($V=\rho$, ω, ϕ), must be studied carefully for some uncertain and undiscovered structures.

(2) Direct hadronic decays of J/ψ and ψ' may also help us to understand better both "new" and "old" physics in the spectrum below 2 GeV. In particlular, the decays $J/\psi\to\{\omega,\phi\}$ X are a useful tool in the investigation of the quark and possible gluonium content of given state X.

(3) One needs to measure the polarization structure of the tensor mesons and some possible exotic states (such as J^{PC}-1^{-+}, 0^{-+} mesons) so as to understand and determine some new states.

(4) Detailed study is needed of decay modes of charmed D mesons, as well as high precistion measurements of many preciously Cabibbo-allowed and, in particular, Cabibbo-suppressed nonleptonic D decays, including measurements of weak KM mixing matrix parameters, D°-$\bar{D}^\circ$ mixing and the CP violation parameter.

(5) The best energy region to find D_s in e^+e^-

annihilation might be between the D_sD_s and $D_sD_s^*$ thresholds (4.0 - 4.14 GeV). Up to now, the study for D_s in e^+e^- annihilation has been unsatisfactory. Further measurement of D_s decay will gain insight into the dominant decay mechanisms of the charmed mesons.

(6) One needs to make precise measurements of the τ non-leptonic branching fractions, to get more stringent limit on the ν_τ mass, and to search for unusual couplings or decays of τ. Measurement of D, $D_s \to \tau\nu_\tau$ can help to determine the ν_τ mass.[9]

(7) With high statistics we may be able to contribute to the study of the charmonium spectrum and the QCD potential. In particular, we could search for the only predicted charmonium state below charm threshold that has not yet been discovered, the $'p_1$ ($J^{pc}=1^{+-}$) state.

(8) We can contribute to the study of charmed baryons, (e.g. Λ_c, Σ_c, ...) which is now only at the beginning stage.

(9) The behavior of $R = \dfrac{\sigma(e^+e^- \to \text{hadrons})}{\sigma(e^+e^- \to \mu^+\mu^-)}$ ... should be studied, especially in the region 4-4.5 GeV where clarification of the $c\bar{c}$ spectrum could be useful.

In fact, in this energy region the Mark III has yielded some very interesting results and raised numerous important questions. There is still quite a deal of physics that needs to be studied experimentally. First, we have to raise the level of accuracy. Secondly, we have to solve some unanswered qewstions. Thirdly, we should observe the rare processes.

As an example, we meet a more complicated structure in the 1400-1500 MeV mass region.[10] A $K_s^\circ K^\pm\pi^\mp$ resonance was first seen in radiative J/ψ decays at $\eta(1400)$MeV (ι) by the Mark II in 1980. The E(1420) has been around for over 20 years as a member of the 1^{++} axial vector nonet, primarily containing $s\bar{s}$ quarks. The E(1420) (f_1(1420)) and ι(1450) are both $K\bar{K}\pi$ resonances but they have been regarded as different resonances. This is partly on the basis of the slightly higher mass and width of the ι ,[2,3]

$$M(\iota) = 1456 \pm 5 \pm 6 \text{ MeV} \qquad \Gamma(\iota) = 95 \pm 10 \pm 15 \text{ MeV}$$

$$M(E) = 1424 \pm 3 \text{ MeV} \qquad \Gamma(E) = 60 \pm 10 \text{ MeV}$$

But the principal basis for differentiation comes from the evidence of their spins and decays. The experiments that observe the ι in radiative J/ψ decay agree on assigning it $J^P=0^-$ and it is a pseudoscalar glueball candidate. However, the E seen in haronic reaction is more controversial with $J^{PC}=0^{-+}$ or 1^{++}. In 1983-1984 we have learned from the Mark III study of 2.7×10^6 J/ψ decays that the situation may be more complicated still: the structure of the $K\bar{K}$ Dalitz plot as a function of the $\bar{K}K\pi$ mass and observation of a $\rho\gamma$ signal raise the possibility that the iota region might contain more than one state. Even now 5.8 million J/ψ decays have not been sufficient to give us a clear understanding of its structure. A recent high statistics partial wave analysis claimed the spin-parity of the f_1(1420) to be 0^{-+}.[11] Therefore, with the possibility of the f_1(1420) being a pseudoscalar, it has been suggested[12] that the f_1(1420) is the same as, or part of, the η(1440)(ι). As mentioned earlier, while the radiative decays could be the ideal place to look for glueballs, ordinary $q\bar{q}$ mesons are also observed here. So the ι/E has been puzzling us for seven years now and it is clear that more detailed studies are needed, requiring higher statistics. If we want to understand gluonium and gluonic coupling to hadrons, radiative J/ψ decays have much to teach us, and a high statistic study of J/ψ decays at BEPC offers one of the best places to look.

Secondly, in 1983 the Mark III collaboration reported a ~ 5σ signal in radiative J/ψ decays for a heavy narrow resonance decaying into $K\bar{K}$, which they named the ξ . However, the DM 2 (Orsay) collaboration has analysed 8.7 million J/ψ decays, and they see no evidence for the $\xi(2.2)$. Results in the Mark III data, which were presented at the SLAC and Kyoto,[13] confirm once again the existence of the $\xi(2.2)$. Both experiments cannot be right. With higher resolution detector may be able to clear up this experimental conflict. A possible explanation might be due to the different TOF measurements.[14]

TABLE 1. Difference in TOF

Time of Flight system	Mark III	DM2
time resolution	190 PS	540 PS
π/K separation	2σ at 1.0 GeV	3σ at 0.45 GeV
solid angle	80%	79%

DM 2 requires only that one of the two charged tracks be compatible with TOF data expected of a kaon. One would expect this to give them more background above 2 GeV than the Mark III has. Even at the Mark III, because of the low level of the $K\bar{K}$ background, the ξ could be seen in the mass histogram with a signal of only ~ 30 K^+K^- events. To measure its spin will require $(2\text{-}5)\times10^7$ J/ψ decays; it might take 25-65 days of operation on BEPC at the design luminosity. We need a search for other decay modes ($J/\psi \to \gamma$ ($K^*\bar{K}$, $K^*\bar{K}^*$, $\eta\eta$, $\eta\eta'$ $\phi\phi$, and $\phi\omega$)). It could not only help to confirm its existence but also determine its properties.

Thirdly, the Mark III data sample in the $\psi''(3770)$ region corresponds to 21700 produced $D^{\circ}\bar{D}^{\circ}$ and 16800 produced $D^{+}D^{-}$ pairs. Even for a Cabibbo-favoured process $D^{+}\to\bar{K}^{\circ}\pi^{+}$, only 141 events are observed.[2] Cabibbo-suppressed channels will be about 5% of Cabibbo-favoured channels and Cabibbo-double-suppressed channels will be much smaller, about 1/400. Hence, it would be a significance step forward in this research area to understand the weak decay mechanism of charm particles. It is clear that much new information will be gained with higher statistics, if a total of 10^{6} $\psi''(3770)$ events is produced, which will take 30 weeks of operation on BEPC at the design luminosity.

There will be two detectors at BEPC: one is the Beijing Spectrometer (BES) another one is Crystal Ball. BES is similar to the Mark III. Since BES will follow seven years of operation by the Mark III, it is necessary to meet the following conditions:[15]

(1) We must collect many more events. If BEPC reaches the design luminosity, we will have the necessary high statistics to do significant physics with BES.

(2) We need high detection efficiency. In particular, we require good charged particle identification of π, K, p, e and μ; good charged particle momentum resolution; high detection efficiency for low energy photon with good position resolution and energy resolution.

(3) We need enough computing power to do the data analysis.

II. THE INSTALLATION AND THE INITIAL CHECK OUT OF BES

According to the schedule, BEPC will have beams colliding by the end of 1988. The check-out of the Beijing Spectrometer will take place in 1989 at the J/ψ, during which BEPC will also be tuned to have luminosity increasing to the design value. It is expected that more than 2×10^6 J/ψ events will be accumulated during 1989,[15] allowing us to repeat measurements made by the Mark III.

The luminosity at SPEAR is listed in table 2. The luminosity of BEPC is optimized at 2.8 GeV, where the peak luminosity is $1.7\times10^{31}/Cm^2S$. Between 1.5 and 2.8 GeV, the design luminosity is expected to be E^2 (not $\propto E^4$), because of the installation of a system of wiggler magnets. The RMS bunch length (σ_z) is 5.8 cm and the

TABLE 2. Luminosity vs Energy at SPEAR

E_{Cm}	3.1	3.77	4.4	5.6	6.0
SPEAR $10^{30}/Cm^2S$	0.6	1.4	2.5	6.6	8.7

energy spread is $\Delta E(MeV) = 2.67\times10^{-7} E^2$. The expected luminosity at the J/ψ is

$$L_{peak} \text{ (3.1 GeV) - } 5 \times 10^{30}/Cm^2S$$

and $$\bar{L}_{av} \text{ (3.1 GeV } = \frac{1}{4} \times 5\times10^{30}/Cm^2S \quad ,$$

where the factor 1/4 reflects actual running experience at installations without full energy injection, including the time needed to fill the ring. Since the J/ψ is a narrow resonance, Γ= 63 KeV$<<\Delta E_{beam}$, where ΔE_{beam} is the energy spread of the beam. The natural RMS value for the BEPC design is $\Delta E_{beam}(J/\psi) \cong 0.6$ MeV. The theoretical cross section

$$\sigma_\psi = \frac{12}{M_\psi^2} Br(J/\psi \to e^+e^-) \cong 1.13 \times 10^{-28} Cm^2$$

must be smeared by the beam spread to find the observed cross section,[16]

$$\bar{\sigma}_{\psi} = \sigma_{\psi} \frac{\Gamma_{\psi}^{tot}}{\sqrt{2}\ \Delta E_{beam}} = 7.4 \times 10^{-30} Cm^2$$

Thus, for a run of $T=10^7$ sec $= \frac{1}{3}$ year we get the event yield

$$N_{\psi} = \bar{\sigma}_{\psi}\ T\ \bar{L}_{av} = 9 \times 10^7$$

In fact, at the beginning of the BES check-out in 1989 the luminosity is not likely to reach the design value, 5×10^{30} $/Cm^2S$; it might start lower, and increase step by step. Then, in the first year we might get an event number,[15,16]

$$N_{\psi} > 2 \times 10^6 \text{ events, for the first year,}$$

which can be compared to the number of events obtained by the Crystal Ball and the Mark III in 1982.

After BES is installed in Sep. 1988, our schedule will be as follows:

(1) The installation of BES Sep.1988-Dec.1988
(2) The initial check out of BES Jan.1989-June 1989
(3) The first data run at the J/ψ July 1989-Dec.1990
(4) The second data run at the $\psi''(3770)$ Jan.1990-Dec. 1992

In order to test our detector the various properties of J/ψ will be studied and checked. The basic analysis technique is to kinematically fit events using the four constraints of energy and momentum conservation. Kinematic fitting substantially improves the energy resolution for the γ, and will be successfully applied to events with up to five γ's and four charged tracks. If the decay includes a π° or η , then an additional mass constraint can be applied to the $\gamma\gamma$ system. TOF and $\frac{dE}{dX}$ identifications will be required when topologies include K or P. Hence, we make some suggestions for the check-out of BES:

(1) Measurement of Bhabha scattering cross section.

(2) Measurement of the cross section, decay width and $Br(J/\psi \to e^+e^-)$ of J/ψ.

(3) Separation of $\pi/K/P$, which will be displayed in a β vs momentum scatter plot.

(4) The reaction $J/\psi \to \pi^+\pi^-\pi^\circ$ will be studied; $\rho^\circ\pi^\circ$, $\rho^+\pi^-$ and $\rho^-\pi^+$ bands will be shown in the Dalitz plot.

(5) The reaction $J/\psi \to \rho^\circ\pi^\circ$ is very useful to study the shower counter γ detection efficiency and the $\gamma\gamma$ mass resolution. The π° from this decay is produced essentially monochromatically at 1.45GeV, and the γ's from these decays have an energy spectrum in the Lab frame that is flat from nearly 0 to 1.45 GeV.

(6) The decay $J/\psi \to \gamma\eta'$ will be studied in two decay modes, $\eta' \to \gamma\rho^\circ$ and $\eta' \to \eta\,\pi^+\pi^-(\eta \to \gamma\gamma)$.

These measurements will demonstrate the performance of the BES detector and in particular the efficiency for low energy γ detection.

After one year of testing BES at the J/ψ energy region. it is hoped that the machine will work satisfactorily and will reach the design luminosity. Within the next two years the first experiments with BES will mainly be done at the J/ψ energy range. In the first year (1989), $(2\text{-}5)\times10^6$ events will be collected at the J/ψ. After continuous running for another year (1990), it is expected to obtain (5-7) x 10^7 events. This will reduce the statistical error of the measurements of the J/ψ decay channels and will make a qualitative difference in the ability to see new structure. Also, such heated problems as the particle and the ι/E puzzle will benefit from significantly more experimental information. It is not excluded that new phenomena might be unearthed with (5-7) x 10^7 J/ψ events. In 1991, it is planned to switch to the energy range for D meson physics. Integrated luminosity of 40 pb^{-1} is expected. This means greater accuracy for the Cabibbo-favored decay channels of the D meson, as well as quantitative improvements for Cabibbo suppressed decay channels.

III. J/ψ PHYSICS AT BEPC

The first data will run at the J/ψ mass region. It is estimated in lowest order perturbation theory that the partial width of the radiative decay of J/ψ is about 8% of the total width:

$$\frac{\Gamma(J/\psi \to \gamma\, gg \to \gamma + \text{hadrons})}{\Gamma(J/\psi \to ggg \to \text{hadrons})}$$

$$= \frac{36}{5}\, e_q^2\, \frac{\alpha}{\alpha_s} \left[1 + \frac{2.2^{\alpha_s}}{\pi} + \ldots\right]$$

$$= 8\%$$

The sum of the partial widths of the now observed radiative decay channels is only 30% of the total radiative decay width. This means that there are still many more radiative decay channels to be searched experimentally, among which there may be interesting new particles and new phenomena.

Direct hadronic decays of J/ψ may also teach us a

great deal about the spectrum between 1 and 2 GeV. They may be especially helpful for distinguishing the "old"($\bar{q}q$) states from "new" gluonic states. The Mark III has just begun a systematic analysis of these decays, which BES will be able to extend with higher statistics.

Here we mention some of the interesting problems requiring experimental study.

(1) The 1400-1500 MeV region[2,10]

The "old and the "new spectroscopy came to meet in a bitter feud in the 1400-1500 MeV mass region. For example, the ι/E puzzle continues today in this region. $\eta(1440)$ (ι) is certainly the most viable candidate of a glueball. In 1983, the Mark III accumulated 340+402 iota events from the analysis of a total sample of 2.7 x 10^6 J/ψ events. In 1985 many reactions have been studied using a sample of $5.8\text{x}10^6$ produced J/ψ decays. The branching ratio is[2]

$$Br(J/\psi\to\gamma\eta(1440))Br(\eta(1440)\to K^{\pm}K^{\circ}_{S}\pi^{\mp})=(16.5\pm1.0\pm2.7)\text{x}10^{-4}$$

$$Br(J/\psi\to\gamma\eta(1440))Br(\eta(1440)\to K^{+}K^{-}\pi^{\circ})=(8.2\pm0.4\pm1.4)\text{x}10^{-4}$$

However, the hadronic experiments[11] support $J^P= 0^-$ for E, while the results from tagged two photon events provide evidence for a 1^+ state.[17] There is a growing suspicion that there may be three states (or more) in this region: one with $J^P= 1^+$ and two with $J^P= 0^-$. So the interpretation of this mass region is unclear due to the complicated structure and $f_1(1420)$. To address this question, the Mark III has compared the radiative decay $J/\psi\to\gamma K\bar{K}\pi$ with the hadronic decays, $J/\psi\to\omega K\bar{K}\pi$ and $J/\psi\to\phi K\bar{K}\pi$, since the quantum numbers J^{PC} of ω or ϕ are same as the photon and the flavor of the ideally mixed vector mesons is known,

$$\phi = |s\bar{s}\rangle$$

$$\omega = \frac{1}{\sqrt{2}}(|u\bar{u}\rangle + |d\bar{d}\rangle)$$

Hence, if a ϕ or an ω is identified in the J/ψ decay, because of the continuation of the quark lines, the recoil system will contain the $s\bar{s}$ or the $u\bar{u}$, $d\bar{d}$ quark content of the respective mesons. This provides a very useful technique to determine the quark contents of various resonances by observing their production in association with a ϕ and an ω in the J/ψ hadronic decays. Neither the $f_1(1420)$ nor the $\eta(1440)$ is observed in the $K\bar{K}\pi$ mode recoiling against a ϕ. If the double OZI contributions could be neglected, this would strongly suggest that the $f_1(1420)$ does not have any substantial $s\bar{s}$ quark content.[18] One of several explanations is that the $K\bar{K}\pi$ peak observed in association with an ω in the J/ψ decay, as well as that observed in the 2-γ reaction, is a hybrid $q\bar{q}g$ 1^{-+} exotic state,[19] and the state observed in $\eta\pi\pi$ in association with an ω is a different state. So a deeper investigation of this mass region is very interesting. More accurate data are needed.

(2) Glueball searches

Since $J/\psi \to \gamma X$ is dominated perturbatively by $J/\psi \to \gamma gg$, radiative J/ψ decay is the preferred channel for glueball searches. The dominate quantum numbers of the digluon, i.e. perturbatively, are 0^{++}, 0^{-+} and 2^{++}, which coincides with the expected quantum numbers of the lightest glueballs. Theory predicts that the mass of the lightest glueballs is likely to be in $\sim$1-2 GeV range. The search for glueballs is one of the most important questions in the experimental study of QCD. It will be a triumph for QCD if a strongly interacting gluonic bound state or a hybrid state is observed. Identifying such a state will be an experimental success.

Although radiative decay of J/ψ is the ideal laboratroy to search for glueballs, it is hard to be sure that any particlar state that is found is a glueball. There are two reasons. First, the mass spectrum of glueballs is not yet reliably calculated. Secondly, the energy range from 1 to 2 GeV is very complicated. There can be as many as 26$(q\bar{q})$ nonets in that region, and we have only identified some of them. More are undiscovered. Their energy spectrum overlaps the glueball spectrum, and there may be mixing between the $(q\bar{q})$ states and the glueball states with the same quantum numbers. Furthermore, there may also be $(q\bar{q}g)$ and $(q\bar{q}\ q\bar{q})$ states in the same energy range, which are also difficult to distinguish. All these make the search for glueballs a very difficult task, thus many more J/ψ events are needed.

The $\eta(1440)(\iota)$ is a glueball candidate, as mentioned before. $f_2(1720)$ is another glueball candidate that was discovered in J/ψ with $M=1640\pm50$ MeV and $\Gamma=220^{+100}_{-70}$ MeV.[20] That is a lower mass and larger width than $f_2(1720)$ later seen in the higher statistics $\pi K\bar{K}$ measurements[5]

$$M = 1720 \pm 7 \text{ MeV} \qquad J^{PC} = 2^{++}$$

$$\Gamma = 132 \pm 15 \text{ MeV}$$

The Mark III has observed the $f_2(1720)$ in radiative decays of the J/ψ into K^+K^-, $K_S^\circ K_S^\circ$ and $\pi^+\pi^-$, with large branching ratios. The analysis of the radiative decay mode, $J/\psi \to \gamma\eta\eta$, is in progress.[18]

Beside $\eta(1440)$ and $f_2(1720)$, there are three wide

resonances observed in hadronic reactions, which may possibly be glueballs.[21]

$$g_T \quad M = 2050 \pm {}^{90}_{50} \qquad \Gamma = 200 \pm {}^{160}_{50} \text{ MeV}$$

$$g_{T'} \quad M = 2300 \pm {}^{20}_{100} \qquad \Gamma = 200 \pm {}^{60}_{50} \text{ MeV}$$

$$g_{T''} \quad M = 2350 \pm {}^{20}_{30} \qquad \Gamma = 270 \pm {}^{270}_{130} \text{ MeV}$$

They are all 2^{++} states. These are not observed by the Mark III and the DM 2 detector, which give only upper limit, perhaps because their detector efficiency is poor for the low momentum K mesons, which they decay into. One of the key problems of BES is to raise the reconstruction efficiency of $\phi\phi$ states, so that g_T, $g_{T'}$ and $g_{T''}$ states can be verified, if they do exist.

A notable problem is that no scalar gluonium has been confirmed or indicated. More precise measurements and more data may produce some valuable results.

(3) $\xi(2200)$ particle

This particle was found by the Mark III after the accumulation of 2.7×10^6 events at SPEAR. The $\gamma K^\circ_S K^\circ_S$ and γK^+K^- analyses are performed on additional statistics of $3.1\text{x}10^6$ J/ψ from the 1985 data, and confirmed the previous observation.[13] The Mark III has shown that[13,18]

$$M(\xi) = 2230 \pm 6 \pm 14 \text{ MeV}$$

$$\Gamma(\xi) = 26\ {}^{+20}_{-16} \pm 17 \text{ MeV}$$

for the charged mode, and

$$M(\xi) = 2232 \pm 7 \pm 7 \text{ MeV}$$

$$\Gamma(\xi) = 18 \, {}^{+23}_{-15} \pm 10 \text{ MeV}$$

for the neutral mode. The measured branching ratios of the two decay modes were

$$Br(J/\psi \to \gamma\xi) \cdot Br(\xi \to K^+K^-) = (4.2^{+1.7}_{-1.4} \pm 0.8) \times 10^{-5}$$

and

$$Br(J/\psi \to \gamma\xi) \cdot Br(\xi \to K^\circ_s K^\circ_s) = (3.1^{+1.6}_{-1.3} \pm 0.7) \times 10^{-5}$$

As the $K^\circ_s K^\circ_s$ system can have only even spin and positive parity, i.e. 0^+, 2^+, 4^+, ..., the spin-parity determination by a maximum likelihood fit to the helicity amplitude favors $J^P \geqq 2^+$.[22] There are altogether about 70 events. Due to the meager statistics a full partial wave analysis (PWA) was not performed.

There are several theoretical speculations, as to the nature of $\xi(2230)$, e.g. a glueball[21] or a hybrid state,[23] an ordinary high spin mesonic state, and others.

If 2×10^7 events are collected by BES at BEPC, then by extrapolation there will be about 240 events. This will be helpful for the investigation of the ξ particle and the mass region in 2200 to 2400 MeV, which shows interesting structures.

The DM2 group at Orsay has also accumulated 8.7×10^6 events, but has not observed the $\xi(2200)$.[24] The upper limit is

$$Br(J/\psi \to \gamma\xi) \, Br(\xi \to K^\circ_s K^\circ_s) < 2.0 \times 10^{-5}$$

$$Br(J/\psi \to \gamma\xi) \, Br(\xi \to K^+ K^-) < 1.2 \times 10^{-5}$$

A possible reason for their negative result is the poorer resolution of π and K from their time-off flight detector compared to the Mark III. BES should be able to verify the existence of ξ. If it does exist, BES will be able to measure its spin and the branching ratios of its decay channels (e.g. $\phi\phi$, $\phi\omega$, $\eta\eta'$, $K^* \bar{K}$, $K^* \bar{K}^*$ besides $K^+ K^-$, $K^o_S K^o_S$).

(4) Four-quark states and hybrid states

In 1982 the process $J/\psi \rightarrow \gamma\rho\rho$ was first discovered by the Mark II. The mass distribution is predominantly under 2GeV, and there is some kind of structure at 1.65 GeV. Later, the enhancement of $\rho\rho$ final states under 2 GeV was also observed in hadronic reactions and in photon-photon scattering.[25]

TABLE 3. $J/\psi \rightarrow \gamma VV$

VV	Mass Range in GeV	J^P	$B(\psi\rightarrow\gamma VV)$ x10³	Peak (MeV)	Experiment
$\rho^o\rho^o$	2	all	(1.3±0.5)	1650±50	Mark II
$\rho^+\rho^-$	1500-1900		mostly 0⁻ or 2⁺	1700±40	Crystal Ball
	1700 peak	all	(3±1)		
$\rho\rho$	< 2	all	(6.2±1.5)	1550,1800	Mark III
	< 2	0⁻	(5±1)	~1600	
$\omega\omega$	all	all	1.8±0.1±0.5	~1800	Mark III
	< 2	all	1.2±0.1±0.3		
	< 2	0⁻	1.0±0.4		
$\omega\omega$	all	all	0.8±0.2±0.2	~1800	DM2
$\phi\phi$	<2.8	all	0.31±0.04±0.06		DM2

The measurement by the Mark III indicates that the vector-vector resonance peak is predominantly pseudoscalar. This resonance might be a four-quark state, a hybrid state

$|q\bar{q}\ g\ \rangle$, or a mixture of glueball and four-quark. To verify these hypotheses, it is necessary to have further measurement and analysis of those new particle states in J/ψ radiative decay products, and to decide their properties precisely, particularly J^{PC} and the branching ratios of their various decay channels.

(5) Polarized production of tensor mesons in J/ψ radiative decay

For the 2^{++} channel, the $f_2(1525)$ and $\theta(1720)$ have been observed in two pseudoscalar mesons decay modes by the Crystal Ball, Mark II and Mark III experiments. The spin analyses are performed by studying the angular distribution of the radiated photon and one of the two pseudoscalars,

$$f_2(1270)\ \left\{ \begin{array}{l} x = 0.96 \pm 0.12 \\ y = 0.96 \pm 0.13 \end{array} \right.$$

$$f_2'(1525)\ \left\{ \begin{array}{l} x = 0.63 \pm 0.10 \\ y = 0.17 \pm 0.20 \end{array} \right.$$

$$\theta(1720)\ \begin{array}{l} x = -1.07 \pm 0.20 \\ y = -1.09 \pm 0.25 \end{array}$$

where x and y are the ratios of the amplitude with helicity one and two to helicity zero. Comparison between some model's predictions and experimental values is helpful to clear up the particle's property.

For some exotic states (e.g. 1^{-+} especially 3^{-+} glueballs) one has derived special x and y behaviour[26] using the perturbative QCD. We believe that a careful experimental measurement may give some new exciting results.

IV. THE CHARMED MESON PHYSICS AT ψ'' ENERGY REGION

The Mark III has collected 9.25 nb^{-1} integrated luminosity at $\psi''(3770)$ peak, which corresponds to 4×10^4 $D\bar{D}$ events. $\psi''(3770)$ is one of the ways to study D meson physics, because

$$\psi''(3770) \to D^+D^-, D^\circ\bar{D}^\circ$$
$$\nrightarrow D\bar{D}^*$$

The Mark III has given the production cross-section, which is smaller than the value by other groups,[2]

$$\sigma_{D^\circ} = 4.9 \pm 0.3 \pm 0.4 \text{ nb}$$
$$\sigma_{D^+} = 3.6 \pm 0.4 \pm 0.4 \text{ nb}$$

The events, which correspond to σ_{D° and σD^+ are separately

$$N_{D^\circ\bar{D}^\circ} = 22700 \pm 1600 \pm 1600$$
$$N_{D^+D^-} = 16800 \pm 2000 \pm 1600$$

It may be seen that the measured $D\bar{D}$ production cross-section is 8.5 nb, which is 75% of the theoretical value, $\sigma_{\psi''} = \frac{12}{M^2{}_{\psi''}} Br(\psi'' \to e^+e^-) \cong 11.3$ nb. This raises a problem that perhaps $\psi''(3770)$ does not decay exclusively into $D\bar{D}$ pairs. So what are the other decay modes?

The recent measurements of two body decays of charm mesons D^+, D°, D_s have obtained some interesting results and raised many important questions. The BES promises to get some solutions to these issues.

(1) The D meson lifetime measurement

The simple spectator model of the decay mechanism predicts

$$\tau(D^+) / \tau(D^\circ) = 1$$

But experimental result of this ratio deviates from 1 substantially. The latest data gives[28]

$$\frac{\tau(D^+)}{\tau(D^\circ)} = \frac{Br(D^+ \to e^+ + X)}{Br(D^\circ \to e^+ + X)} = 2.3 \, {}^{+0.5}_{-0.4} \pm 0.1$$

In 1985, the world average of D meson lifetimes are

$$\tau(D^+) = (8.02 \pm 0.06) \times 10^{-13} \text{ Sec}$$

$$\tau(D^\circ) = (3.97 \pm 0.3) \times 10^{-13} \text{ Sec}$$

These experimental results indicate that there is some kind of mechanism that enhances the D° decay or suppresses the D^+ decay. Further precise measurement of D^+ and D° lifetimes will be significant for the understanding of this kind of mechanism.

Recently, there have been several new neasurements of the $D_s(F^+)$, lifetime: $(3.2\,{}^{+3.0}_{-1.3}) \times 10^{-13}$ sec by the NA 11 experiment,[29] $(3.5\,{}^{+2.4}_{-1.8} \pm 0.9) \times 10^{-13}$ sec by the HRS group,[30] and $(2.6\,{}^{+1.6}_{-0.9}) \times 10^{-13}$ sec by the new E531 experiment.[31] All the new measurements are substantially larger than the old E531 result,[31] $(1.9\,{}^{+1.3}_{-0.7}) \times 10^{-13}$ sec, but still smaller than the world average D° lifetime. However recent E691 experiment,[32] $(0.48\,{}^{+0.05}_{-0.04} \pm 0.02) \times 10^{-13}$ sec show that the life-time of D_s is equal to one of D°. The study of the lifetime of D_s can help to sort out the possible explanations for the $D^\circ - D^+$ lifetime difference.

(2) Improve the measurement accuracy of the Cabibbo favored modes

Up to now, with the 4×10^4 $D\bar{D}$ events, the Mark III has been able to give the branching ratios of a dozen or so

Cabibbo-favored D decay channels. Experimental data of the Cabibbo-favored decay modes are listed in Table 4.[5,7,28]

TABLE 4. Experimental data of the Cabibbo-favored decays

D°	Br(%)	D^+	Br(%)
$K^-\pi^+$	$4.9 \pm 0.4 \pm 0.4$	$\bar{K}^\circ\pi^+$	$3.5 \pm 0.5 \pm 0.4$
$\bar{K}^\circ\pi^\circ$	$2.2 \pm 0.4 \pm 0.2$	$\bar{K}^{*\circ}\pi^+$	$3.0 \pm 1.9 \pm 1.7$
$\bar{K}^\circ\eta$	$1.8 \pm 0.8 \pm 0.3$	$\rho^+\bar{K}^\circ$	$12.2 \pm 2.8 \pm 1.9$
$\rho^\circ\bar{K}^\circ$	$1.3 \pm 0.4 \pm 0.3$		
$\bar{K}^{\circ *}\pi^\circ$	$2.1 \pm 0.9 \pm 0.6$		
$K^{*-}\pi^+$	$7.8 \pm 1.2 \pm 0.9$	D_s^+	Br(%)
	$7.1 \pm 1.6 \pm 1.3$		
ρ^+K^-	$13.7 \pm 1.3 \pm 1.5$		
$\phi\,\bar{K}^\circ$	1.4 ± 0.5[33]	$\bar{K}^{*\circ}K^+$	4.8 ± 2.4
	$1.1^{+0.7+0.4}_{-0.5-0.2}$	$\phi\,\pi^+$	3.3 ± 1.4
$\omega\,\bar{K}^\circ$	$3.8 \pm 1.5 \pm 1.0$		

From Table 4 one can see that some of the decays do not have very good statistics. E.g. for the Cabibbo favored process $D^+ \to \bar{K}^\circ \pi^+$, it has only 141 events, and the branching ratio is (3,5 ± 0.5 ± 0.4)%. If the collected events are to be quadrupled, the accuracy of the number can be improved by reducing the error by a factor of 2.

Although non-leptonic decay of charmed mesons is very complex, current experimental results have already provided much information on non-leptonic decay mechanisms.

The experimental results for exclusive two-body decays of charmed mesons can be analyzed in a model-independent way within the framework of the quark diagram formulation.[34] They can be described in terms of six different quark diagrams: the external W emission a, the internal W emission b (a and b just are those simple spectator diagrams), the W exchange c, the W annihilation d, the horizontal W-loop diagram e, and the vertical W-loop diagram f. Diagram e and f are also called penguin diagrams. (See Figure 1)

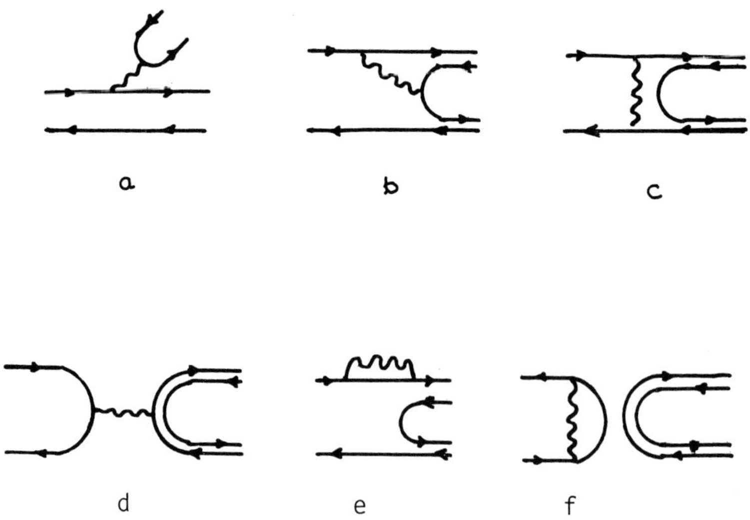

FIGURE 1. Quark diagrams of charm meson decays.

Such a scheme can incorporate any specific strong interaction model calculations, i.e. all quark diagrams include strong-interaction gluon clouds.[35] The Mark III data of charm decaying into two pseudoscalar cannot be fitted if all quark-diagram amplitudes are real and if SU(3) symmetry is assumed. This means that one has to incorporate the effects of SU(3) breaking and final-state interactions in the quark diagrams to account for inelastic final-state interactions due to resonances and rescattering effects.[35]

The large number of D branching ratios measured by the Mark III can now account for about 85% of the D° and D^+ decays. Therefore, the difference between D° D^+ lifetimes should be understandable at the level of two body decays, since three and four body decays are dominated by quasi-two-body channels. The longer lifetime of D^+ compared to D° may be attributed to the suppression of D^+ rate and the enhancement of D° rate.

In fact only a few channels are open to D^+ owing to identical-particle effects in the Cabibbo-allowed decay modes of D^+. More precisely there are 16 channels for Cabibbo-favored D° decays, but only 4 channels for D^+. There is another reason for the suppression of D^+ rate. D° decay amplitude is proportional to a or b, whereas the Cabibbo-allowed D^+ amplitude is always of the form (a+b) due to the identical-particle effect. This implies that the interference between a and b is always destructive. Hence, there is a severe destructive interference in all Cabibbo-favored D^+ decays.[35]

The W-exchange diagram, which does not exist in D^+ decays, may enhance the D° decay rate. Evidence for the decay $D^{\circ} \to \bar{K}^{\circ}\phi$, which occurs only through nonspectator processes, has been presented in the Mark III,[36] with a large branching ratio 1%. This is a signal for the W-exchange diagram.[37]

To quantify these arguments requires additional measurement of decays that are sensitive to interference and

W-exchange. We hope the BES will help to further test these decay mechanism.

A dynamical calculation given by Stech and Bauer,[38] which uses the factorization approach and QCD theory, has explained the observed lifetime ratio of D° and D^+, and the color-mismatched channels. They use only quark decay diagrams, i.e. a and b that contain the QCD radiative corrections and final state interactions. There are basically just two free parameters that can be obtained from a perturbative QCD calculation.[39] But the data of $D^\circ \to K^\circ\bar{K}^\circ$, $\bar{K}^\circ\ \omega$, cannot be explained by their framework. The $\bar{K}^\circ\ \phi$ channel provides evidence for the presence of nonspectator amplitudes in D° decays. This would be expected if the W-exchange amplitude governed the decay. There is a possible contribution to these channels which comes from hair-pin diagrams.[40] In fact, a hairpin diagram only appears when one of the final product particles is a SU(3) singlet. So by introducing the hairpin diagram, only those channels with ϕ, ω or η' in the final products have to change their theoretical expectations.

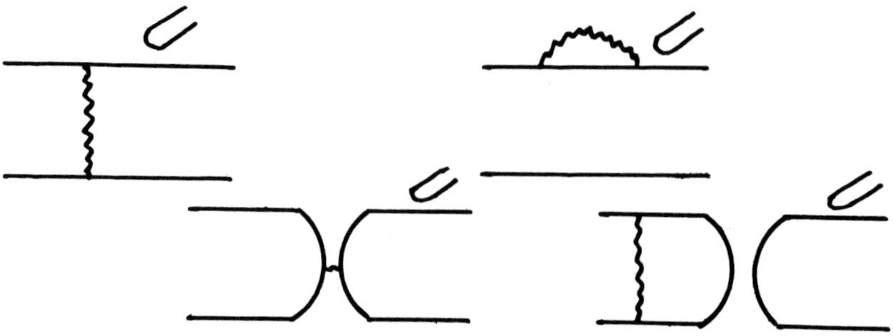

c_h d_h e_h f_h

FIGURE 1. Hair-pin diagram of charm meson decays.

(3) The measurement of Cabibbo-suppressed decay modes

The Cabibbo-suppressed decay modes shed new light on the mechanism of D decay; theoretical interest is particularly intensitive because the penguin diagram (e) may enter the process.[41] This is the diagram that now people assume to explain the $\Delta I = \frac{1}{2}$ rule in the strange particle decays.[42]

The Mark III has got six channels data of Cabibbo-suppressed decays, although the statistics are not very high. They are listed in Table 5[5,7,28]

TABLE 5. Experimental data of the Cabibbo favored decays

Decay Modes	Br (%)
$D^\circ \to K^+ K^-$	0.60 ± 0.10 ± 0.08
$\pi^+ \pi^-$	0.16 ± 0.05 ± 0.03
$K^{*-}K^+$	1.02 ± 0.47 ± 0.21
$D^+ \to \bar{K}^\circ K^+$	1.11 ± 0.34 ± 0.21
$\bar{K}^{*\circ}K^+$	0.53 ± 0.24 ± 0.14
$\pi^+ \phi$	0.93 ± 0.26 ± 0.17

It is well known that the SU(3) symmetry implies

$$R = \frac{\Gamma(D^\circ \to K^+K^-)}{\Gamma(D^\circ \to K^-\pi^+)} = \frac{\Gamma(D^\circ \to \pi^+\pi^-)}{\Gamma(D^\circ \to K^-\pi^+)} = \tan^2 \theta_c = \frac{1}{20}$$

neglecting the phase-space difference, while experimental data show that the $\pi^+\pi^-$ and K^+K^- decays are unequal.[5,7] We can define

$$R_1 = \frac{\Gamma(D^\circ \to K^+K^-)}{\Gamma(D^\circ \to K^-\pi^+)} = (2.25 \pm 0.42)\tan^2\theta_c > \tan^2\theta_c$$

$$R_2 = \frac{\Gamma(D^\circ \to \pi^+\pi^-)}{\Gamma(D^\circ \to K^-\pi^+)} = (0.61 \pm 0.19)\tan^2\theta_c < \tan^2\theta_c$$

The ratio R_1 and R_2 can hardly be explained due to the W annihilation diagram, because in both channels the W ahhihilation diagram has the same relative sign with respect to the spectator diagram (a). It has been shown that by introducing the penguin diagram contribution,[43] the prediction of the ratio gets even worse, because the penguin amplitude would reduce R_1 and enhance R_2. Furthermore, the penguin contribution is rather small, estimated to be $\frac{e}{a} = 1\%$.[44] There is a possibility that final-state interactions are sources responsible for deviation from the SU(3) predictions. The $\pi\pi$ mode can be converted into $K\bar{K}$ through the inelastic final-state interaction. In particular, there are two known 0^+ resonances S(975) and δ(1300), which couple to both $K\bar{K}$ and $\pi\pi$ channels. Of course, SU(3) breaking is another source, but the SU(3) breaking effect in the spectator quark amplitudes is not sufficient to explain the data. So future measurements for the Cabibbo-suppressed decay modes will help us to clarify the D meson decay mechanism, e.g. the measurement of $D^\circ \to \pi^\circ\pi^\circ$ is helpful.

(4) Cabibbo double-suppressed decay mode

The statistics of the Mark III are too low to obtain Cabibbo double-suppressed decay modes, which are at a rate of $\tan^4\theta_c$ ($\sim 1/400$) of the Cabibbo-favored ones in the SU(3) symmetric limit (neglecting the phase-space difference). If BES can quadruple the event rates, then their observation become feasible. That will give a further test of the non-leptonic decay mechanism of the charmed mesons.

(5) D°—$\bar{D}^\circ$ mixing and CP violation

Mark III analysis of $D^\circ\bar{D}^\circ \to (K^\pm + \pi's)(K^\pm + \pi's)$ transitions has yielded some marginal evidence for $D^\circ\bar{D}^\circ$ mixing with a strength well beyond the standard model expectation.[45] With its larger statistics BES will be able to confirm the Mark III data. But the same final state can be produced by double Cabibbo-suppressed D° decays, so a theoretical understanding of the D meson decay mechanism is required. As discussed in the previous part of the proposal, this can be achieved by the analysis of the many other decay modes of D^+, D° and D_s, of which BES will provide accurate results. Equiped with this knowledge, the D°-$\bar{D}^\circ$ mixing angle can be drawn from the data. It is a very important test of the standard model.

If the D°-$\bar{D}^\circ$ mixing does happen with an unexpected large angle, maybe we will also see an unexpected large CP violation effect in the charmed meson system. The large statistics of BES will give us such a chance.

Botella and Chau[46] have pointed out that, if there are four generations of quarks, then the partial-decay rate differences for D_s can be big to show the CP violation in the charmed mesons. The partial-decay rate differences are defined as

$$\Delta \equiv \frac{|A|^2 - |\bar{A}|^2}{|A|^2 + |A|^2}$$

where A is a decay amplitude of a particle and $\bar{A}$ is a decay amplitude of an anti-particle. It is a convenient way to study decay amplitude CP violations. In particular, there is a more advantageous situation in the B decay. However, if there are four generations of quarks, then such as $(F = D_s)$

$$\Delta_F \equiv \frac{\Gamma(F^+ \to K^\circ \pi^+) - \Gamma(F^- \to \bar{K}^\circ \pi^-)}{\Gamma(F^+ \to K^\circ \pi^+) + \Gamma(F^- \to \bar{K}^\circ \pi^-)}$$

can be big (two order of magnitude), which are independent of the masses of the 4th generation of quarks and are dependent solely on the existence of higher generations. Perhaps BES will get such a chance to see a signal for the fourth generations of quarks.

After working in the J/ψ and $\psi''(3770)$ energy ranges the work will be switched to the 4.14 GeV energy region to study the D_s meson and to 5.6 GeV energy region to study charmed baryons.

(6) The measurement of the decays of charmed baryons
An outstanding problem is to extend our knowledge of charmed meson decays to charmed baryon decays. Eventually, we would like to have a theory of non-leptonic decays of both mesons and baryons. So far we still do not have a reliable way to calculate nonleptonic weak decay amplitudes. Indeed, after almost 30 years of the first observation of $\Delta I = 1/2$ enhancement in $K \to 2\pi$, we do not yet have a reasonable framework to compute the $\Delta I = 1/2$ amplitude. Hyperon nonleptonic decays, especially P-wave, have resisted theoretical analysis more than their mesonic counterparts. Here, experimental data is essential for any theoretical development. The measurement of charmed baryon decays by BES will bring invaluable information.

CONCLUSION

The main points are the following:

(1) It is essential to achieve the needed detection

efficiency and resolution.

(2) It is important to accumulate many more events than the Mark III. This requires the accelerator luminosity to eventually reach the design value. This will allow greater accuracy and the observation of rare processes, and it will allow BES to solve some of the unanswered questions left by the Mark III.

(3) In the first few years the main physics will be J/ψ, D and D_s mesons. τ lepton and charmed baryon will be studied later.

(4) Data processing is a key problem. From the experience of the Mark III, the capability has to be equivalent to at least 11 VAX 11/780 machines. The now planned VAX cluster with a VAX 11/780, a VAX 11/785 and two VAX 8550 will be adequate for the first stage.

ACKNOWLEDGEMENTS

We are grateful to R. M. Barnett, S. J. Brodsky, M. Chanowitz, L. L. Chau, Z. Y. Lu, W. Toki, M. H. Ye and H. Y. Zhu for suggestions and discussions. We also thank colleagues in the Institute of High Energy Physics for discussion. T. Huang is very appreciative of the hospitality of UC Davis in April and May 1987, when this paper was completed.

REFERENCES

[1] R. M. Baltrusaitis et al., Phys. Rev. Lett. 52,(1984), 2126.
[2] R. M. Baltrusaitis et al., Phys. Rev. D33 (1986),620; U. Mallik, SLAC-PUB-4238 (1987).
[3] R. M. Baltrusaitis et al., Phys. Rev. D32 (1985) 566, 2883.
[4] W. Stockhausen, SLAC-PUB-4026 (1986); U. Mallik, SLAC-PUB-4238, (1987).
[5] D. M. Coffman et al., SLAC-PUB-3720(1986); R. M. Baltrusaitis et al., Phys. Rev. Lett, 56,(1986) 2136, 2140.
[6] R. M. Baltrusaitis et al., Phys. Rev. Lett. 54(1985), 1976; R.M.Baltrusaitis et al., Phys. Rev. Lett. 55 (1985),150.
[7] R. M. Baltrusaitis et al., Phys. Rev, Lett. 55 (1985), 1842.
[8] G. T. Blaylock et al., SLAC-PUB-4158, (1987).
[9] K. T. Chao et al., Chinese Phys. Lett. 2(1985,177.
[10] S. Cooper, SLAC-BUB-4139 (1986).
[11] S. U. Chung, Phys. Rev. Lett. 55 (1985), 779.
[12] H. Lipkin, Phys. Lett. 171B,(1986),298.
[13] W. Toki, Proceedings of Int. Symp. on Lepton and Photon Interactions at High Energies, Kyoto, (1985); R. M. Baltrusaitis et al., Phys. Rev. Lett. 56(1986) 107, SLAC-PUB-3786 (1986).
[14] S. Cooper, SLAC-PUB-3819 (1986).
[15] T. Huang et al., BIHEP-REPORT-86-1 (1986).
[16] M.Chanowitz, Proceedings of the workshop on Colliding beam Physics, Beijing, (1984),52.
[17] H.Aihara et al., Phys. Rev. Lett. 57 (1986),51,2500.
[18] U. Mallik SLAC-PUB-4238 (1987).
[19] M. Chanowitz, LBL-22611.
[20] C. Edwards et al., Phys. Rev. D25,(1982)3065.
[21] A. Etkin et al., Phys. Rev. Lett. 49 (1982) 1620; S. J. Lindenbaum, Proc. Int. Conf. on HEP, Bar, (1985) 311.
[22] J. J. Becker et al., Contributed Paper to the XIII Int. Conf. on HEP, Berkeley, (1986).
[23] M. Chanowitz & S. R. Sharpe, Phys. Lett. 132B (1983), 413.
[24] B. Jean-Marie, Talk at the XXIII Int. Conf. on High Energy Physics, Berkeley, (1986).
[25] R. M. Baltrusaitis et al., SLAC-PUB-3683 (1985).

[26] Shen Qi-xing et al., Physica Energiae Fortis et Physica Nuclearis, 8 (1984),537.
[27] R. M. Baltrusaitis et al., SLAC-PUB-3861 (1985).
[28] R. H. Schindler, SLAC-PUB-3799 (1985).
[29] R. Bailey et al., Phys. Lett. 139B (1984),320.
[30] HRS Collaboration, C. Jung et al., Phys. Rev. Lett. 56 (1986) 17757.
[31] N. Ushida et al., Phys. Rev. 56 (1986), 1767;
N. Ushida et al., Phys. Rev. Lett. 56 (1983),2362.
[32] M. Witherell, talk at this workshop.
[33] C. Darden et al., DESY 84-04-3; H. Albrecht et al., Phys. Lett. 158B (1985),525.
[34] L. L. Chau, Proc. of the 1980 Guangzhou Conf. on Theoretical Particle Physics, China; Phys. Rep. 95 (1983),1.
[35] L. L. Chau & H. Y. Cheng UDC-86-32.
[36] R. M. Baltrusaitis et al., Phys. Rev. Lett. 56(1986), 2136.
[37] I. I. Bigi & M. Fuhugita, Phys. Lett. 91B(1980),121
[38] M. Bauer & B. Stech, Phys. Lett. 152B (1985),380;
M. Bauer et al., HD-THEP-86-19.
[39] I. I. Bigi, SLAC-PUB-4067 (1986).
[40] X. Y. Li. X. Q. Li & P. Wang, BIHEP-TH-87-1, (1987).
[41] A. I. Sanda, Phys. Rev. D22 (1980),2814; L. L. Chau, H. Y. Cheng, Phys. Rev. Lett. 56 (1986)1655
[42] P. I. Vainshtein, V. I. Zahharov & M. A. Shifman, Nucl. Phys. B120 (1977),316.
[43] M. Gluck, Phys. Lett. 88B (1979),145; G. Eilam & J. P. Leveille, Phys. Rev. Lett. 44 (1980), 1648; J. Finjord, Nucl. Phys. B181 (1981),74.
[44] L. F. Abbott, P. Sihivie, M. B. Wise, Phys. Rev. D21, (1980),768.

Panel Discussion (June 15, 1987)

Chairman: Ye Minghan

Ye Minghan:

We are very glad to have Dr. I. Bigi, Prof. M. Chanowitz, Prof. F. Gilman, Prof. D. Hitlin, and Prof. W. Toki to join our panel discussion. It is intended that this panel discussion should concentrate on the physics that should be done on the Beijing Electron-Positron Collider (BEPC) and related aspects of the detectors. The Beijing spectrometer (BES) is building and right now we are trying very hard to have the Crystal Ball to be moved to another interacting point of BEPC.

D. Hitlin

It's well-known among graduate students that when you are taking an examination, the best tactic is to get the members of your exam committee to argue among themselves. So I thought, I would say some provocative things, which will start some arguments among the members of the panel, and, I hope, among yourselves.

§.1. Specific Physics goals for BEPC/BES

First let's enumerate some physics issues, with some specific physics goals. It is generally accepted that acquiring something.

a) 10^8 J/ψ's is quite importart. You have heard Chanowitz talk at length about a very large number of specific topics in this area, so I will not repeat them.

b) 10^2 pb^{-1} $\psi(3770)$.

Let's turn to charm physics here I will be specific. First

let's talk about D studies at the $\psi(3770)$. The next large step should be an order of magnitude beyond the 9 pb^{-1} Mark III sample, say 100 pb^{-1}.

c) $D^{\circ}\bar{D}^{\circ}$ Mixing and DCSD Decays.

I will talk tomorrow about the three Mark III events which have the same strangeness for both D mesons. You may have heard a lot discussion on whether these are due to D°-$\bar{D}^{\circ}$ mixing or to double Cabibbo suppressed decays (DSSD). There are different ways of answering the question: One is to do with the time dependence; the other has to do with the specific final states which you reconstruct simultaneously. We now have 3 event. If there were 30 such events, perhaps more if your efficiency is higher, you would clearly see combinations of reconstructed channels which would indicate whether you were seeing mixing or doubly cabibbo-suppressed decays.

While seeing mixing would be interesting, Bigi has shown us that an unusually large rate of doubly-Cabibbo-suppressed decays could also indicate the existence of new physics.

d) f_D measurement

With an order of magnaitude more data, or even somewhat less, you should be able to measure f_D. You clealy have the statistical power to make that measurement. The issue will come down to backgrounds. The Mark III was in a situation in which Monte Carlo studies indicated that we expected one background event and we saw none. What our result means for extrapolating to higher statistics, I don't know. You will have to very carefully study the question of backgrounds.

e) Vcd/Vcs measurement

The measurement of this KM matrix element has up to now

always be done in neutrino production. As you saw, Mark III has three examples of semileptonic decays involving a pion. These three events yield you a measurement with, not quite, but almost the same precision on the KM matrix that you get from neutrino scattering with an order of magnitude increase in statistics, will become the standard method of measuring this matrix element, since it has far fewer assumptions than muon production by neutrino's.

f) $D^+ \rightarrow \pi^+ \pi^\circ$

You have heard several of us stress the importance of this exotic Cabibbo-suppressed decay as the key to whether there are important final state ineractions, this can be done with no background. Mark III has one event, with no background. If you had 10 events, that would be very interesting.

g) ψ(3770) Resonance Parameters (Require scan)

I have shown you in some detail problems with the measurement of the resonance parameters of ψ''. A new measurement requires a substantial investment and time. It is unlikely that it will be done by Mark III. We certainly could do it. but I think we will not, because people want to do other things. This is an important measurement which must be done well. It must be done to a precision which is far higher than current experiments.

h) $\psi(3770) \rightarrow \gamma X$
$\pi\pi\psi$

These decays of ψ'' to the states other than D $\bar{D}$, to X states and OZI suppressed decay have been seen with a small number of events by Mark III. A detailed study with higher statistics is certainly warranted.

These are the most interesting topics in my estimation. Other people will have their own list. Remember that as you do these experiments, you also do very precise job on branch ratios, Dalitz plots, etc. These are important physics goals. That's probably not controversial.

i) ~50 pb^{-1} (or more) in the 4 GeV region (D_s^+) Let's turn to the study of the D_s^+.

Again by the criterion that one wants few times the data you need 50 pb^{-1} in the 4 GeV region. It is interesting to think about whether you want to run at $D_s^+ D_s^-$ or $D_s^+ D_s^{-*}$.

j) Additional D_s^+ hadronic braching ratios, especially those involving $\pi°$

You have seen that the study of D_s^+ decay is very much in its infancy. We know far less about the hadronic decays of D_s^+ than we do about D_s. Just understanding a number of additional hadronic branching ratios of the D_s^+, especially these involving $\pi°$, which are essentially unknown, is very interesting.

k) Absolute normalization of D_s^+ branching ratios

There is a question of absolute normalization. The normalization at the moment, is a bit arbitrary, in my estimation. If I enumerate what I think most likely decays and I try to normalize them to brarching ratio, 4%. I can not come anywhere close to accounting for every thing expected. I am therefore suspicious of the normalization. This can be checked, with quite a lot of data. With the data you acquire here, you could probably get 25% precision on the absolute normalization. Mark III has not got the time or luminosity to do this.

l) D_s^+ inclusive semileptonic branching ratio

This is an important number which the lecturers referred to many times. Mark III can probably do this very crudely. It's something you should do.

m) D_s^+ exclusive semileptonic branching ratio (esp. $D_s^+ \to (G)\ \ell\nu$)

This is much harder, but very interesting, especially in view of the fact that you might produce gluon-rich mesons together with the leptons.

n) $D_s^+ \to \tau\nu_\tau$ for f_F

Very hard, but you should always keep it in mind.

o) $D_s^+ \to p\bar{n}$

This decay which has a very small branching ratio, would be very interesting to find.

p) $5 \times 10^1 - 10^2$ pb^{-1} above $\sqrt{s} = 5$ GeV

for charmed baryon studies

Absolute branching ratio

Iuclusive $p\bar{p}$, $\Lambda\Lambda$ production

An absolute normalization measurement here would be very useful. Because there is a problem. When you try to use the $\Lambda_c^+ \to pk\pi$ branching ratio to extract the hadronic production cross section of Λ_c^+ or the $e^+e^- \to \Lambda_c^+$ cross section at such high energy. This requires a precise measurement of inclusive production of $\Lambda\bar{\Lambda}$, $p\bar{p}$ or the simultaneous reconstruction of exclusive Λ_c^+ pairs.

q) Hadronic branching ratios

Very few hadronic branching ratios are measured for charmed baryons. Any increase in the number of known channels is useful for understanding decay mechanisms. These mechanisms are more complicated because weak annihilation is not helicity suppressed. There is thus more complex phenomenology here.

r) Accurate semileptonic branching ratio

There has been a measurement of the Λ_c^+ semileptonic braching ratio from Mark II. It is very important to improve this. This is something Mark III could do in the next few years.

s) Higher charmed baryons

These have never been seen in the SPEAR energy region. They have been seen in hadronic and ν production. The SPEAR region is a very good place to look for them, because the multiplicity is low and the combinative problem is not bad. Whether a run which emphasizes Λ_c^+ should be done at the same energy or at a higher energy is a question work studying.

t) τ decay studies

Decays with π°, η

Rare decays

m_{ν_τ} limits

These have been discussed in detail by Gilman. Physics prospects at BEPC are very bright, extremely interesting. It would be very useful for you to begin some study groups on specific physics topics, for example, D physics, D_s^+, J/ψ, τ, etc. small groups of 3-5 people could get together and stimulate each other to look at the problems in doing physics. What the background is, what the analysis strategy are, etc. That will help you to make more rapid progress in your Monte Carlo. I recommend that you to do this first, it is interesting. Second, it makes progress I know all of you are very busy with detectors and softwares, that goes without saying. Nevertheless you should begin to study how to do the physics. It's not too soon to do that, and it's fun.

§2. Strategies for Data-taking

The decision on the order in which data samples are acquired is of crucial importance. First I will talk about criteria, and then I will give you my personal list of the order of topics.

2.1 Considerations

a) BEPC luminosity

No accelerator reaches its design luminosity on the first day; it takes hard work and time. This has always been the case at SLAC, DESY or Cornell. It will take you some time to reach good luminosity. When you turn on you are not going to accumulate data at very rapid rate. You will gain luminosity as you going on. Therefore, you should not base your strategy on the assumption that you will have very high luminosity at the beginning.

b) BES Data acquisition capability

Data acquisition capability has to be matched to the event rate which you produce, or you will lose luminosity.

c) Data analysis capability

You must match the kind of physics attempt and your ability to do analysis. That involves, of course, the amount of computer power you have as well as the status of pattern recognition and data analysis programs.

d) Monte Carlo capability

You have begun to develop Monte Carlo capability, but you have to do much more in this area. Working on specific physics problems is a good way to motivate this.

e) Competition from e^+e^-, $\bar{p}p$ and fixed target machines

Another important thing is to recognize that you have competition. You probably don't have competition from e^+e^- machines in your energy region, but you do have competition as Witherell has emphasized on this symposium. He clearly showed you that you have serious competition from fixed target hadron machines. We have also seen that ARGUS and CLEO do some interesting charm physics. So you even have competition there. You even have competition from $\bar{p}p$ colliders. With high resolution vertex detectors they can also do charm physics. This competition will have large samples of charmed particles. You have to consider this in determining your run plan.

f) Physics interest

This area will cause the most arguments. The others are based on fact that can be objectively evaluated. On this one you can argue forever, since it is a matter of taste. I have told you little bit about my taste, and the other lecturers will certinly tell you about theirs.

g) Topicality

This is related to both physics interest and competition. You want people to be interested in what you do and what you say. That is not always entirely within your own control, it may depend on what other experimental results.

For example, theorists have been talking about CP violation in D decay for many years. The reaction of experimentalists was "Sure, but that is impossible to measure". Now that ARGUS observes mixing in B decay, there is renewed interest in CP violation in B and D decay. This does not mean it is now possible, but it's more possible. At least one now thinks, talks about it, and works on it.

Topicality is important; you should worry about the kind of impact your results are going to have in the physics world. If you go to a meeting to talk about BES results you want people to be interested, you have the opportunity in front of you. If you do it right, you will have very topical results that people will take care about.

Taking all these things into consideration, here is my controversial scenario for the order of data taking:

There will be as many scenarios as there are physicists. These should be reduced to a few and subjected to analysis according to the criteria I listed above.

1) The best place to begin is the J/ψ. Suppose you get lots of data, lots of tracks, lots of constraints, and lots

of reactions which you can use to check that what you are doing is correct.

You want to collect enough data so that you can establish credibility. You want to repeat measurements that other people have done. If the first time you went to a meeting you said that you have measured the τ mass and width, that would not get everybody very excited, but it would establish your credibility.

With same data sample, you can look at the ξ, and say yes or no. That would be an important accomplishment, people would be very interested in that.

Whether or not you then continue to acquire $50\text{-}100\times10^6$ J/ψ decays is the point in which I differ with the people who was presented. I think this would be very difficult to do. The physics of J/ψ decays is an interesting subject, but 10^8 is a lot of events. If you gave Mark III this many events, we would have a difficult time to analyze them. You have more to learn than we have.

You don't have competition in this field. This physics is very interesting, it will remain interesting until you do it.

$$\left.\begin{array}{ll} \text{2)} & 50\ \text{pb}^{-1} \quad \psi(3770) \\ \text{3)} & D_s^+ \\ \text{4)} & \Lambda_c^+ \ \text{etc.} \end{array}\right\} \text{competition}$$

There is competition in these fields. D, D_s^+ and Λ_c^+ physics can, I'm sure you are convinced, be done with fixed target experiments begining right now at Fermilab. These have the capability of doing major things in charm physics over the next 3-5 years. You should try to take and analyze charm data in this area on this time scale, so that your weak decay results meet the topicalidy criterion.

Ye Minghan:

Thank you very much for your very interesting and comprehensive talk.. Next I would like to ask a theorist, Prof. Gilman to speak.

F. Gilman

Let me add a few things to what has been said, particularly by David Hitlin.

First I would stress the importance of working groups consisting of one or two theorists and five or so experimentalists, each concentrating on one aspect of charm physics relevant to BEPC. This idea has been used very successfully, by the Mark II collaboration in preparation for their experimental work at the SLC. It allows everyone involved to learn physics, as well as helps to prepare for analysis of the data (possibly also employing much data tapes).

Second, I would agree with everyone else that an initial run at the J/ψ is the best beginning for physics at BEPC. After accumulating a data sample that is comparable within a factor of 2 or so with that obtained by Mark III, though, I would recommend spending some time doing another piece of physics. There is a danger in getting locked into a massive program of J/ψ physics. It is good in itself to step back for a time, analyze the data, make improvements in the detector, the trigger or data acquisition system, the analysis programs, etc. while doing other interesting physics. Exactly what physics to do will only be clear at that time, I suspect there will be very strong competition by that time in D physics from fixed target experiments at Fermilab, so my guess is that experiments just above $D_s \bar{D}_s$

threshold or above charm baryon threshold will be most topical. In any case on the basis of the state of physics at that time, one will want to choose one area of physics above open charm threshold and do it well before returning to do the definitive study of J/ψ decays.

Ye Minghan:

Thank you very much next I would like to ask an experimentalist, Prof. Toki to speak.

W. Toki

Let me just put down a few things that might be interesting. Some overlaps with David's. I think it's more interesting to worry about next 3-4 years. So 1 would say it's important to first run at J/ψ resonance, take the order of$(5-10)\times 10^6$.

There is a few things I will say here. The easiest thing would be for J/ψ, at least, at the level$(5-10)\times10^6$ events to check your Monte Carlo, to check your detector. I discuss a little bit about what we do when we start out at J/ψ resonance. Other thing at first run is there may be problems on your detector, you must stand back and look at them. As we first run at spring 1982, we run a little bit at J/ψ, it was a debugging run. Then we more efficiently run at FALL 1982, we had a lot of problems, the runs were not very good ones. Only afterwards at 1983, we had very good runs. There were many problems, when you get this point, you have to work on them and solve them. May be you don't get a lot of physics immediately with these runs because you have problems with detector. The drift chamber had high currents, the shower counter leaks, you have problems with masking synchrotron radiation background. There are a lot of problems you must first face and look at, depending on how things go here. But at this level, there is a number of things you can do.

a) $\xi(2.2) \to K\bar{K}$

to verify if $\xi(2.2)$ is there.

b) $\eta_c \to VV$

to verify whether or not η_c goes to two vectors as Mark II sees and DM II sees.

These are very simple and very straitforward things you can study at this level.

c) $\psi \to \gamma X$

The next level, Mark III has talked about, would require a large amount of statistics. At this point at least you can begin the radiative decays to see what is involved for spin-parity test, to understand your detector. For us, to study radiative decay with spin-parity test really requires a lot of studies. In some case we have not done certain spin-parity measurement, because we don't think we understand our detector as well as we like to.

The next set of runs. I would say is again the taste whether or not you want to go on to 10^8 of J/ψ or to do charm physics. I would say if you want to do 10^8 J/ψ and if the theory that Chanowitz is proposing turns out to true, there is a possibility of spectacular discovery. If you can verify conclusively there is an exotic state, such as 1^{-+}. This will require tremendous studies to understand these decays. We thought Mark III began to study this, perhaps the only place is here to do completely right.

d) F decay will be easier to do than D decay Absolute branching ratio is probably something you can do.

May be by then Mark III will have a very approximate number, say in the order of 30-50%, very rough measurement. May be you can bring it down to 10%, that will be very important.

Other area will be semileptonic measurement. Again Mark III may get a number, you can push down errors to much better, much more significant number. I dont think you have too much competition from other experiments.

Other numbers, probably you will be lucky enough to do, is $F \to \tau\nu$, Mark III doesn't have statistics to do that. Nobody else can do that. You have the best opportunity to do that.

In D decay, you can try to measure $\pi e\nu/K e\nu$, which is important for KM matrix element. Mixing will be important. The problem here is, as Fred say, there could be competition from hadronic machine. Particularly they look at $D^{+*} \to \pi^+ D^\circ$. May be by then they will have much better data.

Λ_c measurement is difficult for them to do, because Λ_c life time is very short, as they make decay length measurements, they lost of their most events. These may be possible for you at higher energy to study, very few measurement so far had been made.

An other area which hasn't been touched is Ξ°, Ξ^+ measurement, only very few from them having been made.

This is what we are looking at Mark III. Again without statistics it's not clear what impact we might do.

I think all things I talk here require extremely high statistics.

Before you actually decide, you should first run here to see how you stand, that is, how much data you are really going to get, how will the machine works, also you must ask yourself whether or not there are some repairs you have to do for your detector.

I would not decide now, I would listen to the arguements carefully.

There is one thing, that is Mark III experience. It is useful to first take data, to think about where you stand. It is not a good idea to keep switching back and forth. When you understand your detector, you should decide to study one or two years at one area. Don't jump back and forth. Let people get a complete data set then finish it analyze it and do publication.

Ye Minghan:
Thank you very much. Now I would like to ask Dr. Bigi to give his opinion.

1. I agree with the remarks made by Gilman. It is essential to work on concrete problems in groups with one or two theorists in each group.
2. There are 4 fields of research:
   a) $D\bar{D}$
   b) $F^{(*)}\bar{F}^{(*)}$, $D^{(*)}\bar{D}^{(*)}$
   c) $\Lambda_c + X$
   d) 10^8 J/ψ

I feel very strongly that you should turn your attention to open charm physics after the initial J/ψ run. However I do not suggest that you should cover all three topics (a),(b),(c) first before you return to a very high statistics J/ψ run since this would delay the latter by at least five to six years. Plan to cover one (at most two) of the first three topics before concentrating on (d). Which of the first three fields you should choose, depends-as already stressed by Hitlin-on the competition which develops in the meantime. Therefore you have to allow for some flexibility in your planning.

If you force me to make a prediction, which you should not do, my present feeling is to favour option (b). I guess most, but not all, physics issues in (a) might be settled in the next two or three years. Now why do I say "not all of them"? I mean there are two decay modes which are generic in the problems they pose in the sense that it is not clear to me that, for example, fixed target machines can do a sufficient measurement:

$$D^+ \to \pi^+ \pi^\circ, \quad D^+ \to \ell\nu$$

All these techniques on fixed target machine need two prongs. That's why I leave a loophole open. The above mentioned decay modes are single prong decay, they are very important.

3. Last comment is on τ physics. One interesting task consists of establishing the V-A structure of the τ-decay. It is, as far as we know, consistent with the data. However, if you turn the argument around and say how much can we rule out V+A or other combinations, then it is very unimpressive. Repeating measurement of the mixing parameter is not very decisive. I'm not sure what the situation really is for your accelerator configuration. But I would like to mention spin effects, at least for the transeverse polarization. It does not work for ψ, but it might work for ψ'', one could consider rotating the transverse polarization into the longitudinal direction. Then you can make polarization measurements and correlation measurements between the two sides. You have gained much more sensitivity in measurements about V-A versus V+A and so on. That is the kind of analysis people did for muon decay, which is sometimes

referred to as the polarization parameter ξ. To repeat: Such studies are very important, in particular, for τ-decay.

Ye Minghan:

Thank you very much. I would like to ask Prof. Chanowitz to speak.

M.Chanowitz

First a suggestion for the theorists about one way to become familiar with physics topics relevant to BEPC. In Berkeley we have a theory journal club on four dimensional physics that meets once a week. Everyone takes a turn to speak including students. The speaker for each week reports on an interesting paper, either theoretical or experimental, often on a topic in which he is not an expert. Sometimes the topic is a very recent result, sometimes an interesting older result. We try to have a lot of questions and discussion. Experimenters may also wish to attend some of the sessions. It's a nice way to discuss and become familiar with a range of topics that could be important at BEPC. In Berkeley we meet at lunch time. We eat while we listen so the talks don't get too long because the speaker starts to get hungry.

Regarding the physics plan, you should be happy to see us debating between J/ψ and open charm physics, like two dogs fighting for a bone. There is a lot of interest in doing both kinds of physics at BEPC.

It is clearly correct to begin by tuning up at the J/ψ. After that it is not clear how to proceed. That decision depends both on BEPC performance and on physics interest. I see from the BES checkout proposal that the design calls for almost a factor of 10 better luminosity at J/ψ for BEPC relative to SPEAR, compared to an advantage of only about $2\frac{1}{2}$ above $\bar{D}D$ threshold. If actual running approximats this design, then high statistics J/ψ data can be accumulated more rapidly for J/ψ studies then for open charm.

I therefore have a concern about the suggestion that high statistics J/ψ running be delayed until after completion of the open charm studies for D, F and charmed baryons. This could mean a considerable delay. It would be nice to have high statistics J/ψ data at about the same time that the next generation of lattice computation begins to provide more reliable QCD calculations of the spectrum, since that will be a moment of maximum physics interest. It could be an active and interesting period of collaboration with theorists and experimenters working togather to understand the spectrum of gluonic states.

I do not agree with the remark that the experiment should not be done until after the theorists have made their "predictions". The interplay between theory and experiment is more complex and interesting than that remark implies, an interactive period of collaboration is the best way; experiment guiding theory to an understanding of what calculations and approximations are adequate while theory guides experiment by pointing in a procise way to interesting experimental questions and interpretations.

There is also a human time scale to physics that we should not forget. We need to maintain momentum or enthusiasm is lost. That is why we want the SSC in the U.S. in the 90's rather than in the next century. BEPC is potentially a tremendous resource for studying gluonic states but it does not have a monopoly on the subject. It needs to contribute in a timely fashion, at the same time that exciting contributions are being made by other experiments and by theory.

I am not arguing against the possibility of doing some open charm physics after the initial J/ψ physics check-in period, but I am concerned about a long delay before trying for high statistics J/ψ data. Fortunately we don't have to commit to a definite plan today. It's clear from Hitlin's hep1ful remarks that the data acquisition problem at design luminosity is at least as severe for open charm physics as for the J/ψ since a major component of the trigger rate is from backgrounds. That are proportional to or independent of beam current. Therefore for either the J/ψ or open charm, the data acquisition system will need an up grade to take advantage of the design luminosity. As Hitlin said, it's not too early to begin to think about this.

The real difference between the requirements for J/ψ and open charm physics seems to be in off-line computing. This represents a problem and an opportunity. It seems that the way to go is a parallel processor farm, as demonstrated by LASS at SLAC (140,000,000 KP scattering events recorded and reconstructed) and at Fermilab as Witherell discussed, several experiments are already building on the Fermilab system, which is VAX based. Besides the immediate benefit to particle physics, it could be a more general benefit to bring such technology to China, we should also remember that the computing horizon is expanding at a rate that it is difficult to anticipate. In 1984 we were discussing 2 VAX 780's and now we are already anticipating a much more powerful system with a 785 and one or more 8550's.

The lead-time for the off-line computing problem is also considerable. A commitment to solve it would probably have to be two or more years in advance.

So I have just expressed some concerns and problems and no solutions. Fortunately the decisions can be made with the benefit of some experience, though keeping in mind the long lead-times needed to create a system that can realize the full potential of the machine.

Ye Minghan

Time is running out.

Thank you very much. We are very grateful to all there valuable opinions, they are very helpful to us. Everyone agrees, the first run on BEPC should be at J/ψ. The cross-section is very large. There are plenty of well known physics processes, they could be used to test and calibrate the detector. It is also easier for the accelerator, since the output power of klystrons is higher than 20 MW, full energy injection at 1.55 GeV is possible, it is good for the operation of the accelerator and higher average luminosity. Synchrotion radiation background should be low at this energy. Above all, there are certainly new physics there, such as the search for glue ball, we are very exciting to enter this energy region.

It is important to think about the data acquisition problem with the designed high luminosity, thanks to Prof. Hitlin and others for this helpful remark.

As to what happens after the first J/ψ run, it is still open.

Let us thank the speakers again for all these very valuable and helpful remarks and interesting discussions.

List of Participants

Name		Affiliation
Bigi, Ikaros	比　基	Stanford Linear Accelerator Center
Chanowitz, Michael S.	卡诺维基	Lawrence Berkeley Laboratory
Chen, Chaoqing	陈朝清	Institute of High Energy Physics
Chen, Fengzi	陈风至	Institute of High Energy Physics
Chen Guomin	陈国民	Institute of High Energy Physics
Chen, Hesheng	陈和生	Institute of High Energy Physics
Chen, Shihao	陈世浩	Northeast Normal University
Chou, Kuangchao	周光召	Institute of Theoretical Physics
Cui, Huachuan	崔化传	Institute of High Energy Physics
Dai, Yuanben	戴元本	Institute of Theoretical Physics
Ding, Yibing	丁奕兵	Graduate School, University of Science and Technology of China
Dongfang, Xiao	东方晓	Institute of High Energy Physics
Gao, Chongshou	高崇寿	Peking University
Gao, Shuqi	高树琦	Institute of High Energy Physics
Gao, Wenxiu	高文绣	Institute of High Energy Physics
Gilman, Frederick J.	吉尔曼	Stanford Linear Accelerator Center
Gou, Liang	勾　亮	Institute of High Energy Physics
Gu, Weixin	顾维新	Institute of High Energy Physics
Gu, Yifan	顾以藩	Institute of High Energy Physics
Guo, Xinheng	郭新恒	Institute of High Energy Physics
Guo, Yanan	过雅南	Institute of High Energy Physics
Han, Shiwen	韩世温	Institute of High Energy Physics
Han, Shurong	韩树荣	Institute of High Energy Physics
He, Jingtang	何景棠	Institute of High Energy Physics
He, Yudong	和煜东	Yunnan University
Hitlin, David	希特林	California Institute of Technology
Hu, Haiming	胡海明	Southwest Jiaotong University
Hu, Ning	胡　宁	Peking University
Huang, Tao	黄　涛	Institute of High Energy Physics
Huang, Yeqin	黄业勤	Tsinghua University
Huang, Zheng	黄　峥	Institute of High Energy Physics
Jiang, Xiangdong	江向东	Institute of High Energy Physics
Ju, Changsheng	鞠长胜	Institute of High Energy Physics
Lee, T. D.	李政道	Columbia University, New York
Li, Bingan	李炳安	Institute of High Energy Physics
Li, Jia	李　佳	Institute of High Energy Physics

Li, Jin	李 金	Institute of High Energy Physics
Li, Peiqin	李佩琴	Institute of High Energy Physics
Li, Rubai	李如柏	Institute of High Energy Physics
Li, Tiezhong	李铁忠	Institute of High Energy Physics
Li, Xiaoyuan	李小源	Institute of Theoretical Physics
Li, Xueqian	李学潜	Institute of Theoretical Physics
Li, Yuekuan	李悦宽	Institute of High Energy Physics
Lianng, Xiancui	梁仙翠	Hangzhou University
Lin, Shuzi	林树子	Institute of High Energy Physics
Lin, Zhongjin	林仲金	Xia-Men University
Liu, Dongsheng	刘东胜	Institute of Theoretical Physics
Liu, Feng	刘 峰	Huazhong Normal University
Liu, Tiehui	刘铁辉	Nankai University
Lu, Changguo	陆昌国	Institute of High Energy Physics
Lu, Gongru	鲁公儒	He-Nan Normal University
Lu, Huaqing	吕华庆	Huazhong Normal University
Lu, Zuyin	陆祖荫	Academia Sinica
Ma, Zhongqi	马中骐	Institute of High Energy Physics
Panofsky, Wolfgang K. H.	潘诺夫斯基	Stanford Linear Accelerator Center
Qian, Yukun	钱裕昆	Peking University
Qin, Danhua	秦旦华	Peking University
Qiu, Zhongping	裘忠平	Huazhong Normal University
Ruan, Tongze	阮同泽	Institute of High Energy Physics
Shen, Qixing	沈齐兴	Institute of High Energy Physics
Shi, Huanzhang	史焕章	Institute of High Energy Physics
Sun, Changpu	孙昌璞	Northeast Normal University
Tao, Zhijian	陶志坚	Institute of Theoretical Physics
Tian, Weihua	田卫华	Institute of High Energy Physics
Tian, Xu	田 旭	Huazhong Normal University
Toki, Walter H.	托 基	Stanford Linear Accelerator Center
Wang, Jianxiong	王建雄	Institute of Theoretical Physics
Wang, Linzhou	王临州	Institute of High Energy Physics
Wang, Luyu	王鲁豫	Xing-jing University
Wang, Ping	王 平	Institute of High Energy Physics
Wang, Qinmou	王勤谋	Anhui Normal University
Wang, Shaosuen	王韶舜	University of Science and Technology of China
Wang, Shuqin	王淑琴	Institute of High Energy Physics
Wang, Taijie	王泰杰	Institute of High Energy Physics
Wang, Xueying	汪雪瑛	Institute of High Energy Physics
Wang, Zhengben	王正本	Yunnan University
Wen, Jiaru	闻家如	Institute of High Energy Physics
Witherell, Michael	韦斯瑟尔	University of California, Santa Barbara
Wu, Jimin	吴济民	Institute of High Energy Physics
Wu, Yueliang	吴岳良	Institute of Theoretical Physics

Xiong, Chuansheng	熊传胜	Institute of Theoretical Physics
Xu, Dezhi	徐德之	Institute of High Energy Physics
Xu, Mingkang	许明康	Hangzhou University
Xue, Shengtian	薛生田	Institute of High Energy Physics
Xue, Xun	薛 迅	Institute of Theoretical Physics
Yan, Wuguang	严武光	Institute of High Energy Physics
Ye, Minghan	叶铭汉	Institute of High Energy Physics
Yu, Hong	郁 宏	Institute of High Energy Physics
Zhang, Changchun	张长春	Institute of High Energy Physics
Zhang, Ge	张 戈	Institute of High Energy Physics
Zhang, Liangsheng	张良生	Institute of High Energy Physics
Zhang, Meiman	张美曼	Institute of High Energy Physics
Zhang, Shiluan	张世鸾	Yunnan University
Zhang, Xiaofeng	张晓峰	Institute of Theoretical Physics
Zhang, Yingping	张英平	Institute of High Energy Physics
Zhang, Zhaoxi	张肇西	Institute of Theoretical Physics
Zhang, Ziping	张子平	University of Science and Technology of China
Zhao, Guangda	赵光达	Peking University
Zhao, Peiying	赵佩英	Institute of High Energy Physics
Zheng, Linsheng	郑林生	Institute of High Energy Physics
Zheng, Min	郑 敏	Institute of High Energy Physics
Zheng, Zhipeng	郑志鹏	Institute of High Energy Physics
Zhou, Yongsheng	周永参	Institute of High Energy Physics
Zhou, Yuehua	周月华	Institute of High Energy Physics
Zhu, Guosheng	朱国胜	Institute of High Energy Physics
Zhu, Hongyuan	朱洪元	Institute of High Energy Physics
Zhu, Shangen	朱善根	Institute of High Energy Physics
Zhu, Xuetian	朱雪天	Zhejiang University
Zhu, Yongsheng	朱永生	Institute of High Energy Physics
Zhu, Yucan	祝玉灿	Institute of High Energy Physics

Index

C (1480) 168
θ (1730) 168
ξ (2230) 168
Axial vector current 27
B Meson 9, 227, 371
B-semileptonic decay 285, 403
BEPC 2, 149, 163, 166, 169, 170, 196, 475-503, 508, 510, 535, 540
BES 149, 171, 475-502, 509, 511-512, 517, 519, 529, 531, 541
Bose statistics 210
Boson masses 7
Bottom decay 21, 24
Bottomonium 46, 59, 63, 66, 74
Breit-Fermi potential 57
Breit-Wigner 131-132, 194, 266, 278
Bremsstrahlung 350, 430
η_C Decays 111
CLEO 477
CP violations 416
Cabibbo-allowed decays 228, 234, 240, 246, 272, 309, 348, 430, 433, 505, 525
Cabibbo-depressed decays 109
Cabibbo-suppressed decays 11, 21, 23, 28, 137, 228, 240, 244, 246, 252, 262, 272, 309, 341, 351, 353- 354, 368- 369, 396, 399, 433, 439, 453, 479, 505, 509, 527, 536
Charm-anticharm pairs 411
Charm decays 12, 340, 356, 370
Charm threshold 76-77
Charm vertex 428
Charmed baryons 295, 395, 460, 470, 480, 530, 539
Charmonium spectroscopy 4, 39, 46, 59, 63, 66, 70, 72, 76, 109, 380, 427, 465-467, 471, 480, 506
Cornell potential 41, 48
Coulomb interaction 40, 44, 46, 63
Crystal ball 89-159, 509, 511, 535
D Meson11, 245, 260, 262, 264, 344, 350, 386, 427, 464, 504, 505, 513, 521, 522
D-hadronic branching ratios 256, 539
D-hadronic decays 246, 252, 256, 266
DM 2 89-159, 477
Daughter quark tag method 318
Decays 111
Double tags 260, 262, 264, 444
Doubly Cabibbo-suppressed D decays (DCSD)402, 454, 456, 480, 509, 528, 536
Eigenstates 8, 174, 183, 313, 409, 416, 419, 453, 469
Eigenvalues 179
Exotic D decays 406
Fierz transformation 15, 239, 295
Fixed target experiment 470, 471, 471
G-parity 99, 108
Gauge bosons 6, 205, 383
Gauge theory 1, 38, 161
Gell -Mann-Nishijima formula 178
Glueball 109, 146, 161, 162, 167, 180, 185, 186, 189, 203, 478, 504, 507, 515, 516, 517
Gluonic coupling 507
Gluonic states 184, 188, 191
Gluons 11, 17, 39, 57, 72, 101-102, 107-108, 163, 165, 173, 178, 190, 306, 342, 349, 368, 385, 414
Ground state vector 60
HRS collaboration 36
Hadron 6, 11, 16, 20, 23, 37, 72, 102, 161-162, 219, 227
  charmed 274, 345

Hadron spectroscopy 129
Hadronic decays 103-120, 137, 166, 180, 231, 272, 504-505, 513
Hadronic state 27
Hadroproduction 460-463, 470
Hamiltonian 15-16, 20, 73, 176, 178, 234, 241, 243, 295, 312
Helicity applitudes 135
Helicity suppression 345, 349, 433
Isodoublets 174
Isoscalars 148, 181, 193, 208
Isospin 99, 367
J/decays 89, 90, 122, 161, 184, 191, 192, 196
J/ψ spin parity98, 107, 131
J/ψ width 97
J/ψ decays 161, 544, 545
Langrangian 342, 356, 371, 385
Lepton tag 318
Leptonic decays 26, 32, 34, 228, 284
Leptons1, 2, 12, 25, 102, 222, 246
Leptoquark 286
Lorentz Invariant Phase Space
(LIPS) 165
Lorenz scalar 58
Mark II 89-159, 477
Mark III 89-159, 457- 458, 460, 477, 503, 511, 514, 517, 519- 520, 522, 524, 529, 536, 537, 543, 546- 547, 550
Martin 42, 48
Meiktons 161, 190, 198, 200, 202, 212, 213
Mesons 101, 110, 120
 charged 361
 charmed 233, 245, 429, 479
 gluonic 478
 ordinary 172
 tensor 520
 vector 140
Monte Carlo 536, 540, 541, 545
Muon 14, 24, 26
 decay 226
Muon neutrinos 25
Non-leptonic decays14, 243, 366, 393, 421, 433, 479, 524, 529, 530
Non-relativistic quark model 165, 173
Nonet 165, 174, 181, 183, 204
Open charm spectroscopy 467, 468 - 470, 471
Penguins 20-21, 23, 398, 405, 524, 528
Perturbation theory 202
Prongs 33, 444
Pscaudoscalar 60
Pseudoscalar 62, 100, 103- 104, 108-109, 120, 131, 148, 168, 182, 204, 272, 298, 317, 383, 414, 478, 520, 524
Quantum chromodynamics (QCD) 6, 15, 17, 20, 37-39, 102, 109, 126, 161-162, 166-167, 175, 211, 239, 243-245, 295, 300, 339, 361, 379, 385, 387, 390, 392, 433, 461, 478, 526, 551
Quark masses 2
Quarks 1-2, 8, 10, 12, 14, 17, 23, 40-41, 57
 charm 340
 heavy 11, 12, 16, 37, 59, 61,
 222, 231, 285, 312
 light 60-61, 90, 161, 222
Quarks charm 102, 219, 227, 293
Radiative decays 105, 122, 132, 148, 163, 198, 504, 505, 506, 508, 516
Richardson 41, 42, 48
Rosner formulation 104
SU (3) classification 99, 100
Schindler 46

Schrodinger equation 60
Semi-hadronic decays 27, 31
Semileptonic D decays 272, 282, 321, 404, 504
Semileptonic branching ratio 274, 282, 295, 354, 391, 538
Semileptonic decays 12, 14, 23-24, 228, 231, 239, 243, 245- 246, 289, 350, 400, 457
Semileptonic hyperon decay 226
Single tags 260
Spectator model 23-24
Spin dependent potential 50, 64
Spin independent potential 39, 51
Spin orbit 62-63
Spin-parity204, 208, 465, 469, 504, 507, 518
Spin-spin interactions 57, 58
Standard model 6, 8, 222, 317, 341, 407, 411, 413, 421
Stickiness 110, 131, 165, 193, 194, 212, 213
Tau 14, 26, 27
    decay27, 34
Tau lepton 24
Tensor interaction 57, 62
V-spin 347, 352, 359, 378
Valence gluons 211
Vectar current 27
Vector bosons 6, 383
Vector-pseudoscalar decays 121
W-boson 342
Weak decays 11, 219
Weak doublet 25